INSECT MOLECULAR SCIENCE

INSECT MOLECULAR SCIENCE

Edited by

Julian M. Crampton and Paul Eggleston

Wolfson Unit of Molecular Genetics, Liverpool School of Tropical Medicine, Liverpool, UK

16th Symposium of the
Royal Entomological Society of London
12–13 September 1991
at
Imperial College, London

ACADEMIC PRESS

Harcourt Brace Jovanovich, Publishers
London   San Diego   New York
Boston   Sydney   Tokyo   Toronto

ACADEMIC PRESS LIMITED
24/28 Oval Road, London NW1 7DX

United States Edition published by
ACADEMIC PRESS, INC.
San Diego, CA 92101

Copyright © 1992 by
The Royal Entomological Society of London

All rights reserved. No part of this book may be reproduced
or transmitted in any form or by any means, electronic or
mechanical, including photocopy, recording, or any
information storage and retrieval system without permission
in writing from the publisher

A CIP record for this book is available from the British Library

ISBN 0-12-195210-X

Typeset by Keyset Composition, Colchester
Printed in Great Britain by TJ Press (Padstow) Ltd, Padstow, Cornwall

Contents

Contributors ix
Preface xi

Part I. Potential

1. Potential Application of Molecular Biology in Entomology
   J. M. CRAMPTON

   I. Introduction 3
   II. The Requirements for Genetic Manipulation 5
   III. The Potential Application of Transgenic Technology
        in Insect Systems 14
   IV. Transgenic Insects: the Future 18
       References 18

2. DNA Analysis in Relation to Insect Taxonomy, Evolution and Identification
   R. J. POST, P. K. FLOOK AND M. D. WILSON

   I. Introduction 21
   II. The Major Features of Genome Organization and
       Evolution 22
   III. Ribosomal RNA Genes in Evolution and Systematics ... 24
   IV. Uncharacterized Repetitive DNA Sequences for
       Species Identification 30
       References 34

3. Transposable Elements and their Biological Consequences in *Drosophila* and Other Insects
   D. J. FINNEGAN

   I. Introduction 35
   II. Biological Consequences of Transposition 38
   III. Experimental Exploitation of Transposable Elements ... 41
   IV. Can *P* Elements be Used in Other Species? 43
   V. How to Find a New Transposable Element 44
      References 46

Part II. Genome Mapping, Development and the Immune System

4. Mapping Insect Genomes
   M. ASHBURNER

I.	Introduction	51
II.	Genetic Maps	51
III.	Polytene Chromosome Maps	56
IV.	Microdissection Libraries – "Polytene Chromosomes on a Filter"	59
V.	Maps of Restriction Fragment Length Polymorphisms	60
VI.	Mini- and Micro-satellite DNA Mapping	62
VII.	Whole Genome Molecular Mapping	64
VIII.	The Next Stage	69
	References	71

5. Molecular Aspects of Sex Determination in Insects
   M. BOWNES

I.	Introduction	76
II.	The Genetics of Sex Determination in *Drosophila*	77
III.	The Genetics of Sex Determination in Diptera	80
IV.	The Genetics of Sex Determination in Haplodiploid Hymenopterans	84
V.	Is a Single Model for Sex Determination in Insects Possible?	85
VI.	The Molecular Basis of Sex Determination in *Drosophila*	89
VII.	Molecular Analysis of Sex Determination in Other Insects	95
	References	99

6. Gene Control Systems Affecting Insect Development
   M. G. PETITT AND M. P. SCOTT

I.	Introduction	101
II.	Pattern Formation in the *Drosophila* Embryo	102
III.	Current Issues and Strategies	110
IV.	Summary and Outlook	121
	References	122

7. Insect Immune Systems
   Y. ENGSTRÖM

I.	Introduction	125
II.	Background	126

	III.	Insect Immune Proteins	128
	IV.	*Drosophila* as an Insect Immune Model System	131
		References	135

Part III. Interactions with the Environment

8. Semiochemicals: Molecular Determinants of Activity and Biosynthesis
J. A. PICKETT AND C. M. WOODCOCK

	I.	Introduction	141
	II.	Molecular Determinants of Activity	141
	III.	Interactions Between Different Semiochemical Types	144
	IV.	Analogues of Semiochemicals	146
	V.	Semiochemical Biosynthesis: New Targets for Molecular Genetics	147
		References	149

9. The Perception of Semiochemicals
L. J. WADHAMS

	I.	Introduction	152
	II.	Perception of Odours	152
	III.	Peripheral Coding of Odours	154
	IV.	Conclusion	161
		References	161

10. Semiochemically Mediated Behaviour
G. M. POPPY

	I.	Introduction	163
	II.	The Role of Behavioural Studies in Chemical Ecology	164
	III.	Aphid Sex Pheromones	165
	IV.	Host Plant Volatiles	167
	V.	Alarm Pheromone	169
	VI.	Conclusion	170
		References	171

11. Molecular Biology of Insecticide Resistance
A. L. DEVONSHIRE, L. M. FIELD AND M. S. WILLIAMSON

	I.	Introduction	173
	II.	Target Site Resistance	174
	III.	Metabolic Resistance	179
		References	182

Part IV. Physiological Regulation

12. Locust Adipokinetic Hormones: A Review
    M. O'SHEA AND R. C. RAYNE

I.	Introduction	187
II.	Localization and Site of Synthesis	189
III.	Release and Functions	191
IV.	Inactivation and Metabolism	191
V.	Prohormone and Precursor Biosynthesis	194
VI.	Processing *In Vivo* and *In Vitro*	198
VII.	Conclusions and Future Directions	202
	References	203

13. The Structure and Functional Activity of Neuropeptides
    G. GOLDSWORTHY, G. COAST, C. WHEELER,
    O. CUSINATO, I. KAY AND B. KHAMBAY

I.	Introduction	205
II.	Neuropeptides that Affect the Metabolism of the Fat Body	206
III.	Attempts to Predict and Determine Secondary Structure	210
IV.	N-terminal Modification of Peptides	218
V.	Insect Diuretic Peptides	221
VI.	Conclusions	222
	References	223

14. Molecular Approaches to Insect Endocrinology
    L. M. RIDDIFORD

I.	Introduction	226
II.	Contributions of Molecular Biology to the Study of Developmental Insect Neuropeptides	227
III.	Hormone Receptors, Transcription Factors, and the Regulation of Moulting and Metamorphosis	231
IV.	Conclusions	237
	References	237

Appendix. DNA Transformation of Non-Drosophilids (Workshop Abstracts)

I.	Introduction	241
II.	Abstracts	243
III.	Overview	258

Index 263

Contributors

Numbers in parentheses indicate the page numbers on which the author's contributions begin.

M. ASHBURNER (51), Department of Genetics, University of Cambridge, Cambridge CB2 3EH, UK

M. BOWNES (76), Institute of Cell and Molecular Biology, Division of Biological Sciences, University of Edinburgh, King's Building, Mayfield Road, Edinburgh EH16 6HJ, UK

J. CARLSON (241), Department of Microbiology, Colorado State University, Fort Collins CO 80523, USA

G. COAST (205), Department of Biology, Birkbeck College, University of London, Malet Street, London WC1E 7HX, UK

J. M. CRAMPTON (3), Wolfson Unit of Molecular Genetics, Liverpool School of Tropical Medicine, Pembroke Place, Liverpool L3 5QA, UK

O. CUSINATO (205), Department of Biology, Birkbeck College, University of London, Malet Street, London WC1E 7HX, UK

A. L. DEVONSHIRE (173), AFRC Institute of Arable Crops Research, Rothamsted Experimental Station, Harpenden, Herts AL5 2JQ, UK

P. EGGLESTON (241), Wolfson Unit of Molecular Genetics, Liverpool School of Tropical Medicine, Pembroke Place, Liverpool L3 5QA, UK

Y. ENGSTRÖM (125), Department of Molecular Biology, University of Stockholm, S-106 91 Stockholm, Sweden

L. M. FIELD (173), AFRC Institute of Arable Crops Research, Rothamsted Experimental Station, Harpenden, Herts AL5 2JQ, UK

D. J. FINNEGAN (35), Institute of Cell and Molecular Biology, University of Edinburgh, Mayfield Road, Edinburgh EH9 3JR, UK

P. K. FLOOK (21), Vector Biology and Control Group, Liverpool School of Tropical Medicine, Pembroke Place, Liverpool L3 5QA, UK

G. J. GOLDSWORTHY (205), Department of Biology, Birkbeck College, University of London, Malet Street, London WC1E 7HX, UK

I. KAY (205), Department of Biology, Birkbeck College, University of London, Malet Street, London WC1E 7HX, UK

B. KHAMBAY (205), Department of Insecticides and Fungicides, AFRC Institute of Arable Crops Research, Rothamsted Experimental Station, Harpenden, Herts AL5 2JQ, UK

M. O'SHEA (187), Sussex Centre for Neuroscience, School of Biological Sciences, University of Sussex, Falmer, Brighton, East Sussex BN1 9QG, UK

M. G. PETITT (101), Department of Developmental Biology, Stanford University School of Medicine, Stanford, California, USA

J. A. PICKETT (141), AFRC Institute of Arable Crops Research, Rothamsted Experimental Station, Harpenden, Herts AL5 2JQ, UK

G. M. POPPY (163), AFRC Institute of Arable Crops Research, Rothamsted Experimental Station, Harpenden, Herts AL5 2JQ, UK

R. J. POST (21), Animal Taxonomy Section, Department of Entomology, University of Wageningen, PO Box 8031, 6700 EH Wageningen, Netherlands

R. C. RAYNE (187), Sussex Centre for Neuroscience, School of Biological Sciences, University of Sussex, Falmer, Brighton, East Sussex BN1 9QG, UK

L. M. RIDDIFORD (226), Department of Zoology, University of Washington, Seattle WA98195, USA

M. P. SCOTT (101), Department of Developmental Biology, Stanford University School of Medicine, Stanford, California, USA

L. J. WADHAMS (152), AFRC Institute of Arable Crops Research, Rothamsted Experimental Station, Harpenden, Herts AL5 2JQ, UK

C. WHEELER (205), Department of Biology, Birkbeck College, University of London, Malet Street, London WC1E 7HX, UK

M. S. WILLIAMSON (173), AFRC Institute of Arable Crops Research, Rothamsted Experimental Station, Harpenden, Herts AL5 2JQ, UK

M. D. WILSON (21), Department of Biological Sciences, University of Salford, Salford M5 4WT, UK

C. M. WOODCOCK (141), AFRC Institute of Arable Crops Research, Rothamsted Experimental Station, Harpenden, Herts AL5 2JQ, UK

Preface

This volume contains the scientific papers presented at the Royal Entomological Society's symposium on Insect Molecular Science held at Imperial College, London, 12–13 September 1991. The Symposium was preceded by a workshop on DNA transformation of non-drosophilids and the abstracts of the papers presented at the workshop are given at the end of this volume.

The aim of the meeting was to provide an outline and update of the use and application of molecular biology in relation to entomology. As is clear from the contents, a broad spectrum of topics was covered in the symposium in order to stimulate interest in the application of this technology to further our understanding of insect biology. The field of insect molecular science is of increasing interest to a broad spectrum of researchers and commercial concerns. This interest is primarily due to the enormous potential that the approach offers for pest control through transgenic technology, pesticide application and management, the design of new attractants and insecticidal compounds and novel pest surveillance strategies. We hope, therefore, that this volume will provide a useful overview of research in insect molecular science and be of interest to informed entomologists, those interested in entering the field of molecular entomology and particularly students starting their careers in this exciting field.

The financial support of the Wellcome Trust and ICI is gratefully acknowledged. We would also like to thank the Royal Society for providing bursaries to encourage young scientists to attend the Symposium.

JULIAN M. CRAMPTON AND PAUL EGGLESTON
Convenors and Editors
Liverpool 1992

Part I.
Potential

1. Potential Application of Molecular Biology in Entomology
   J. M. CRAMPTON
2. DNA Analysis in Relation to Insect Taxonomy, Evolution and Identification
   R. J. POST, P. K. FLOOK AND M. D. WILSON
3. Transposable Elements and their Biological Consequences in *Drosophila* and Other Insects
   D. J. FINNEGAN

1

Potential Application of Molecular Biology in Entomology

J. M. CRAMPTON

    I. Introduction 3
        A. Transgenic Technology in Insects 4
   II. The Requirements for Genetic Manipulation 5
        A. Genome Organization and Complexity 6
        B. Introducing DNA into Mosquito Embryos 7
        C. The Search for a Mosquito Transformation Vector
           System 11
  III. The Potential Application of Transgenic Technology in
      Insect Systems 14
        A. Transgenic Technology as an Analytical Tool 14
        B. Transgenic Technology in Insect Populations of
           Economic Significance 15
        C. Potential Target Genes for Manipulation 15
        D. Transgenic Insects in Natural Populations 17
  IV. Transgenic Insects: the Future 18
      References 18

I. INTRODUCTION

Over the last few years it has become increasingly clear that the use of molecular techniques can provide remarkably detailed descriptions and insights into many aspects of insect biology. The processes being examined using a spectrum of molecular techniques (used here in its broadest sense and not, as in many people's minds, confined to recombinant DNA techniques and the analysis of DNA and its products) cover the range from basic molecular biology of insect genomes and how they evolve through to the study of molecules involved in communication between insects in a population and between cells and tissues within the insect itself. It seems an appropriate time, therefore, to consider how this technology may be applied to economically important insects. Such insects include not only those of direct commercial value, such as the honey bee and silkworm moth, but also those that have a profound impact on agriculture and human health.

This chapter seeks to explore how molecular biology may be used in the future as a means to manipulate insect populations of economic significance. The use of transgenic technology for the genetic manipulation of insect genomes will be particularly highlighted. First, the major requirements for producing transgenic insects will be described together with the progress which has been made to date with particular reference to the work we have done on the development of transgenic technology in mosquitoes. Subsequently, the types of genetic manipulations of economically important insects which may be envisaged will be considered. Clearly, there are considerable practical and ethical problems associated with the release of genetically manipulated insects into natural populations. Some of the questions posed by the need to assess the biological consequences of such a release are therefore discussed, particularly in relation to highlighting the areas of research which may need to be emphasized in the future. Such research will be essential if we wish to make a fully informed appraisal of the net benefit against the potential risks of the approach.

A. Transgenic Technology in Insects

Recombinant DNA technology and transgenic techniques provide the means for the controlled genetic manipulation of vector genomes by the direct introduction of DNA into the insect germ line. Such manipulation of economically important insects therefore provides the opportunity to introduce and express foreign genes and/or disrupt existing gene functions so that the desirable modifications may be inherited by subsequent generations, removing the need for frequent mass releases. Two particular advantages of using transgenic technology over classical genetics for future manipulation are evident and should be emphasized. One is the potential to exploit genes and gene constructs across species barriers, and the other the ability to introduce particular, defined sequences without the genome disruption of a conventional cross.

In any consideration of transgenic technology and how it may be applied to economically important insects, the following factors should be considered: (1) the practical requirements for creating transgenic insects; (2) how best to apply the technology and thus, (3) what gene systems from insects or other organisms need to be defined in order to undertake the desired manipulations; (4) once transgenic insects incorporating the desired characteristics have been created, what further research needs to be carried out prior to their release into natural populations. Each of these factors is considered in turn below.

II. THE REQUIREMENTS FOR GENETIC MANIPULATION

Despite the dramatic advances made recently with respect to genome manipulation in *Drosophila melanogaster* there is an urgent need for a much greater understanding of the molecular biology of other insects. This must include an analysis of the complexity and organization of their genomes and an understanding of the distribution of coding and repetitive sequences. These details form an integral part of the design and interpretation of cloning and hybridization experiments. In addition, the exploitation of transgenic technology requires methods for the introduction of DNA both into living insects and into cultured insect cells. Ideally, this will involve a transformation vector which is capable of directing efficient and stable integration into the chromosomes of the recipient. The introduced DNA has not only to be expressed but should also carry a selectable marker for the identification of transformed individuals or cells. Finally, there will be a need to study alternative promoter and enhancer sequences so that the spatial and temporal expression of the introduced DNA may be controlled. Each of these aspects will be considered and illustrated by considering our own work on the molecular biology of medically important insects.

Although there is a trend towards integrated pest management, in which several approaches are used in combination to suppress insect populations, the main emphasis for insect vector control has been the elimination of breeding sites and the application of chemical agents. There are now considerable problems associated with the use of synthetic insecticides. The most important of these is the seemingly inevitable evolution of resistance and there are now populations which are multiply resistant to all four classes of insecticidal compound (organophosphates, organochlorines, carbamates and pyrethroids). Coupled with this is the high cost of developing and registering new insecticidal variants, increasing legislation over their use and growing environmental awareness over their toxic residues. These problems have stimulated us to investigate the potential that transgenic technology has for developing novel strategies for controlling vector populations or their ability to transmit pathogens. We have concentrated on the mosquito *Aedes aegypti*, the major urban vector of arboviral diseases, such as yellow fever, dengue and dengue haemorrhagic fever (Rudnick, 1967). These acute diseases affect millions of people and result in substantial mortality. *Aedes aegypti* is therefore an important disease-carrying vector and this, coupled with its suitability for laboratory research and its potential for studies in genetics (Knipling *et al.*, 1968), has made it the focus of our research effort.

A. Genome Organization and Complexity

If the genomes of insect disease vectors are to be manipulated in a controlled and directed fashion, it is important to determine the size of the genomes involved. Genome organization, that is the nature and dispersion pattern of repetitive sequences and how they are organized in relation to the coding sequences, is also important. This is because it will have a profound influence on the types of manipulation which can be envisaged and the approaches to be adopted in order to identify and to clone sequences of interest. Until recently, very little was known about the size and organization of mosquito genomes. However, the DNA of all higher eukaryotes is conveniently subdivided into three components, namely highly repetitive, moderately repetitive and unique or single copy sequences, and the organization of these components can be determined through experiments which involve denaturing the DNA and measuring the reassociation of complementary strands over time. Generally, highly repetitive sequences will reassociate most quickly since there are many copies whereas unique sequences will reassociate more slowly. The complexities of these components and of the total genome are determined by comparing the reassociation values with that of a single copy sequence of known complexity, namely *E. coli* DNA. Of equal importance, however, is the way in which the different components are distributed throughout the genome. Black and Rai (1988) suggest that the existence of two basic organization patterns throughout all higher eukaryotes is indicative of a set of rules governing the establishment and spread of repetitive elements. The first pattern is known as *short period interspersion* (SPI), in which 1–2-kb segments of single copy sequence alternate regularly with short (0.2–0.6-kb) or moderately long (1–4-kb) repetitive sequences. This pattern is characteristic of the majority of animal species. The second pattern, *long period interspersion* (LPI), is characterized by long (5–6-kb) repetitive sequences alternating with very long (>13-kb) uninterrupted stretches of unique sequence DNA. Clearly, the evolution of such organization is an interesting phenomenon in itself and the family Culicidae (to which the mosquitoes belong) may be of particular interest. According to Black and Rai (1988), this is the only family so far shown to contain species exhibiting both patterns of organization and it may, therefore, explain how the transition from one to the other occurs.

Black and Rai (1988) have analysed the genomic DNA from four species of mosquito, namely *Anopheles quadrimaculatus*, *Culex pipiens*, *Aedes albopictus* and *Aedes triseriatus*. More recently, Cockburn and Mitchell (1989) have shown that the genomes of anopheline mosquitoes are generally relatively small and exhibit the LPI pattern of repetitive sequences. This is in marked contrast to *Aedes aegypti* which has the more complex SPI pattern (Warren and Crampton, 1991). The *Aedes aegypti* genome is also relatively large, being five times the size of the *Drosophila* genome or, to place this in context, one third the size of the human genome. Thus, almost every clone isolated from *Aedes aegypti* genomic libraries will contain repetitive DNA and this may well mask the hybridization characteristics of sequences of interest (particularly single copy sequences) when using complex genomic probes. This is in direct contrast to organisms like *Drosophila* and *Anopheles*, which display the LPI pattern, where cloned sequences are more likely to consist entirely of either repetitive or unique DNA. In insect studies to date, *Anopheles quadrimaculatus*, *Drosophila melanogaster*, *Apis mellifera*, *Sarcophaga bullata* and *Chironomous tentans* have been shown to exhibit the LPI pattern. Conversely, *Aedes aegypti*, *Culex pipiens*, *Aedes triseriatus*, *Aedes albopictus*, *Lucilia cuprina*, *Musca domestica*, *Bombyx mori* and *Antherea pernyi* exhibit the SPI pattern (Black and Rai, 1988; Crampton et al., 1990b; T. Howells, personal communication). Clearly, there appears to be no simple relationship between genome size and organization.

B. Introducing DNA into Mosquito Embryos

1. Microinjection of mosquito embryos

Here, we describe the system which we have developed in our own laboratory for transformation of the mosquito *Aedes aegypti*. However, similar techniques have been used elsewhere both for *Anopheles* and for other *Aedes* species. Unlike that of *Drosophila*, the rigid, opaque endochorion of the mosquito embryo cannot be removed, and the embryos are extremely sensitive to desiccation. However, glass capillaries with tips of $100-300 \times 4-10 \mu m$ can be used to puncture the rigid endochorion without tearing it and deliver the DNA solution without damage to the embryo. The slightly viscous DNA solution cannot be expelled manually from such a fine needle and is therefore injected by means of a two-phase nitrogen supply. The lower pressure prevents backflow and the higher pressure delivers 160–800 pl of DNA solution (corresponding to 1–5% of the embryo volume) into the posterior pole of the embryo at the syncitial blastoderm stage, before cell partitioning occurs. This is where the pole cells, which are the germ line primordia, develop. Experience from *Drosophila* suggests that injection of DNA close to the site of pole cell formation is not critical to germ line incorporation, but the timing is clearly important if the DNA is to be taken up by the developing germ line cells. All of our injections are normally completed within 2 hours of oviposition. After injection, the embryos are covered with a water-saturated halocarbon oil, which permits the normal uptake of water until they are returned to standard insectary conditions. In this way transformation vector sequences have been introduced into the embryos, with survival rates comparable to those obtained with *Drosophila* (Spradling and Rubin, 1982).

2. Use of the P element transformation vector in mosquitoes

Germ line transformation of insect cells requires the establishment of appropriate techniques for the introduction of a eukaryotic DNA transformation vector. This should be capable of directing the efficient and stable integration of DNA sequences into the chromosomes of the recipient, as well as providing some means for the identification of transformants. The transposable genetic P element from *Drosophila melanogaster* has been used in a variety of insect species in an attempt to create transgenic individuals and appropriate transformation vector constructs are now used as a routine research tool in *Drosophila*. P elements (Engels, 1988) are a distinct class of transposable element discovered when a syndrome of correlated genetic traits (hybrid dysgenesis) occurred if male flies from P strains (normally from wild populations) were mated with M strain females (normally from laboratory populations). A series of abnormalities resulted, which included high rates of mutation, male recombination, chromosomal rearrangements, sterility, abnormal germ line development and meiotic drive. Genetic observations showed that a number of these mutations were unstable and had high reversion frequencies, all of which led to the hypothesis that P strains carry a family of transposable sequences called P elements. The abnormalities only arise in hybrids because of the absence of a repressive factor which, although not fully understood, is normally found in P strains. It is now thought that P factors have arisen in wild *D. melanogaster* only within the past 30 years. The P element family consists of intact elements (2.9 kb) and a heterogeneous group of smaller elements (0.5–2.5 kb) which are derived from the intact element by internal deletion. The intact element encodes a transposase enzyme which catalyses excision and transposition. The smaller elements lack this function and are incapable of transposing themselves. They can, however, transpose when supplied with functional transposase *in trans* by an intact element. All P elements carry perfect 31-bp inverted repeat sequences at their termini that are absolutely required for excision and transposition. This process results in an 8-bp duplication at the target site. P elements are capable of precise excision which leaves behind a single copy of the 8-bp sequence. More frequently they will excise imprecisely, leaving part of the element behind or removing flanking sequences.

The pUChsneo transformation vector carries only the inverted terminal repeats of the P element. Between these termini it has been engineered to carry a neomycin resistance gene driven by the *Drosophila* heat shock promoter HSP70, which serves as a marker for the identification of transformed individuals, and a site where foreign DNA can be ligated. The inverted repeats are joined by 500 bp from the *white* locus of *D. melanogaster* which may function as a hybridization probe to distinguish random integration events from precise P-mediated transposition (Miller et al., 1987). The latter would involve only those sequences within, and including, the inverted terminal repeats of the P element, but excluding the *white* locus DNA. In our experience, however, this diagnostic feature is limited, perhaps because of a low frequency of precise excisions. The site of integration of foreign DNA within the genome, rather than the precision of the mechanism, is likely to be of greater importance to normal gene expression.

Transposition of vector sequences and integration into the genome will only be mediated by the terminal repeats if they are supplied with a functional transposase. A helper plasmid, pUChsπ(Δ2-3), which carries the entire P element transposase coding region, is therefore introduced with the transformation vector. Although the helper carries most of the P element sequences, it is unable to transpose and integrate into the genome because of a specific deletion in one of its terminal repeats (Steller and Pirrotta, 1985). The sole function of the helper is therefore to supply transposase and, in *D. melanogaster*, further injections of helper have been used to switch transposition on in those generations following the initial transformation. Normally, P element transposition occurs only in the germ line of *D. melanogaster* because the intron between open reading frames 2 and 3 of the coding sequence is not removed in somatic tissue. However, transposase expression from constructs which have been modified *in vitro* to remove this intron does allow transposition in somatic tissue. There is now good evidence that similar processing problems prevent normal P element transposition in non-drosophilids (O'Brochta and Handler, 1988) although excision has been observed in mammalian and yeast cells (Rio et al., 1988). If there is a requirement for non-P-element-encoded proteins to achieve transposition in Drosophila, then the distribution of homologous genes in related genera may explain the divergent results obtained in other insects.

3. Creating transgenic mosquitoes

The basis of the experimental design to create transgenic mosquitoes is such that G_0 individuals which survive microinjection with the P element vector/helper DNA are mated to wild type and allowed to produce progeny. These G_1 individuals are the first which might be expected to express antibiotic resistance throughout all tissues and larvae are therefore subjected to selection with the neomycin derivative G418. The molecular nature of any transformation events is determined by DNA analysis using radioactively labelled transformation vector DNA to probe Southern blots of genomic DNA extracted from the putative transformants and their progeny (Morris et al., 1989; Miller et al., 1987). Intact vector P elements have been detected in 5–10% of adults that have developed from injected embryos (G_0), confirming that the introduced DNA is not immediately broken down by the mosquito. Furthermore, we have detected the chromosomal integration of vector DNA in several G_0 individuals. This probably reflects direct incorporation into a proportion of the somatic cell nuclei since it is only in the following generation (G_1) that we might expect a germ line integration event to have been transmitted to every nucleus. More promisingly, vector DNA has been identified in the chromosomes of the G_1 and G_2 progeny of injected embryos, suggesting that integration has occurred in the germ line of the mosquito and that this DNA shows normal Mendelian inheritance. Some of these events, however, appear to be unstable from one generation to the next and this phenomenon, together with the molecular basis of the transformation events, awaits further investigation.

As indicated above, chromosomal integration of the introduced P element DNA has been observed in both *Anopheles* and *Aedes* mosquitoes and the integration events appear, in some cases, to be heritable and clearly involve the germ line of the transgenic mosquitoes (Miller et al., 1987; McGrane et al., 1988; Morris et al., 1989). Although these events did not result from normal P element transposition, some functional role of the P sequences cannot be excluded. This is particularly true since similar experiments in *Lucilia cuprina* (T. Howells, personal communication) and *Ceratitis capitata* (M. Ashburner, personal communication) have failed to produce any integration of vector sequences. Research in other laboratories is now being directed towards the identification of these accessory *Drosophila* proteins and the cloning of the genes involved may facilitate high-efficiency P transposition in non-drosophilids. It is clear, however, from this work and from other experiments involving the transfection of the same DNA into cultured mosquito cells (Lycett *et al.*, 1989) that the P element system in its present form is not suitable for routine use in the mosquito. Thus, whilst the means are currently available for introducing DNA into both mosquito embryonic germ lines and cultured cells, a major stumbling block is the lack of an appropriate DNA transformation vector system for manipulating the mosquito genome.

C. The Search for a Mosquito Transformation Vector System

It is clear that the germ line integration events so far observed in mosquitoes do not involve normal P element transposition. The absence of this controlled mobility poses certain limitations, for example with respect to transposon tagging for functional cloning (see Section III.A). Research elsewhere is concentrating on the precise mechanism of P transposition, and attempts are being made to modify the P element system for more general use (O'Brochta, 1990). Such research may yet lead to the "universal vectors" originally envisaged. At the same time, there remains the possibility that P elements may never function as efficient transposition-mediated transformation vectors in non-drosophilids. To this end we are investigating a number of alternative transformation vector systems which may improve the overall efficiency of the system. These include fully processed cDNA copies of both the *Ac* and *Spm* elements from *Zea mays* which have been shown to transpose actively in a number of evolutionarily disparate organisms and may prove to act autonomously in mosquitoes (Kunze and Starlinger, 1989). By using cDNA clones it is hoped to overcome the problems of transposase processing seen with existing constructs. As an alternative to the above, we are actively searching for endogenous transposable elements which may yet prove to be the most suitable transformation vectors.

1. Transposable genetic elements in the mosquito genome

The isolation of endogenous transposable genetic elements may ultimately prove central to the development of efficient transformation and transposon tagging systems in mosquitoes. A number of approaches have been taken to identify such mobile elements and one of these was to analyse specific gene systems, such as the ribosomal DNA of mosquitoes, in an attempt to isolate variants of these genes which may have arisen from the insertion of transposons. No such insertions have, as yet, been identified in *Aedes aegypti* DNA (Gale and Crampton, 1989) but insertion events have been detected in the rDNA of *Anopheles gambiae* and these elements are being fully defined (Paskewitz and Collins, 1989). The elements appear to resemble a particular class of mobile element known as non-viral retroposons. It is unlikely, however, that these elements will prove useful as transformation vectors because of the ill-defined nature of their mode of transposition. We have adopted an alternative strategy to identify directly a specific class of mobile elements, known as retrotransposons, in the mosquito DNA. The approach relies on utilizing the characteristic biochemical and structural properties of these elements. This work has led to the successful isolation of several retrotransposon-like elements from the *Aedes aegypti* genome (Crampton *et al.*, 1990a,b). More recently, we have used the polymerase chain reaction (PCR) to develop a particularly rapid methodology for identifying endogenous retrotransposon-like elements in mosquito DNA (Warren and Crampton, 1992). Once such elements have been isolated and fully characterized, and their ability to transpose autonomously established, they may be engineered to form the core of a transformation vector system.

2. Markers for the selection/identification of transformed individuals

A very important feature to be incorporated into a DNA transformation vector for use in mosquitoes is an improved selectable marker system for the identification of transformed individuals. A number of research laboratories have now reported problems with the existing neomycin resistance technique. In our experience, the main problem is the low activity of the neomycin phosphotransferase produced by the Tn5 *neo* gene in the transformation vector. The system is also less than satisfactory in that mosquitoes exhibit a spectrum of sensitivities to G418 and the survival of transformed mosquitoes relies on high levels of expression of the resistance gene. In addition, *Aedes aegypti* cells appear to have a tendency to retain intact vector plasmids which do not integrate into the recipient chromosomes but which transiently express antibiotic resistance. Alternative selectable markers, including the Tn903 neomycin resistance gene, which has been reported to show much higher phosphotransferase levels in yeast cells (Langhinrichs *et al.*, 1989), and hygromycin resistance, together with alternative promoters to drive their expression, are now being investigated. A considerable body of work now suggests that the *Drosophila* heat shock promoter currently incorporated in the P element transformation vector constructs is not entirely satisfactory for driving the expression of selectable marker gene systems in mosquitoes (Lycett *et al.*, 1992). Ultimately, therefore, cloned phenotypic markers available for use in transgenic mosquitoes, such as the eye colour mutations used routinely in *Drosophila*, would appear to offer the most efficient selection system for incorporation into a mosquito DNA transformation vector. However, although eye colour mutations exist in both *Aedes* and *Anopheles*, such markers have yet to be isolated and characterized at the molecular level.

3. Stage- and tissue-specific promoter/enhancer sequences

At some stage it will be desirable to express defined genes in mosquitoes in a tissue- or stage-specific fashion. For this to be envisaged, mosquito stage- and tissue-specific promoters have to be defined. None are, as yet, available but attempts to characterize the DNA sequences responsible for expressing certain genes in mosquitoes are underway by identifying genes which are expressed in a tissue-specific fashion and then defining the upstream, putative tissue-specific promoter sequence. One example of this approach has been the identification of an *Aedes aegypti* sequence which is only expressed in the female salivary gland (James *et al.*, 1989). The expectation is, therefore, that this will allow the definition of a salivary-gland-specific promoter sequence which may eventually allow the controlled expression of an introduced gene sequence in this tissue. However, it will be necessary to develop methods for establishing the functionality of putative promoter sequences. To this end, we have begun to develop the methodologies for transfecting mosquito cells in culture. The simple, controlled environment of cultured cells allows one to follow the expression of cloned genes, and so delineate promoter and enhancer sequences. Genetically manipulated cell cultures are also able to overproduce specific proteins, which facilitates their isolation and purification. Cultured mosquito cells have been used to examine different transfection techniques and vectors and to help establish a suitable system for germ line transformation of *Aedes aegypti* (Lycett, 1990). Initially, experiments involved introduction of the P element vector and helper constructs into several cell lines by a variety of techniques devised to generate transient cell membrane pores, including calcium phosphate precipitation (Wigler *et al.*, 1977), dextran sulphate (Lopata *et al.*, 1984), polybrene (Durbin and Fallon, 1985), electroporation (Chu *et al.*, 1987) and lipofection (Felgner *et al.*, 1987). Much of this work has concentrated on the immortal Mos20 fibroblast cell line which was derived from minced, trypsinized, neonate larvae of the *Aedes aegypti* London strain in 1969. Polybrene- and electroporation-mediated transfection have proved to be most successful for these cells, producing approximately 30 and 4000 transformants per 10^6 cells, respectively. Subsequently, constructs incorporating the chloramphenicol acetyl transferase (CAT) reporter gene system have been utilized to optimize expression of the CAT gene under the control of the *Drosophila* heat shock promoter, HSP70, in the Mos20 mosquito cultured cells (Lycett et al., 1992). This type of approach, in addition to experiments utilizing a range of reporter genes and microinjection of embryos, will eventually allow fully functional constitutive stage- or tissue-specific mosquito promoters to be defined.

III. THE POTENTIAL APPLICATION OF TRANSGENIC TECHNOLOGY IN INSECT SYSTEMS

Once the systems necessary to create transgenic insects have been developed, how may this technology be applied? Two aspects will be discussed in order to illustrate the potential of the technology. The first deals with the use of the technique for analytical purposes and the second with applying transgenic technology to economically significant mosquito populations.

A. Transgenic Technology as an Analytical Tool

The introduction and insertion of a mobile genetic element at or near a particular locus can cause that allele to mutate, producing a structural or developmental effect. In *Drosophila* in particular, transposable genetic elements (TGEs) have been used as mutagens in order to clone genes or gene clusters of interest via transposon tagging (Bingham et al., 1981). In essence, the TGE is introduced into the germ line of the insect by microinjection of the embryo, and the progeny scored for mutants in the phenotype of interest. Subsequently, cloned TGE probes are used for *in situ* hybridization to chromosomes of mutant and wild type individuals. This identifies a TGE "newly" integrated at or near the genetic locus of interest. DNA clones are then retrieved from a genomic library prepared from the mutant stock using the TGE DNA as a probe. DNA sequences adjacent to the TGE in such clones represent the gene for the locus of interest. This approach is an extremely powerful application of the technology, as it allows the cloning of genes purely on the basis of their function.

B. Transgenic Technology in Insect Populations of Economic Significance

In the long term, perhaps the most exciting applications of transgenic technology will be in insects of economic significance. Clearly, the type of manipulation to be envisaged will depend on the target insect and the scale of its economic impact. For example, insect pest populations have perhaps the most significance in terms of their commercial importance and here the aim may be either to suppress or control specific pest populations by transgenic means. Also in the agricultural sphere, there are an increasing number of beneficial insect populations. Such insects may be predators of pest insects and so form an important means of biological control. In this instance, the beneficial insects may be manipulated in such a way as to confer resistance to insecticidal compounds which may be applied as part of an integrated pest management programme.

Insects are also important vectors of disease to both man and agricultural animals. In this way, vector populations may have a profound impact on the economy of regions which, in many cases and particularly in the tropics, are the most fertile and potentially productive areas. Transgenic technology may eventually have a role to play in controlling vector-borne disease by providing the means to suppress vector populations by rendering them vulnerable to subsequent control measures, such as insecticide susceptibility, temperature sensitivity or ability to survive diapause. A second possibility and, perhaps, a more exciting approach, would be to alter the ability of the insect to transmit the disease. Clearly, such possibilities are for the future but it is quite feasible to consider genetic manipulation of insect populations of direct commercial value, such as the honey bee or silkworm moth. Here, transgenic technology may be employed to confer a number of beneficial characteristics to these insects to create novel and highly productive strains. For example, the insect may be manipulated to increase the yield of the product by increasing the growth rate or by enhancing the resistance of the insect to infection, temperature shock or other detrimental factors.

C. Potential Target Genes for Manipulation

Having discussed the types of manipulation of insect genomes which may be beneficial in economic or health terms, it is worth considering the types of gene systems which may be potential targets for manipulation to achieve these aims. In this respect, there are a number of obvious targets for manipulation, including the genes involved in the insect immune system, development control genes and insecticide resistance genes. As discussed in subsequent chapters, genes influencing all of these factors have now been characterized at the molecular level for a number of different insects and it is now feasible to consider manipulating them in the germ line of these insects. In addition, a number of genes are of particular interest because they are directly implicated in the ability of insects to transmit disease-causing organisms. Examples include the filarial susceptibility (f^m) and *Plasmodium* susceptibility (*pls*) loci of the mosquito *Aedes aegypti*. The f^m locus is genetically well defined and there are good data on its linkage relationships. Refractoriness to infection is due to a partially sex-linked, dominant gene (Macdonald and Ramachandran, 1965). There is marked variation in the susceptibility of this mosquito to different filarial worms, although all of the alleles concerned map at about the same place on the sex chromosome. Also of particular interest is a strain of *Anopheles gambiae* which has been selected for refractoriness to the malaria parasite and characterized genetically (Collins *et al.*, 1986). Attempts are currently underway to clone these genes but it is difficult to undertake such a cloning exercise in the absence of any knowledge of the gene product. Clearly, the use of transgenic technology through transposon tagging will assist in the characterization of refractory genes and their products.

An important genotypic characteristic not possessed by the majority of genes encoding refractoriness is that any such gene introduced into the insect would have to be capable of altering the phenotype through the expression of a single gene copy. Unfortunately, at present, there is no gene or gene product defined at the molecular level which is known directly to affect phenotype in relation to pathogen development in, or transmission by, any insect. However, in the mosquito system a number of molecules are known to affect the transmission of malaria by anophelines. Foremost among these are the so-called transmission-blocking vaccines, which can achieve a total transmission blockage (Winger *et al.*, 1987). These vaccines attack antigens present on the gametes and ookinetes of the malaria parasite and antibodies which recognize these antigens are able to block the development of the parasite in the mosquito midgut. It may therefore be feasible to create a transgenic mosquito incorporating an antibody gene which will be expressed in the insect midgut in response to a blood meal and which blocks the transmission of malaria. If successful, transgenic mosquitoes expressing antimalarial antibodies may represent a potential strategy for controlling malaria and may establish a precedent for a wide range of new anti-disease strategies.

D. Transgenic Insects in Natural Populations

Once transgenic insects with the necessary characteristics have been created, there remains the question, what next? Clearly, if the manipulated insects are themselves to be cultivated for production it may be possible directly to apply novel strains created by transgenic means. However, where this is not the case, it is necessary to consider the problems likely to be faced in applying the technology in experimental and natural populations. It may well be that such a situation would disrupt the normal adaptive process and therefore be opposed by natural selection. If this is so then some form of drive mechanism may be needed to force the desired gene through the population. This is not an alien concept to those who have worked on the genetic control of insect populations. However, the testing of such mechanisms has been limited since, in reality, they have awaited the advent of recombinant DNA technology to provide the necessary raw material.

Two types of drive mechanism have been suggested. One is meiotic drive, where a given chromosome is transmitted to more than the expected 50% of offspring. Any desirable genes linked to the driven chromosome would eventually approach fixation even with the release of relatively few individuals. There is experimental evidence to support the use of meiotic drive in *Aedes aegypti*. This mechanism, driven by the M^D locus, has been used to force the marker gene *re* (red eye) into a laboratory cage population (Wood *et al.*, 1977). Interestingly, meiotic drive also occurs during hybrid dysgenesis and it might, therefore, also be possible to exploit this phenomenon by using either the P element itself, or a mobile element with properties similar to P, as an efficient mechanism to drive a specific gene construct through an insect population. The second type of drive mechanism is the exploitation of genetic traits that reduce heterozygote fitness (Curtis and Graves, 1988). For example, the gene to be driven could be introduced into a translocation chromosome such that viable and fertile homozygotes were formed, whereas heterozygotes would display reduced fertility or viability. In this way, translocations, pericentric inversions, inter-racial hybrid sterility, cytoplasmic incompatibility and compound chromosomes all have potential since, in each case, hybrids have reduced fitness. Such mechanisms require larger release numbers since there is no exponential increase in the frequency of the driven chromosome as with meiotic drive. However, fixation of desirable genes would occur more quickly than with meiotic drive because of the reduced fitness of heterozygous combinations. Efficiency could be improved by providing the released individuals with some form of temporary advantage. For example, insecticide resistance could be incorporated into the genome and then insecticide applied (Whitten, 1970). Ideally, the insecticide resistance gene would be fused to the desirable gene and introduced as a unit to prevent disruption of useful combinations by meiotic recombination. Certainly in the case of vector populations, the most useful end result of such programmes would be the progressive replacement rather than the eradication of disease-transmitting populations since an emptied ecological niche might be colonized rapidly by migration of wild types.

IV. TRANSGENIC INSECTS: THE FUTURE

Eventually, embryo transformation will provide the raw material to test the proposed drive mechanisms in laboratory and natural populations. The questions posed by considering the release of transgenic insects emphasize the need to assess the biological consequences of such a release. It is, however, difficult to gauge the possible hazards of such a release in the absence of experimental evidence and these ethical and safety considerations need to be faced at an early stage. In order to undertake an informed appraisal where the possible net benefits may be balanced against the potential hazards, considerable effort will have to be devoted to utilizing caged populations and the controlled release of molecularly tagged individuals together with mathematical modelling of these populations. There is clearly some way to go before any release of transgenic insects can be considered. The power of the technology is, however, so enormous that it must be explored and there is every indication that over the next few years the potential of transgenic technology in insects will be fully exploited.

Acknowledgements

I would like to thank Alison Morris, Ann Warren, Gareth Lycett, Iain Comley, Teresa Knapp and Paul Eggleston, all of whom have been involved in the work described in this chapter. I would also like to thank the Wellcome Trust, Medical Research Council, Wolfson Foundation and Liverpool University for providing financial support. The author is a Wellcome Trust Senior Research Fellow in Basic Biomedical Sciences.

REFERENCES

Bingham, P. M., Levis, R. and Rubin, G. M. (1981). *Cell* **25**, 693–704.
Black, W. C. and Rai, K. S. (1988). *Genet. Res. (Camb.)* **51**, 185–195.

Chu, G., Hayakawa, H. and Berg, P. (1987). *Nucleic Acids Res.*, **15**, 1311–1326.
Cockburn, A. F. and Mitchell, S. F. (1989). *Arch. Insect Biochem. Physiol.* **10**, 105–113.
Collins, A. F., Sakai, R. K., Vernick, K. D., Paskewitz, S., Seeley, D. C., Miller, L. H., Collins, W. E., Campbell, C. C. and Gwadz, R. W. (1986). *Science* **236**, 607–610.
Crampton, J. M., Morris, A. C., Lycett, G. J., Warren, A. M. and Eggleston, P. (1990a). *Parasitology Today* **6**, 31–36.
Crampton, J. M., Morris, A. C., Lycett, G. J., Warren, A. M. and Eggleston, P. (1990b). *In* "Molecular Insect Science" (H. H. Hagedorn, J. G. Hildebrand, M. G. Kidwell and J. H. Law, eds), pp. 1–11. Plenum Press, New York.
Curtis, C. F. and Graves, P. M. (1988). *J. Trop. Med. Hyg.* **91**, 43–48.
Durbin, J. E. and Fallon, A. M. (1985). *Gene* **36**, 173–178.
Engels, W. R. (1988) *In* "Mobile DNA" (D. Berg and M. Howe, eds), pp. 437–484. ASM Publications, Washington, D.C.
Felgner, P. L., Gadek, T. R., Holm, M., Roman, R., Chan, H. W., Wenz, M., Northrop, J. P., Ringhold, G. M. and Danielson, M. (1987). *Proc. Natl. Acad. Sci. USA* **84**, 7413–7417.
Gale, K. and Crampton, J. M. (1989). *Eur. J. Biochem.* **185**, 311–317.
James, A. A., Blackmer, K. and Racioppi, J. V. (1989). *Gene* **75**, 73–83.
Knipling, E. F., Laven, H., Craig, G. B., Pal, R., Kitzmiller, B., Smith, C. N. and Brown, A. W. A. (1968). *Bull. WHO* **38**, 421–438.
Kunze, R. and Starlinger, P. (1989). *EMBO J.* **8** (11), 3177–3185.
Langhinrichs, C., Berndorff, D., Seefeldt, C. and Stahl, U. (1989). *Appl. Microbiol. Biotechnol.* **30** (4), 388–394.
Lopata, M. A., Cleveland, D. W. and Sollner-Webb, B. (1984). *Nucleic Acids Res.* **12**, 5707–5717.
Lycett, G. J. (1990). *Insect Molecular Genetics Newsletter* **4**, 1–3. Royal Entomological Society of London Publication.
Lycett, G. J., Eggleston, P. and Crampton, J. M. (1989). *Heredity* **63**, 277.
Lycett, G. J., Eggleston, P. and Crampton, J. M. (1992). *Trans. R. Soc. Trop. Med. Hyg.* **86**, 344.
Macdonald, W. W. and Ramachandran, C. P. (1965). *Ann. Trop. Med. Parasitol.* **59**, 64–73.
McGrane, V., Carlson, J. O., Miller, B. R. and Beatty, B. J. (1988). *Am. J. Trop. Med. Hyg.* **39**, 502–510.
Miller, L. H., Sakai, R. K., Romans, P., Gwadz, R. W., Kantoff, P. and Caon, H. G. (1987). *Science* **237**, 779–781.
Morris, A. C., Eggleston, P. and Crampton, J. M. (1989). *Med. Vet. Entomol.* **3**, 1–7.
O'Brochta, D. A. (1990). *Bull. Ent. Res.* **80**, 241–244.
O'Brochta, D. A. and Handler, A. M. (1988). *Proc. Natl. Acad. Sci. USA* **85**, 6052–6056.
Paskewitz, S. M. and Collins, F. H. (1989). *Nucleic Acids Res.* **17**, 8125–8133.
Rio, D. C., Barnes, G., Laski, F. A., Rine, J. and Rubin, G. M. (1988). *J. Mol. Biol.* **200**, 411–415.
Rudnick, A. (1967). *Bull. WHO* **36**, 528–532.
Spradling, A. C. and Rubin, G. M. (1982). *Science* **218**, 341–347.
Steller, H. and Pirrotta, V. (1985). *EMBO J.* **4**, 167–171.
Warren, A. M. and Crampton, J. M. (1991) *Genet. Res. (Camb.)* **58**, 225–232.
Warren, A. M. and Crampton, J. M. (1992). *Trans. R. Soc. Trop. Med. Hyg.* **86**, 347.
Whitten, M. J. (1970). International Atomic Energy Agency Symposium, "The Sterility Principle for Insect Control or Eradication", IAEA, Athens, pp. 399–410.
Wigler, M., Siverstein, S., Lee, L. S., Pellicer, A., Cheng, Y. and Axel, R. (1977). *Cell* **11**, 223–232.

Winger, L., Smith, J. E., Nicholas, J., Carter, E. H., Tirawanchai, N. and Sinden, R. E. (1987). *Parasite Immunol.* **10**, 193–207.
Wood, R. J., Cook, L. M., Hamilton, A. and Whitelaw, A. (1977). *J. Med. Entomol.* **14**, 461–464.

2

DNA Analysis in Relation to Insect Taxonomy, Evolution and Identification

R. J. POST, P. K. FLOOK AND M. D. WILSON

I. Introduction	21
II. The Major Features of Genome Organization and Evolution	22
III. Ribosomal RNA Genes in Evolution and Systematics	24
IV. Uncharacterized Repetitive DNA Sequences for Species Identification	30
References	34

I. INTRODUCTION

Systematics is the study of phylogeny and taxonomy, which itself divides into descriptive taxonomy and identification. As such, systematics attempts to use a variety of variable traits, which are assumed to have a strong genetic base, to assess the relatedness between populations. DNA analysis is highly suitable for such studies because it is the most direct analysis of the genetic material that is possible, and is generally unlikely to show life-stage or environmentally mediated variation. Furthermore the structure of DNA lends itself to this sort of analysis, being a simple linear sequence of only four different sorts of nucleotide bases. Each nucleotide position is a potential data-point, and so even quite short DNA sequences present a potentially large database.

To use DNA sequence analysis for systematics it is essential to know something about how these sequences vary, and how they are organized. A consequent understanding of genome evolution will help in choosing sequences suitable for systematics, and indicate suitable techniques for analysis.

II. THE MAJOR FEATURES OF GENOME ORGANIZATION AND EVOLUTION

The genome of all higher eukaryotes divides into single copy DNA and repetitive DNA. The single copy DNA (each sequence present only once per haploid genome) includes most of the protein-coding genes, but also non-coding intergenic spacer regions. The repetitive DNA similarly consists of coding sequences, for gene families such as the ribosomal RNA genes, and non-coding sequences such as satellite DNA. The relative proportions of these different classes of DNA varies greatly between species, but in *Drosophila melanogaster*, for example, about 30% and 24% of the genome is single copy genes and intergenic spacers respectively. About 6% consists of repetitive gene families, and the remaining 40% a heterogeneous collection of mostly satellite DNA and transposable elements (John and Miklos, 1988).

Both single copy and repetitive DNA can vary in two basic ways. Firstly, the chromosomal location of a sequence can change by a variety of translocation mechanisms. Secondly, and more importantly for systematics, the exact base sequence can change (point mutation) by small deletions and insertions, but most commonly by substitution, the change of one nucleotide for another. If two populations are isolated from one another they will slowly accumulate independent substitutions and their DNA sequences will diverge. The limit to this divergence will be when the sequences are effectively random with respect to each other, although they will still show 25% similarity because there are only four different sorts of nucleotide (Fig. 1).

As well as evolutionary divergence by substitution, repetitive DNA can also increase or decrease in copy number by a variety of amplification and deletion mechanisms. The interaction of these processes within families of repetitive DNA results in the phenomenon of concerted evolution (Dover and Tautz, 1986). Consider a newly amplified DNA sequence, which forms an array of identical copies. With time these will diverge from each other by substitution to become a heterogeneous array. However, there are continuous turnover processes within the array, such that some copies are amplified whilst others are deleted. The molecular nature of these processes is varied and includes unequal exchange, gene conversion, replicative transposition and others, which are collectively known as molecular drive (Fig. 2). The result of molecular drive is that if we have two isolated populations, with time all the copies of a repetitive DNA sequence will diverge with respect to the other population, but they will diverge more or less in concert. For example, the 360 satellite located on the X chromosome of the *Drosophila melanogaster* species group shows

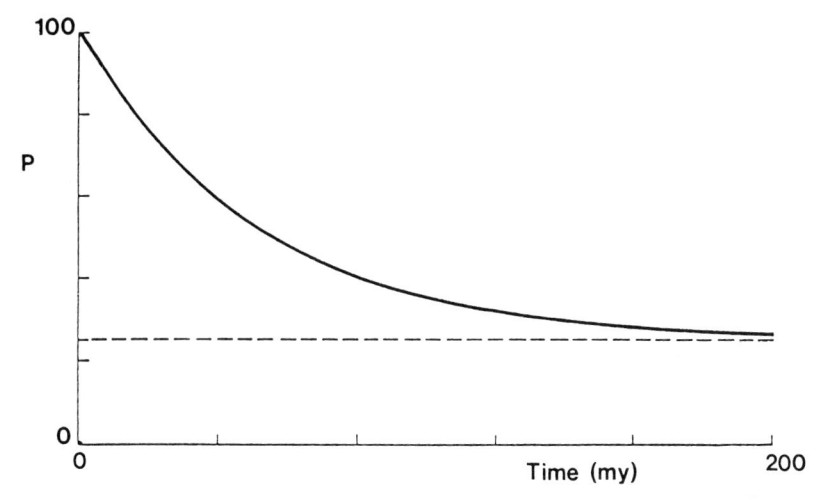

Fig. 1. The evolutionary divergence of a DNA sequence through time with 5×10^{-9} substitutions per site per year, where P = percentage similarity.

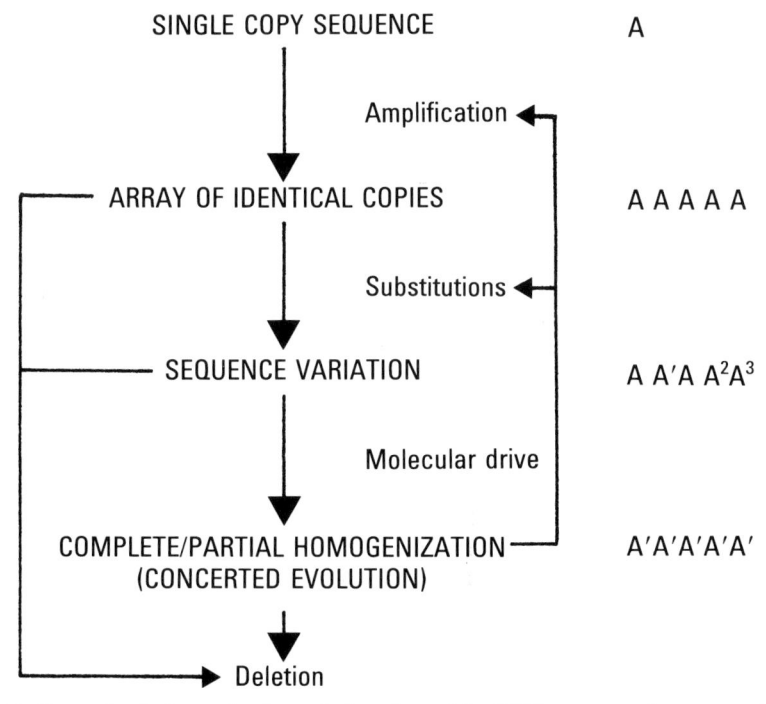

Fig. 2. Generalized scheme for the evolution of repetitive DNA sequences by concerted evolution.

mean variation between copies within species at 3% of nucleotide positions, but between-species variation of 32% (Dover and Tautz, 1986). The limit to this divergence will, of course, be 75% difference.

Given the same mutation rate and base composition, the rate of divergence of the consensus sequence of a repetitive DNA family with unbiased molecular drive will be the same as the rate of divergence of single copy DNA. However, in practice when we consider particular sequences, the rate will be affected dramatically by differences in mutation rate or natural selection, for example. However, one of the perceived advantages of repetitive DNA for systematics is that, being repetitive, there is proportionally more of it, and hence it is easier to detect in small organisms, or parts of organisms.

When choosing a particular repetitive DNA sequence to work with there have been two different approaches generally adopted: firstly, to examine sequences which are already known, from studies of other organisms, to vary in a manner suitable for systematics, and secondly, randomly to search the genome for uncharacterized repetitive sequences which show systematic variation. The first of these approaches is well illustrated by studies of the ribosomal RNA coding genes.

III. RIBOSOMAL RNA GENES IN EVOLUTION AND SYSTEMATICS

In eukaryotes the genes encoding the 18S (small subunit) and 28S (large subunit) ribosomal RNA (rRNA) are clustered as tandem repeats in the nucleolus-organizing regions of the chromosomes (Fig. 3). Different parts of this repeating unit evolve at different rates. Thus, for example, a comparison between the sequence of the complete 28S gene of *Drosophila melanogaster* (Tautz *et al.*, 1988) with the nematode *Caenorhabditis elegans* (Ellis *et al.*, 1986) indicates that some domains are very similar and others are very different. This is simply illustrated in the dot-matrix comparison (Fig. 4). The solid diagonal shows the very slowly evolving domains, known as core segments, which show little divergence between *Drosophila* and *Caenorhabditis*. The gaps in the diagonal are sequences which have diverged more quickly, and are called expansion segments. These differences in sequence conservation are undoubtedly related to structural and functional constraints. Conserved regions are often important for maintaining the characteristic secondary and tertiary structure of rRNA molecules (Simon, 1991).

The slowly evolving core segments are even homologous to parts of the prokaryote large subunit rRNA, and comparisons have allowed attempts to reconstruct very deep phylogenies, to the origin of the kingdoms. An

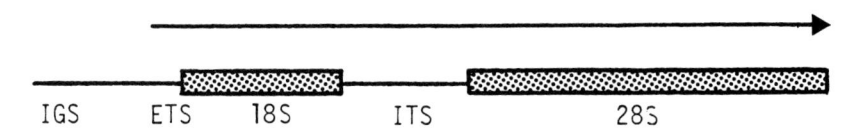

2. DNA Analysis in Relation to Insect Taxonomy, Evolution and Identification

```
         IGS      ETS    18S         ITS              28S
```

Fig. 3. Simplified map of the ribosomal DNA repeat of eukaryotes. IGS, intergenic spacer; ETS, external transcribed spacer; 18S, small subunit rRNA gene; ITS, internal transcribed spacer; 28S, large subunit rRNA gene. Arrow indicates transcription.

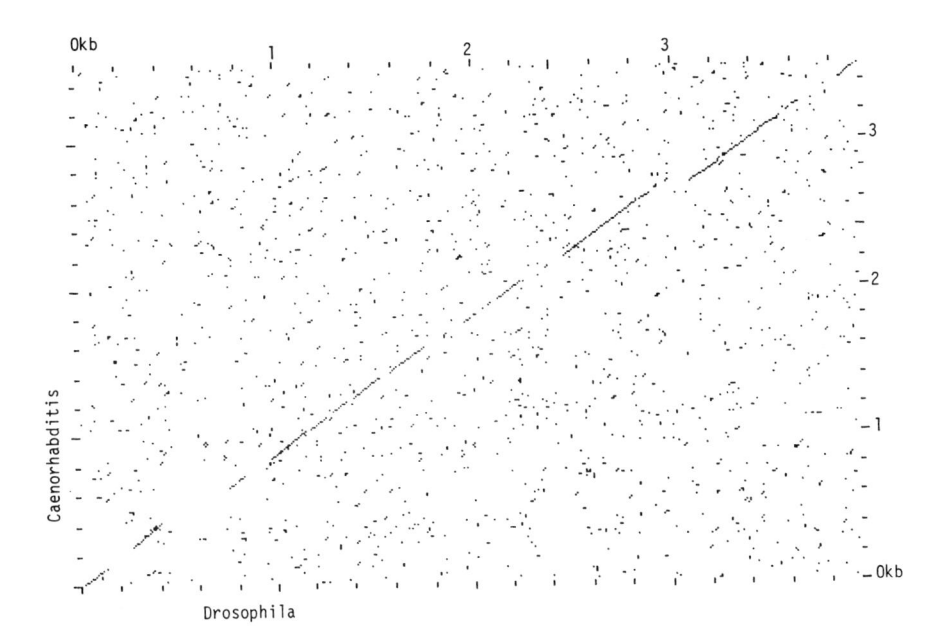

Fig. 4. Dot-matrix comparison between the complete 28S rRNA gene of *Drosophila melanogaster* and *Caenorhabditis elegans*. A moving window of 21 nucleotides along the *Drosophila* sequence is compared with the entire *Caenorhabditis* sequence, and any similarity of 13+ out of 21 is indicated by plotting a dot.

example from Cedergren *et al.* (1988) based on the small subunit rRNA is shown in Fig. 5. This also shows the origin of the mitochondrial DNA from the Eubacteria.

To study the phylogenetic relationship at the family level it is necessary to study the more fastly evolving domains. Vossbrink and Friedman (1989) studied a 324-bp sequence from the second expansion segment of

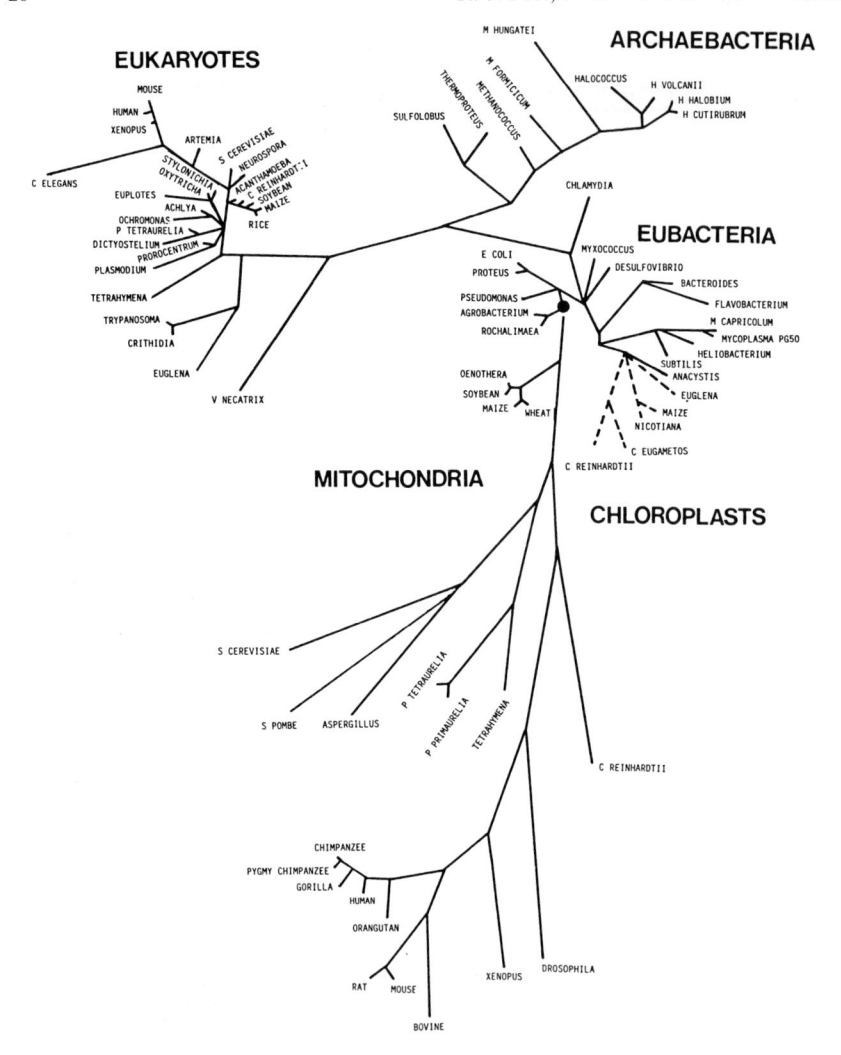

Fig. 5. Unrooted tree from sequence comparison of the small subunit rDNA of eukaryotes, prokaryotes and plastids (redrawn from Cedergren *et al.*, 1988).

the 28S gene to reconstruct a phylogeny of some calyptrate Diptera. The consensus tree (Fig. 6) throws light on a number of long-standing taxonomic problems. For example, adult bot flies (Gasterophilidae) are to some extent morphologically intermediate between the Calyptrata and Acalyptrata (Rohdendorf, 1974), but other authors have placed them with the oestroids (Hennig, 1973) or the muscoids (Borrer *et al.*, 1981).

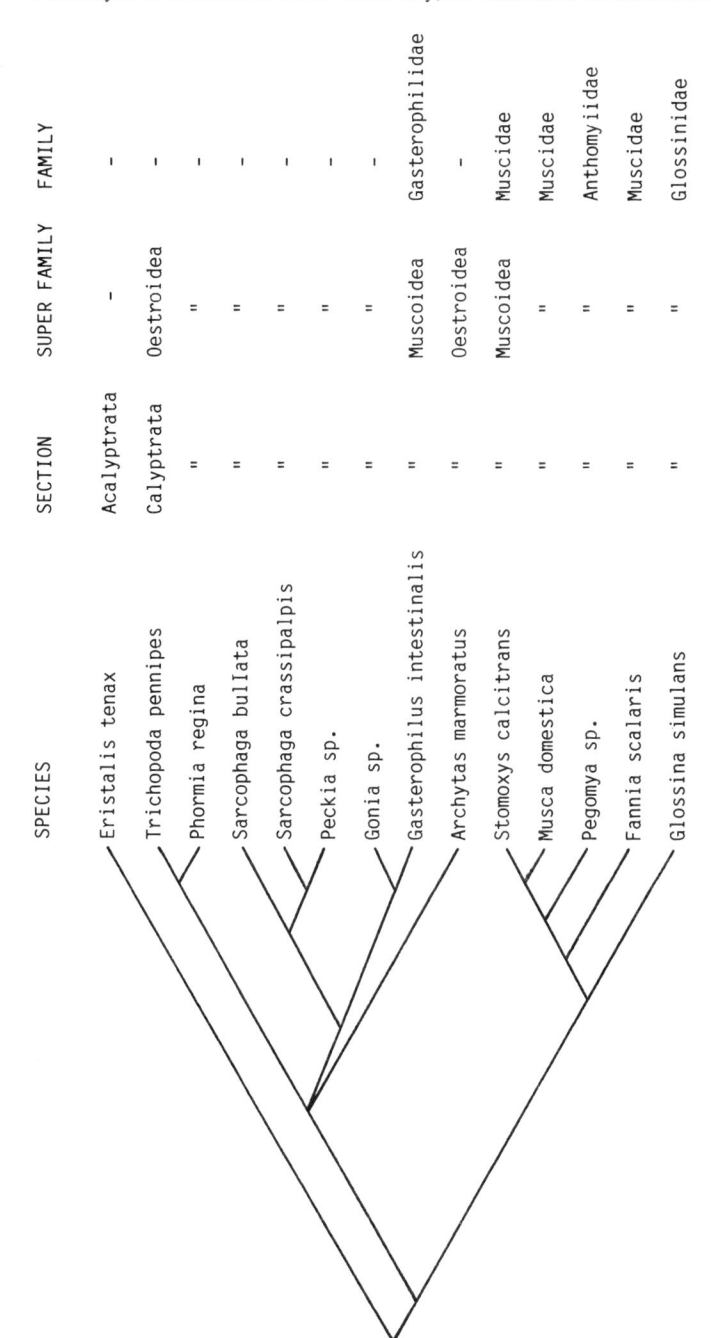

2. DNA Analysis in Relation to Insect Taxonomy, Evolution and Identification

SPECIES	SECTION	SUPER FAMILY	FAMILY
Eristalis tenax	Acalyptrata	–	–
Trichopoda pennipes	Calyptrata	Oestroidea	–
Phormia regina	"	"	–
Sarcophaga bullata	"	"	–
Sarcophaga crassipalpis	"	"	–
Peckia sp.	"	"	–
Gonia sp.	"	"	–
Gasterophilus intestinalis	"	Muscoidea	Gasterophilidae
Archytas marmoratus	"	Oestroidea	–
Stomoxys calcitrans	"	Muscoidea	Muscidae
Musca domestica	"	"	Muscidae
Pegomya sp.	"	"	Anthomyiidae
Fannia scalaris	"	"	Muscidae
Glossina simulans	"	"	Glossinidae

Fig. 6. Consensus tree from sequence comparison of an expansion segment of the 28S rDNA from the calyptrate diptera (redrawn from Vossbrink and Friedman, 1989).

The phylogeny clearly supports an association with the oestroids. Similarly, the latrine fly (*Fannia*) has been variously placed in the Muscidae (Hennig, 1958), Anthomyiidae (Crampton, 1944), or its own family the Fanniidae (Roback, 1951). The phylogeny supports the latter position.

As a general rule mitochondrial DNA evolves faster than genomic DNA (Simon, 1991); hence the markedly longer branch lengths in Fig. 5. Working at the generic taxonomic level Xiong and Kocher (1991) chose the large subunit (16S) rDNA from the mitochondria to attempt a phylogenetic reconstruction of the Simuliidae (Fig. 7). Members of the same genus cluster together, as expected, and the close relationship between *Stegopterna* and *Cnephia* also reflects current taxonomic opinion. However, there are currently two different suprageneric classifications of the Simuliidae (Crosskey, 1985), which are indicated in Fig. 7. The reconstruction is consistent with either at the level of the tribe, but does not support Rubtsov's designation of subfamilies.

The rDNA can also be used for taxonomy at the species level. The most variable part of the rDNA repeat is the intergenic spacer (Fig. 3), and species-specific variation can be used for identification of sibling species of the *Anopheles gambiae* (Diptera) complex using Southern blot (Collins *et al.*, 1988) or polymerase chain reaction (Paskewitz and Collins, 1990).

We have been using polymerase chain reaction (PCR) to study the structure of the adjacent tenth and eleventh expansion segments of the 28S gene of the *Simulium damnosum* complex of sibling species, which transmit human onchocerciasis in West Africa. Oligonucleotide primers were chosen from the conserved core segments on either side, from the published *Drosophila melanogaster* sequence (Tautz *et al.*, 1988). These oligonucleotides amplify a 769-bp fragment from *Drosophila*, as expected, which can be cut with restriction endonucleases to give fragments of sizes also expected from the published sequence. However, using *Simulium* DNA two fragments are amplified which co-migrate at about 830 bp in ethidium bromide agarose gel electrophoresis. These fragments have different restriction enzyme recognition sites from each other, and from *Drosophila*, and one of them shows species-specific restriction fragment length variation (Fig. 8). Both fragments cross-hybridize with the cloned *Drosophila* rDNA repeat, PDM 238 (Glover and Hogness, 1977), but we are currently sequencing them to determine their exact homology.

In summary, the utility of the rDNA is truly remarkable. By carefully choosing different segments, with different evolutionary characteristics, it is possible to use it for systematic analysis of taxonomic groups at any

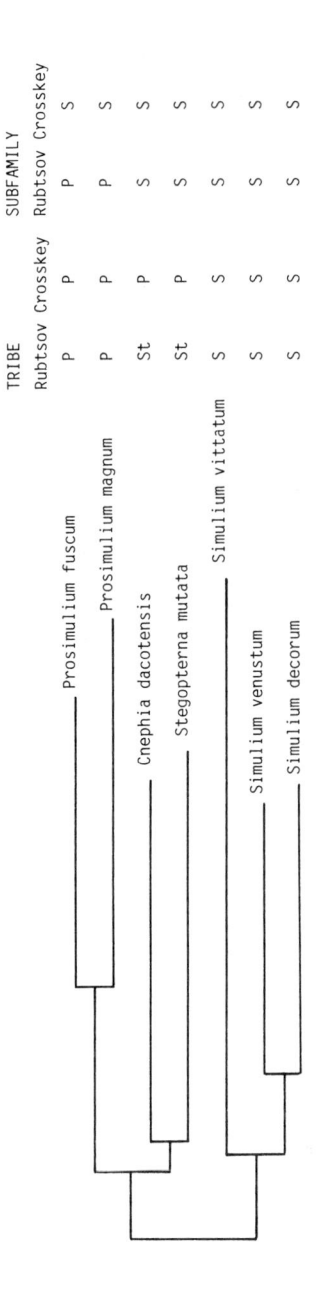

2. DNA Analysis in Relation to Insect Taxonomy, Evolution and Identification

	TRIBE	SUBFAMILY
	Rubtsov Crosskey	Rubtsov Crosskey
Prosimulium fuscum	P P	P S
Prosimulium magnum	P P	P S
Cnephia dacotensis	St P	S S
Stegopterna mutata	St P	S S
Simulium vittatum	S S	S S
Simulium venustum	S S	S S
Simulium decorum	S S	S S

Fig. 7. Unrooted tree from sequence comparison of the large subunit mitochondrial rDNA from seven species of Simuliidae (redrawn from Xiong and Kocher, 1991).

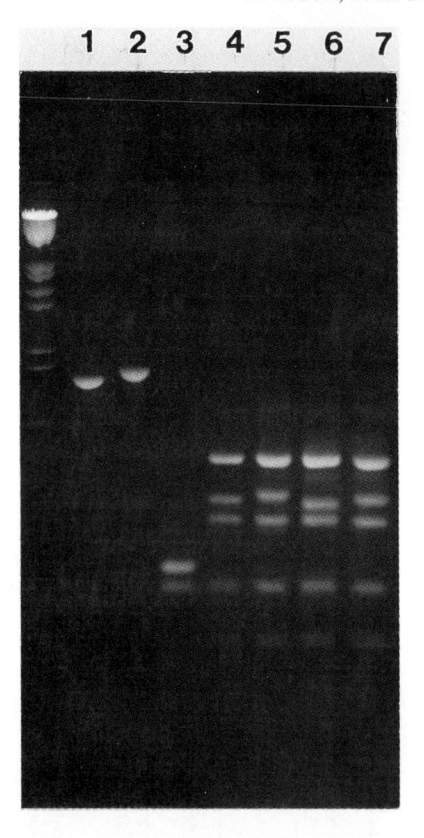

Fig. 8. Agarose gel electrophoresis of the 28S rRNA gene PCR product. Lane 1, 769-bp product *D. melanogaster*; lane 2, 830-bp product *S. sirbanum*; lanes 3–7, Alu-I-digested product of *D. melanogaster, S. santipauli, S. yahense, S. squamosum* and *S. sirbanum*.

level, from the origin of the kingdoms, to the identification of sibling species. Furthermore, the intergenic spacer region is now being used for the analysis of intraspecific population structure (McLain *et al.*, 1989).

IV. UNCHARACTERIZED REPETITIVE DNA SEQUENCES FOR SPECIES IDENTIFICATION

Sequences such as the rDNA and mitochondrial DNA have often been used in systematic studies because they are already known to vary in a suitable way. However, an alternative approach has been widely successful for the identification of difficult insect vectors and the parasites they

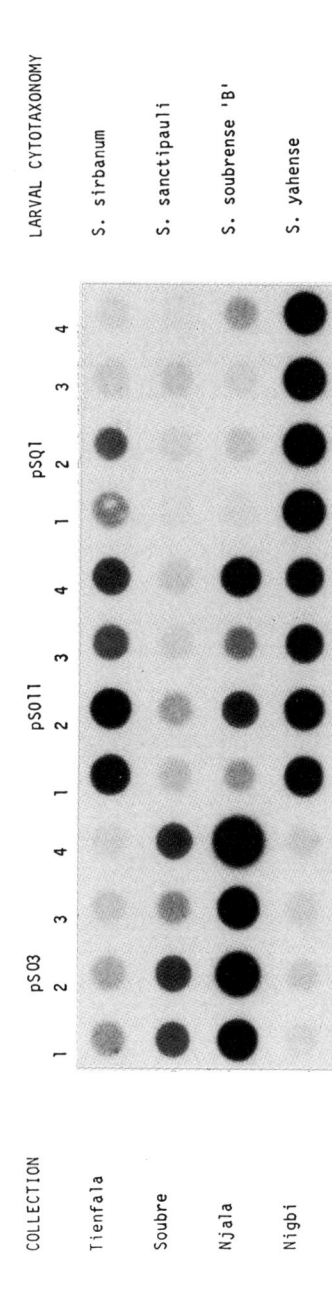

2. DNA Analysis in Relation to Insect Taxonomy, Evolution and Identification 31

Fig. 9. Three replica dot blots of genomic DNA extracted from four female *S. sirbanum*, *S. sanctipauli*, *S. soubrense* and *S. yahense*, probed with radiolabelled pSO3, pSO11 and pSQ1

transmit, such as the *Anopheles gambiae* complex (Hill et al., 1991) and malaria (Hughes et al., 1990), or the *Simulium damnosum* complex and *Onchocerca volvulus* (Post et al., 1991). This approach has been broadly similar for most groups studied, and is detailed by Post and Crampton (1988). A random genomic library is constructed for the species in question. This will randomly contain repetitive and single copy DNA, and it is screened comparatively using radiolabelled genomic DNA from each of the two, or more, species we wish to separate. The sensitivity of the screen is such that the random clones containing single copy DNA are not revealed by autoradiography, but the clones of repetitive DNA can be identified. Clones of repetitive sequences at high copy number in one species but absent or at low copy number in another are revealed by differences in the amount of hybridization in the screen, revealed by autoradiography.

In this way we have isolated three sequences, pSO3, pSO11 and pSQ1, which all hybridize to dot blots of genomic DNA of all species within the West African *Simulium damnosum* complex. However, the relative amount of hybridization varies, and can be used to identify the three

```
   1    GAATTCACGT GGTAAAGATA AACAAATTCA ATTTTATTTT TCAAAAGTCA
  51    GTTTTTGGTC AGAAAACCGA CTTAAAAATT TCAAAACTTC AAATTTGGCA
 101    AAGTTGATTT TGATGAAAAA CTGCTTATAT ACAAATTTTA TTAAGACCAA
 151    CTACTAATAT GAGTAGTATA TTTGTGGGGA TTCTGCGCCA GTGGTGAGGT
 201    GGTGAGTTGA GTGGTGAGGT GTGGTGAGGG AGCTTTTTTA ATGAAAATTT
 251    TTCCGCTTGG TGGCGCTTTG TGAGCTTTTA GCGTGTGTGG TGAGGACTTT
 301    TTAATGAAAG TTTTTTCCGG TAGATGGCGT TTTTGAGATT TTAGTGCTTT
 351    TTCATAAAAA CTTTTGAGCT TTTTGAGCTT TCATCGTGTG GTGAGGCTGA
 401    GTCAGCAATT TTTTAATGAA AGTTAGATGG CGCGGTAGAT AAAGTTTTCA
 451    ATAGGGTAAA CGTGAGCAAG ATGAATCATA TTATCAAAAT ATTTAGTAAA
 501    CATACAAACA AACATATTCA AACTCAAATA CCTGATGAAA TTATTTTAAA
 551    AATTTTCAAG TACTTTTTCT ATCAATGATC CAATTTACCC CAATTTTACC
 601    ATAGACAATT GCGCTTTTCG AGCTTTTTCA TAAAAACTTT TGAGCTTTTT
 651    GAGCTTTCAT CGTGTGGTGA GGCTAAGTCA GCAATTTTTA ATGAAAGTTA
 701    GATGGCGCGG TAGATAAAGT TTTCAATAGG GTAAACGTGA GCAAGATGAA
 751    TCATATTGTC AAAATATTTA CTGACGTATT AAACTCAAAT ACCATATGAA
 801    ATTTCAGTAC TTTTTTCAAT CAATGTTCCA ATTTACCCCA ATTTACCATA
 851    GACAATTGTG CTTTTCGAGC TTCTTTATAA AAACTTTTGA GCTTTTTGAG
 901    CTTTCATCGT GTGGTGAGGC TGAGTCAGCA ATTTTTTAAT GAAAGTTAGA
 951    TGGCGCGGTA GATAAAGTGT GCAATAGGGA AACGGGAGCA AGATGAATCA
1001    TATTGTAAAA ATATTTAGTA AACATACAAA CAAACATATT CAAACTCAAA
1051    TATCTGATGA AATTTTTCAA GTATTCAAGA ATTTTTTCTA TAAATGATCC
1101    AATTTACCCC AATTTACCCT AGACAATTGC ACTTTTCGAG CTTTTTCATA
1151    AAAATTTTTA ATGTAATTTG ATATTATGTC AATACATGTA TGATTTTCGT
1201    CGTGTGGTGA GGCTGAGTCA GCAATTTTTT TTTGTGATAT CATGTTCAAA
1251    CATGTATGAT TTTCATCGTG TGGTGAGCCT GAGTCAGCAA TTTTTTGTGA
1301    TATCATGTTC AAACATGTAT GATTTTCGTC GTGTGGTGAG GCTGTGTCAG
1351    CAATTTTTTA ATAAAAGCTT
```

Fig. 10. Sequence of part of the cloned insert of pSO11, with three near-identical repeats underlined.

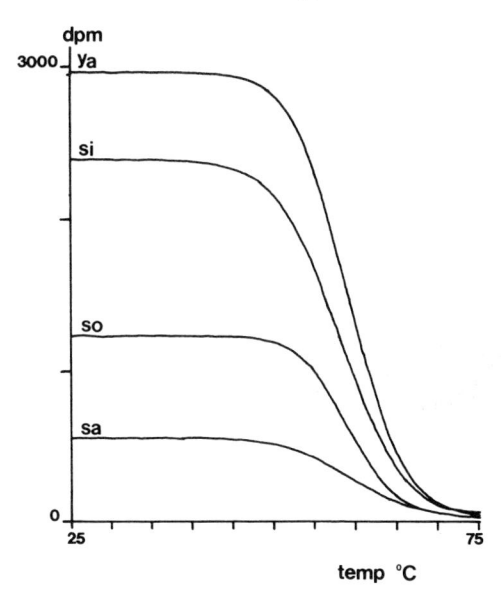

2. DNA Analysis in Relation to Insect Taxonomy, Evolution and Identification

Fig. 11. Melting curves (generalized logistic curves fitted to data) of dot blots of genomic DNA of *S. yahense* (ya), *S. sirbanum* (si), *S. soubrense* (so) and *S. sanctipauli* (sa) to radiolabelled pSO11 and washed at increasing temperatures.

major West African subcomplexes, the *damnosum* subcomplex, the *sanctipauli* subcomplex and the *squamosum* subcomplex (Fig. 9).

To investigate the nature of the species-specific variation we have sequenced 1.4 kb of the complete 3.5-kb cloned insert from pSO11 (Fig. 10). In the sequence there is a major repeat which is not tandem, but is separated by variable lengths of spacer sequence. This repeat was subcloned, and found to show the same pattern of hybridization to the various sibling species as was shown by the whole clone (Fig. 9). It is undoubtedly this major repeat within pSO11 which is providing the species-specific variability.

Furthermore, radiolabelled pSO11 was hybridized to dot-blotted genomic DNA of various species and washed off at increasing temperatures to obtain a melting curve (Fig. 11). In such curves the melting temperature is lower for sequences that have diverged, but for pSO11 the melting temperature was not significantly different between species, showing that the sequence is essentially the same in all the species. The differences in the amount of hybridization therefore reflect differences in the copy number of the sequence, and it is this difference in copy number which provides the species-specific variation. This is in contrast to the rDNA sequences described above, where specificity was provided by sequence divergence.

REFERENCES

Borrer, D. J., de Long, D. M. and Triplehorn, C. A. (1981). "An Introduction to the Study of Insects", 5th edn. Saunders College Publishing.
Cedergren, R., Gray, M. W., Abel, Y. and Sankoff, D. (1988). *J. Mol. Evol.* **28**, 98–112.
Collins, F. H., Finnerty, V. and Petrarca, V. (1988). *Parasitologia*, **30**, 231–240.
Crampton, G. C. (1944). *Bull. Brooklyn Ent. Soc.* **34**, 1–31.
Crosskey, R. W. (1985). *Ent. Mon. Mag.* **121**, 167–178.
Dover, G. A. and Tautz, D. (1986). *Phil. Trans. R. Soc. Lond.* B **312**, 275–289.
Ellis, R. E., Sulston, J. W. and Coulson, A. R. (1986). *Nucleic Acids Res.* **14**, 2345–2364.
Glover, D. M. and Hogness, D. S. (1977), *Cell* **10**, 167–176.
Hennig, W. (1958). *Beitr. Ent.* **8**, 505–688.
Hennig, W. (1973). Handbuch der Zoologie, eine Naturgeschichte der Stamme des Tierreiches, IV. Band: Arthropoda-2. Hafte: Insecta 2. Teil: Spezielles 31. W. Hennig: Diptera (Zweifluger). Berlin.
Hill, S. M., Urwin, R., Knapp, T. F. and Crampton, J. M. (1991). *Med. Vet. Entomol.* **5**, 455–463.
Hughes, M. A., Hommel, M. and Crampton, J. M. (1990). *Parasitology* **100**, 383–387.
John, B. and Miklos, G. (1988). "The Eukaryote Genome in Development and Evolution". Allen & Unwin, London.
McLain, D. K., Collins, F. H., Brandling-Bennett, A. D. and Were, J. B. O. (1989). *Heredity* **62**, 257–264.
Paskewitz, S. M. and Collins, F. H. (1990). *Med. Vet. Entomol.* **4**, 367–373.
Post, R. J. and Crampton, J. M. (1988). In "Biosystematics of Haematophagous Insects" (M. W. Service, ed.), pp. 245–255. Systematics Association Special Volume 37, Clarendon Press, Oxford.
Post, R. J., Murray, K. A., Flook, P. K, and Millest, A. L. (1991). In "Molecular Techniques in Taxonomy" (G. M. Hewitt, A. W. B. Johnston and J. P. W. Young, eds), pp. 271–281. NATO ASI Series, Springer-Verlag, Berlin.
Roback, S. S. (1951). *Annals Ent. Soc. Am.* **44**, 327–361.
Rohdendorf, B. (1974). "The Historical Development of the Diptera". University of Alberta Press, Alberta.
Simon, C. (1991). "Molecular Techniques in Taxonomy" (G. M. Hewitt, A. W. B. Johnston and J. P. W. Young, eds), pp. 33–71. NATO ASI Series, Springer-Verlag, Berlin.
Tautz, D., Hancock, J. M., Webb, D. A., Tautz, C. and Dover, G. (1988). *Mol. Biol. Evol.* **5**, 366–376.
Vossbrink, C. and Friedman, S. (1989). *Syst. Entomol.* **14**, 417–431.
Xiong, B. and Kocher, T. D. (1991). *Genome* **34**, 306–311.

3

Transposable Elements and their Biological Consequences in *Drosophila* and Other Insects

D. J. FINNEGAN

I. Introduction	35
A. Classification of Elements	36
II. Biological Consequences of Transposition	38
A. Transposable Elements as Mutagens	38
B. Hybrid Dysgenesis	38
C. Evolution of Genes and Genomes	39
D. Horizontal Transfer of Genetic Information	40
E. Transposable Elements and Speciation	41
III. Experimental Exploitation of Transposable Elements	41
A. Insertional Mutagenesis and Transposon Tagging	41
B. Germ Line Transformation	42
C. Enhancer Traps	43
IV. Can *P* Elements be Used in Other Species?	43
V. How to Find a New Transposable Element	44
References	46

I. INTRODUCTION

The application of recombinant DNA techniques to the study of eukaryotic genomes led to the unexpected discovery that they contain large numbers of transposable elements, that is DNA sequences that can move from one place in a genome to another. These have been detected in a wide range of eukaryotes and over one hundred have been described so far, about forty of them in *Drosophila melanogaster*. Transposable elements increase in copy number during transposition and are found in the component of a genome known as moderately repetitive DNA. The proportion of a genome that comprises transposable elements varies from species to species but is usually about 10%.

A. Classification of Elements

Nearly all transposable elements can be classified according to their structure and presumed mechanism of transposition (Fig. 1). There are two main classes of element. Class I elements transpose via an RNA intermediate while class II elements transpose directly from DNA to DNA.

Class I.1 elements are related to retroviruses in both structure and sequence and have direct repeats, known as LTRs (long terminal repeats), a few hundred base pairs long at each end. They contain coding regions similar to the *gag* and *pol* genes of a retrovirus and for many years have been thought to transpose by reverse transcribing an RNA intermediate by a mechanism related to a retroviral life-cycle. This has been demonstrated directly for *Ty1* elements of *Saccharomyces cerevisiae* (Boeke *et al.*, 1985; Eichinger and Boeke, 1988, 1990) and *IAP* elements of mice (Heidmann and Heidmann, 1991).

Class I.2 elements have no terminal repeats but also contain open reading frames with similarities to *gag* and *pol* genes. The *pol*-like gene encodes a reverse transcriptase (Ivanov *et al.*, 1991; Gabriel and Boeke, 1991; Mathias *et al.*, 1991) and two elements of this type, *I* elements of *D. melanogaster* and L1 elements of mouse, have been shown to transpose by reverse transcription of an RNA intermediate (Pélisson *et al.*, 1991; Jensen and Heidmann, 1991; Evans and Palmiter, 1991). This is presumably generally true and the RNA transposition intermediates are probably packaged into virus-like particles (Deragon *et al.*, 1990; Martin, 1991).

Many of the 28S ribosomal RNA genes of *D. melanogaster* contain insertion sequences that appear to be transposable elements of this type. There are two families of such sequences that insert at closely spaced sites in a conserved region of the rDNA. They were originally called type I and type II but are now called *R1* and *R2* (Jakubczak *et al.*, 1990). The rRNA genes of 47 species of insects from nine different orders have been screened for these sequences and they have been found in 43 of them (Jakubczak *et al.*, 1991). The proportion of rRNA genes containing insertions varies but can be as high as 60%. Ribosomal RNA genes containing insertions are inactive and are probably maintained by transposition of *R1* and *R2* elements into uninterrupted genes (Xiong and Eickbush, 1988).

Class II.1 elements have short terminal inverted repeats and have one or more open reading frames coding for functions required for their transposition. The best-studied elements of this type are the *P* elements of *D. melanogaster*. Complete *P* elements are 2.9 kb long and have 31-bp

CLASS I

I.1

LTR — gag — RT — ? — LTR

I.2

gag — RT — A_n

CLASS II

II.1

← TRANSPOSASE →

II.2

← ? →

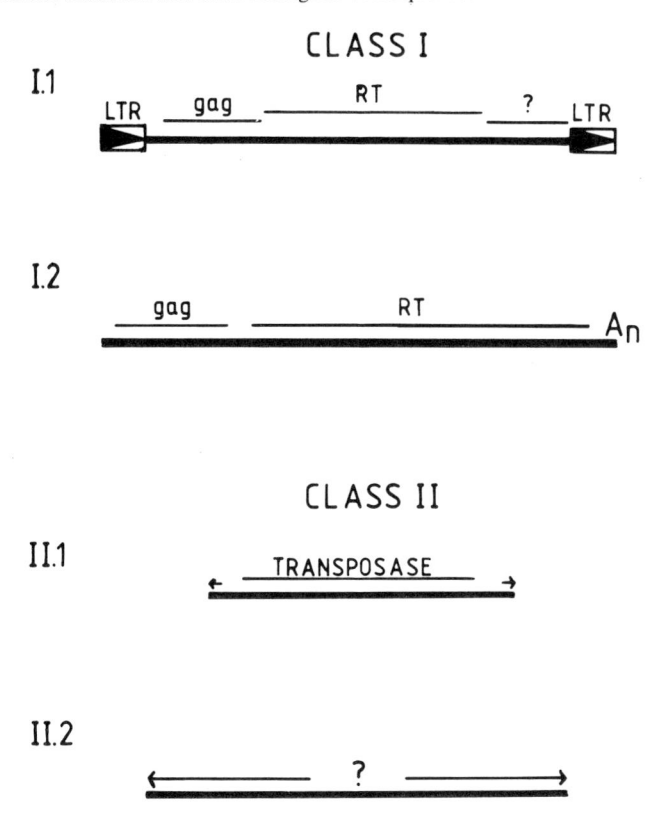

Fig. 1. Transposable elements classified according to their structure and mechanism of transposition. Class I elements transpose by reverse transcription of an RNA intermediate. Class I.1 elements are related to retroviruses. Class I.2 elements have no terminal repeats. Class II elements transpose directly from DNA to DNA. Class II.1 elements have short inverted terminal repeats. Class II.2 elements have long inverted repeats. LTR, long terminal direct repeat sequences; gag, an open reading frame with similarity to a retroviral *gag* gene; RT, an open reading frame encoding a reverse transcriptase; ?, a reading frame the function of which is unknown; A_n, an A-rich sequence found at the 3' end of one strand of Class I.2 elements. Only some Class I.1 elements have three open reading frames.

terminal inverted repeats. There are four open reading frames that are joined by RNA splicing to produce a protein, transposase, required for transposition (Laski *et al.*, 1986). Transposase does not bind to the inverted terminal repeats of *P* elements (Kaufman *et al.*, 1989) but may stimulate transposition by interacting with a host protein that does. Such a protein, inverted repeat binding protein, has been isolated from cells of an M strain (Rio and Rubin, 1988).

The least well-understood transposable elements are those that have long terminal inverted repeats. These make up class II.2. Only one family, the *FB* elements of *D. melanogaster*, has been studied in detail. These have inverted terminal repeats several hundred base pairs long that are themselves made up of short tandem repeats (Potter, 1982). There are no other sequences in many *FB* elements although one particular element has been identified that has a non-repetitive central region that may encode a transposition function (Templeton and Potter, 1989; Harden and Ashburner, 1990).

In some strains of *D. melanogaster* regions of the genome several hundred kilobases long can move from one site to another in large transposable elements called TEs (Ising and Block, 1981). These have *FB* elements at each end that appear to be responsible for their mobility (Paro *et al.*, 1983). It is not clear whether all *FB* elements can mobilize other sequences in this way or just a subset.

II. BIOLOGICAL CONSEQUENCES OF TRANSPOSITION

A. Transposable Elements as Mutagens

Although transposable elements can undoubtedly affect the evolution of genes and genomes they generally have no discernible effect on the lives of individuals within a species. This is because their rate of transposition is low, probably of the order of 10^{-4}–10^{-5} per element per generation (Harada *et al.*, 1990). This is not surprising as any element that continued to transpose at a high rate would so debilitate individuals containing it that it would soon be lost. Several authors have suggested that transposition frequencies are sensitive to conditions such as heat shock and ultraviolet radiation and that this may provide a pool of new mutations that help a species survive substantial changes in its environment (Strand and McDonald, 1985; McDonald *et al.*, 1988; McEntee and Bradshaw, 1988; Paquin and Williamson, 1988). This is an attractive idea although the evidence to support it is rather limited.

B. Hybrid Dysgenesis

There are some circumstances that greatly increase transposition of particular elements in *D. melanogaster* and the same may be true of other species. This gives rise to the phenomenon of hybrid dysgenesis and is associated with reduced fertility and high frequencies of mutations. The best-studied example is known as P–M dysgenesis. This is due to activation of P elements in the progeny of crosses between males of a strain that carries functional P elements, a P strain, and females of a strain that does not, an M strain. This is due to the breakdown of the mechanism that regulates transposition in P strains. This regulation is complex (Rio, 1991). It may be mediated by the products of both complete and internally deleted P elements (Black et al., 1987; Nitasaka et al., 1987; Robertson and Engels, 1989; Misra and Rio, 1990) and is at least in part due to transcriptional control (Kaufman and Rio, 1991; Lemaitre and Coen, 1991).

Although P transposition is greatly increased if M strain females are mated with P strain males this is not true in the progeny of the reciprocal cross. The reason for this asymmetry is not known precisely. The 66-kDa regulator of P activity is found at high levels in the ovaries and oocytes of P strain females but must be absent in M strain females. The activity of P elements from the sperm of P strain males is presumably inhibited in P oocytes but not in M oocytes.

C. Evolution of Genes and Genomes

There are many ways in which transposable elements can affect genes and genomes in the long term (Finnegan, 1989). They are undoubtedly a major source of spontaneous mutations in *D. melanogaster* and about one half of such mutations studied in the laboratory are due to insertions. This may have contributed to the success of genetic studies in this species, particularly in the days when geneticists had to use spontaneous rather than induced mutations. Transposable elements can also cause gross chromosomal rearrangements either as a direct result of transposition or indirectly because of recombination between copies of an element at different sites in the genome. The expression of genes may change because transposable elements brings them under the control of new regulatory sequences (Errede et al., 1987) while the sequence of gene products may change because transposable elements alter pre-mRNA processing (Kim et al., 1987) or because the sequence of a coding region is altered by insertion and/or imprecise excision of an element (Schwarz-Sommer et al., 1985). These effects have been observed as rare mutational events and may well have contributed to genome evolution in the past.

D. Horizontal Transfer of Genetic Information

Another evolutionary consequence of transposition that has come to light recently is the possibility that transposable elements may move horizontally between species. Similar elements have now been isolated from a wide range of species and when their base sequences have been compared they have been found to be related in ways that do not reflect the relationships between the host organisms (Xiong and Eickbush, 1990). This is most clearly demonstrated by *P* elements in *Drosophila*. These appear to have spread through wild populations of *D. melanogaster* during this century. None of the stocks that were isolated from the wild in the 1930s and 1940s contain *P* sequences today and these only start to appear in stocks established in the 1950s (Bingham *et al.*, 1982; Kidwell *et al.*, 1983; Anxolabéhère *et al.*, 1988).

This pattern seems to result from the acquisition of *P* elements by M strains in the wild rather than loss of *P* sequences by P strains in the laboratory. One may ask from where these *P* elements came and the answer seems to be that they have come from a long way away. *P* sequences cannot be found in the species most closely related to *D. melanogaster* but are present in more distantly related species and in particular in members of the *D. willistoni* group (Daniels *et al.*, 1990). A complete *P* element has been cloned from *D. willistoni* itself and has been found to be identical in sequence to a complete and functional *P* element from *D. melanogaster* except for a single base substitution (Daniels *et al.*, 1990). Given the evolutionary distance between *D. melanogaster* and *D. willistoni*, the degree of similarity between complete *P* elements isolated from these species and the discontinuous distribution of *P* elements in species lying between them it is difficult to come to any conclusion other than that *P* elements have moved horizontally between them. The fact that *D. willistoni* is present in Central and North America, the region from which the first P strains of *D. melanogaster* were isolated, is consistent with this.

It is not clear when the horizontal transfer of *P* elements would have taken place. It need not have occurred close to the time at which the first *P*-containing strains of *D. melanogaster* were isolated since few strains have survived from earlier times and gene flow between populations in the wild has probably been accelerated as a result of increases in international trade and travel.

If *P* elements have moved from one species to another then they probably did so with the aid of a vector of some sort. Parasites or pathogens of the species concerned are the most likely candidates for this. Nuclear polyhedrosis viruses (bacculoviruses) can pick up transposable elements from the genomes of cells that they infect (Miller and Miller, 1982; Wang *et al.*, 1989; Cary *et al.*, 1989) and the mite *Proctolaelaps regalis* acquires P sequences if it grows in the same culture vessel as a P strain of *D. melanogaster*. In neither case has it been shown that these vectors can transfer elements to another species and it is not known whether the P sequences in *P. regalis* are integrated into the genome of the mite or remain extracellular. Nevertheless, these data do suggest possible routes for horizontal transmission.

E. Transposable Elements and Speciation

The progeny of crosses between P strain males and M strain females are dysgenic and are less fertile than their parents. If they were completely sterile then this would genetically isolate the two parental strains at least for crosses between P males and M females. This has led to the suggestion that transposable elements might be a driving force in speciation (Bingham *et al.*, 1982; Rose and Doolittle, 1983; Ginzberg *et al.*, 1984). One can imagine that separate populations of the same species might gradually acquire different sets of transposable elements, perhaps by horizontal transfer from other species. Eventually the reduction in the fertility of their hybrid progeny might be so great as to prevent gene flow between them should they ever establish contact again. There is no direct evidence that transposable elements have been involved in a speciation event but it is an attractive possibility.

III. EXPERIMENTAL EXPLOITATION OF TRANSPOSABLE ELEMENTS

Transposable elements can be useful experimental tools if they can be induced to transpose at high frequencies either naturally or artificially. This is particularly true of P elements in *D. melanogaster* which have been used for insertional mutagenesis and transposon tagging, germ line transformation and enhancer trapping.

A. Insertional Mutagenesis and Transposon Tagging

Transposon tagging is a technique that simplifies the molecular cloning of genes of interest. Transposable elements cause mutations by inserting within or adjacent to genes. If a transposable element that has already been cloned mutates a gene of interest then the DNA sequence of that element can be used as a probe with which to screen a library of cloned sequences from the mutant stock. Among the clones that hybridize to the transposable element will be those that carry DNA from the target gene. The progeny of P-M dysgenic flies can be screened for *P*-element induced mutations that can be used in this way. Mutations due to *P* insertions are recognizable because they are unstable in dysgenic individuals and are associated with *P* sequences as assayed by *in situ* hybridization. There will be several *P* elements in the genome of mutant flies isolated in this way and the efficiency of subsequent cloning steps is increased if elements that do not lie near the gene of interest are removed by crossing to an M strain that has no P sequences (Kidwell, 1986).

A variant of this technique has been developed to select flies having a *P* element inserted close to any previously cloned DNA (Ballinger and Benzer, 1989; Kaiser and Goodwin, 1990). Flies carrying the appropriate insertion are identified with the aid of a polymerase chain reaction using one primer from the target sequence and the other from the P element itself. This allows one to recover mutations in genes that have been identified via a cDNA or in a chromosome walk without any reference to their function. The insertions detected in this way may be close to but not within the target sequence. If this is the case it may be possible to induce mutations affecting the gene itself by expressing *P* transposase in flies carrying the insertion. This will stimulate excision of the *P* element and as this will not always be precise some flies may be recovered that have lost all or part of the target gene.

B. Germ Line Transformation

Complete *P* elements produce transposase when injected into early embryos of an M strain and transpose to the chromosomes of host germ cells. They can also stimulate transposition of internally deleted *P* elements that are coinjected with them, providing a means by which other sequences can be introduced into the germ line (Rubin and Spradling, 1982). A sequence to be introduced into the genome is inserted within a *P* transformation vector that has the terminal sequences required in *cis* for transposition and a marker gene that allows transformed flies to be distinguished from non-transformants. The vector is then injected into the posterior pole of M strain embryos together with a *P* element that has its coding region intact but has lost a few bases from one end so that it cannot itself integrate into the target genome.

C. Enhancer Traps

P elements can also be used to identify genes that are expressed in a particular tissue or developmental stage. Many genes have their expression restricted to particular cells or tissues because their transcription is controlled by *cis*-acting regulatory elements known as enhancers. These are DNA sequences that can act over long distances in a manner that is independent of their orientation with respect to a target gene. O'Kane and Gehring (1987) suggested that enhancers could be detected with the aid of a *P* vector that had the Z gene of *E. coli* under the control of a weak promoter with no enhancer of its own. They argued that if this vector were to insert in the vicinity of an enhancer then the *lacZ* gene would be expressed in the tissue in which the enhancer is active. This can be detected histologically using a chromogenic substrate for β-galactosidase.

This strategy has proved to be remarkably successful (Bellen *et al.*, 1989; Wilson *et al.*, 1989). Thousands of strains of flies have been obtained with enhancer trap vectors inserted at sites scattered around the genome. About 65% of these strains express β-galactosidase in a restricted number of cells (Plate 1). One would expect that in these lines the vector would be adjacent to sequences the expression of which is restricted in a similar way. This is often the case (Bellen *et al.*, 1989) and enhancer trapping is a powerful means of identifying and cloning genes that are expressed in particular tissues. In many instances these are genes that could not have been identified in conventional genetic experiments either because their function is unknown or because they have no easily recognized mutant phenotype.

IV. CAN *P* ELEMENTS BY USED IN OTHER SPECIES?

Unfortunately it is unlikely that *P* elements will be useful experimental tools outside the Drosophilidae at least for the moment. *P* elements transpose after injection into embryos of *D. simulans* (Scavarda and Hartl, 1984) or *D. hawaiiensis* (Brennan *et al.*, 1984) and transposase activity can be detected if *P* sequences are injected into embryos of a variety of other drosophilids but not other insects (O'Brochta and Handler, 1988; O'Brochta *et al.*, 1991). Two groups have tried to mimic the *D. melanogaster P* transformation system in mosquitoes (Miller *et al.*, 1987; Morris *et al.*, 1989). In both cases they were able to recover adults with stably integrated vector DNA but this did not require *P* transposase and was presumably the result of non-homologous recombination events. The transposase gene can be expressed in at least some non-drosophilids (O'Brochta and Handler, 1988) so these negative results may indicate that one or more host factors are required for transposition and that these are not found in other genera. The inverted repeat binding protein found in M strains of *D. melanogaster* is a likely candidate for such a factor. Once the gene for this protein has been cloned it may be possible to incorporate it into a transformation system that would supply everything required for transposition in a heterologous organism.

A second transformation system has been developed for *D. melanogaster* based on *hobo*, another transposable element with short inverted terminal repeats (McGinnis *et al.*, 1983; Calvi *et al.*, 1991). A marked *hobo* element can integrate into the chromosomes of a strain that lacks endogenous elements if it is injected into embryos together with a complete element (Blackman *et al.*, 1989). Nothing is known about the biochemical requirements for *hobo* transposition except that these include a *hobo*-encoded protein that has some sequence similarity to the transposase of the maize transposable element *Ac* (Calvi *et al.*, 1991). If there is no requirement for a host protein that is restricted to drosophilids then this system might be used in a wide range of species.

V. HOW TO FIND A NEW TRANSPOSABLE ELEMENT

There are several ways of searching for transposable elements that might be useful in the genetic and molecular analysis of the genomes of insects and other species. The most direct is to look for insertional mutations of genes with easily scored phenotypes. This can be done genetically by screening for unstable mutations although these can also be due to duplications and other chromosome rearrangements, and molecularly by comparing the DNA of mutant and wild type alleles of genes that have

---

Plate 1. An enhancer trap element used to search for a gene expressed in germ cells. The testis (A) and ovary (B) of two different strains of *D. melanogaster* carrying the enhancer trap element shown in (C) have been stained for β-galactosidase activity. The colourless substrate X-gal (5-Bromo-4-chloro-3-indolyl-β-D-galactopyranoside) is converted to a blue derivative in the presence of β-galactosidase. This can be seen in sperm heads in (A) and in the nuclei of nurse cells and oocytes in (B). The map of the P enhancer trap element is shown in (C). The solid shading indicates sequences from the ends of a complete P element and the hatched region indicates the *E. coli lacZ* gene. This is fused to the second intron of the P sequence. The unshaded region contains the *D. melanogaster rosy* gene that is used as a marker with which to detect flies carrying this element. The photographs in (A) and (B) were generously provided by Nian Zhang and Mary Bownes.

A.

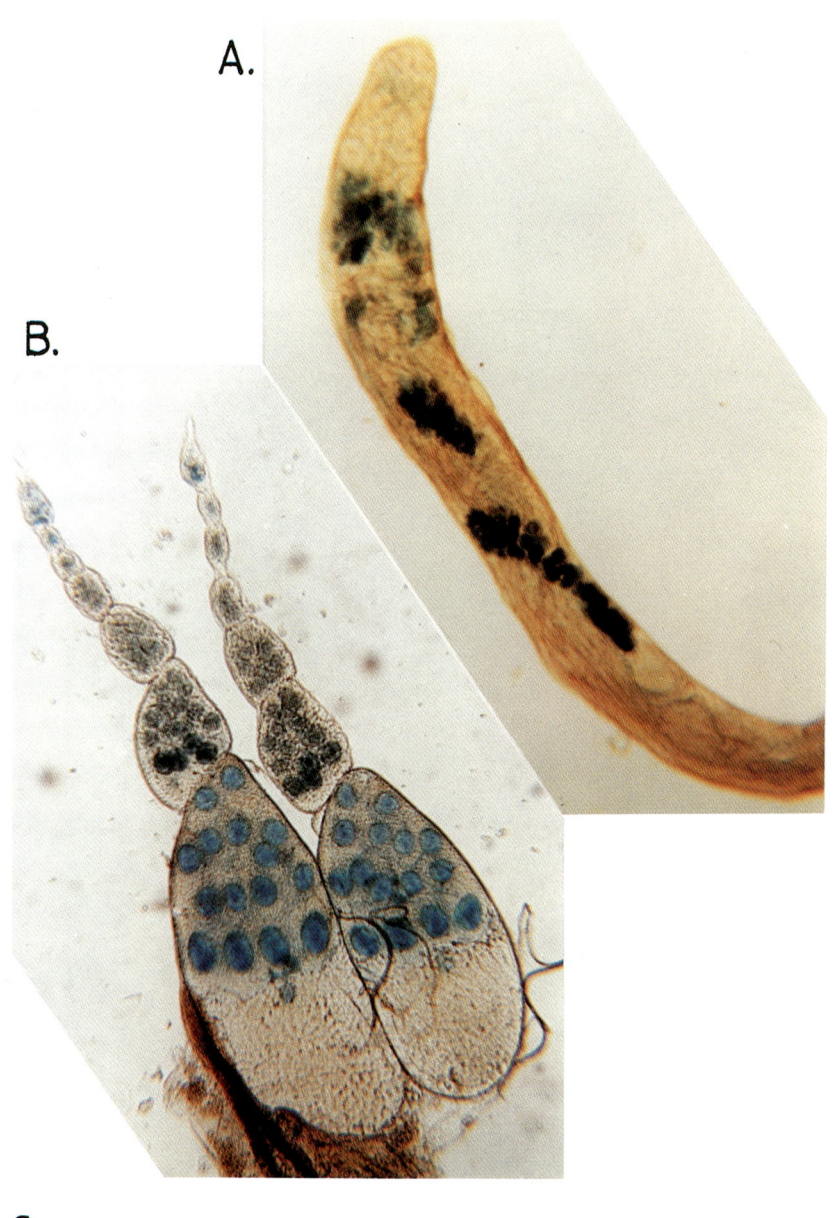

B.

C.

P   lacZ   rosy   P

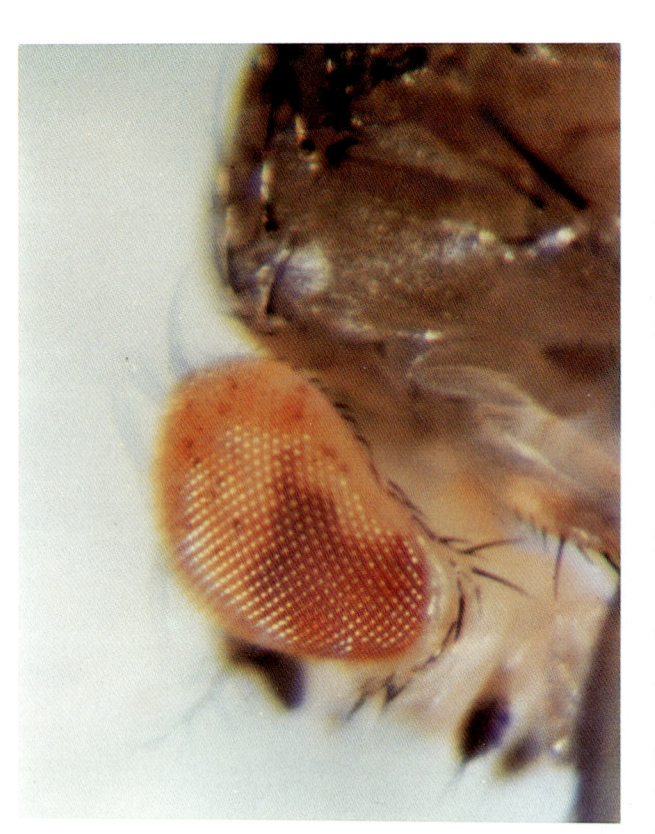

Plate 2. The *mariner* element of *D. mauritiana* is unstable in somatic cells. This shows the eye of a fly with a *mariner* element inserted in the *white* gene of *D. mauritiana*. The pale regions are due to segments of the eye carrying this insertion. The red regions are due to segments of the eye in which the *mariner* element has excised from the *white* gene. This photograph was generously provided by Glenn Bryan.

3. Transposable Elements and their Biological Consequences

been cloned. The genetic screen will only reveal the presence of elements that can excise under the influence of a transposase and will most readily detect elements that excise both somatically and germinally. Most transposable elements appear to be inactive in somatic tissues; however, *mariner* elements were detected in *D. mauritiana* because of a somatically unstable *white* gene mutation (Plate 2; Jacobson *et al.*, 1986). The *Tc1* element of *Caenorhabditis elegans* can also excise both germinally and somatically (Moerman and Waterston, 1989). Both of these elements have short terminal inverted repeats.

A less direct method for detecting transposable elements is to take advantage of the fact that they are moderately repetitive sequences. If they are sufficiently abundant they can be recognized because they give visible bands in ethidium-bromide-stained digests of total genomic DNA. This is how L1 elements were first detected in mammalian DNAs (Rogers, 1983) and how the retroviral-like element *del-1* was detected in *Lilium henryi* (Sentry and Smyth, 1989). Less abundant elements may be recovered by cloning fragments of purified moderately repetitive DNA. In either case a potentially transposable sequence can be checked by comparing its chromosomal distribution in different individuals or populations of the organism concerned. This can be done by *in situ* hybridization to polytene chromosomes if they are suitable or by Southern hybridization experiments.

Elements that transpose by reverse transcription of an RNA intermediate contain open reading frames that encode putative reverse transcriptases. These contain short amino acid sequences that are conserved in all reverse transcriptases, and degenerate oligonucleotides corresponding to these motifs can be used to detect cloned genomic DNA containing retrotransposons or to amplify regions of such elements by the polymerase chain reaction. These methods have been used successfully to clone new transposable elements from several complex genomes (Warren and Crampton, 1991; Flavell *et al.*, 1992).

Transposable elements are most useful if they can be induced to transpose at high frequency. Transposition of the *Ty1* element of *S. cerevisiae* was increased by placing it under the control of an inducible promoter (Boeke *et al.*, 1985) and in some cases transposition can be induced by crossing strains that do and do not contain functional copies of the element concerned. This is done to induce hybrid dysgenesis with P, I and *hobo* elements in *D. melanogaster* and it may be possible to induce any element in this way if it regulates its own transposition through a *trans*-acting factor and if there are strains that lack functional copies of the element concerned. There is probably only a brief period in the history of a species during which such conditions can be found. Any element that can be induced in this way is likely to spread rapidly through all populations of a species into which it is introduced and in which it can transpose.

Acknowledgements

I am grateful to Nian Zhang and Mary Bownes for providing the photographs for Plate 1, and Glenn Bryan for Plate 2, Carol McLean for having read a draft of the manuscript, and Graham Brown for photography.

REFERENCES

Anxolabéhère, D., Kidwell, M. G. and Periquet, G. (1988). *Mol. Biol. Evol.* **5**, 232–269.
Ballinger, D. G. and Benzer, S. (1989). *Proc. Natl. Acad. Sci. USA* **86**, 9402–9406.
Bellen, H. J., O'Kane, C. J., Wilson, C., Grossniklaus, U., Pearson, R. K. and Gehring, W. J. (1989). *Genes Dev.* **3**, 1288–1289.
Bingham, P. M., Kidwell, M. G. and Rubin, G. M. (1982). *Cell* **57**, 171–183.
Black, D. M., Jackson, M. S., Kidwell, M. G. and Dover, G. A. (1987). *EMBO J.* **6**, 4125–4135.
Blackman, R. K., Macy, M., Koehler, D., Grimaila, R. and Gelbart, W. M. (1989). *EMBO J.* **8**, 211–217.
Boeke, J. D., Garfinkel, D. J., Styles, C. A. and Fink, G. R. (1985). *Cell* **40**, 491–500.
Brennan, M. D., Rowan, R. G. and Dickinson, W. J. (1984). *Cell* **38**, 147–151.
Calvi, B. R., Hong, T. J., Findley, S. D. and Gelbart, W. M. (1991). *Cell* **66**, 465–471.
Cary, L. C., Goebel, M., Corsaro, B. G., Wanh, H., Rosen, E. and Fraser, M. J. (1989). *Virology* **172**, 156–169.
Daniels, S. B., Peterson, K. R., Strausbaugh, L. D., Kidwell, M. G. and Chovnik, A. (1990). *Genetics* **124**, 339–355.
Deragon, J. M., Sinnet, D. and Labuda, D. (1990). *EMBO J.* **9**, 3363–3368.
Eichinger, D. J. and Boeke, J. D. (1988). *Cell* **54**, 955–966.
Eichinger, D. J. and Boeke, J. D. (1990). *Genes Dev.* **4**, 324–330.
Errede, B., Company, M. and Hutchison, C. A. (1987). *Mol. Cell. Biol.* **7**, 258–264.
Evans, J. P. and Palmiter, R. D. (1991). *Proc. Natl. Acad. Sci. USA* **88**, 8792–8795.
Finnegan, D. J. (1989). *Trends Genet.* **5**, 103–107.
Flavell, A. J., Smith, D. B. and Kumar, A. (1992). *Mol. Gen. Genet.* **231**, 233–242.
Gabriel, A. and Boeke, J. D. (1991). *Proc. Natl. Acad. Sci. USA* **88**, 9794–9798.
Ginzberg, L. R., Bingham, P. M. and Yoo, S. (1984). *Genetics* **107**, 331–341.
Harada, K., Yukuhiro, K. and Mukai, T. (1990). *Proc. Natl. Acad. Sci. USA* **87**, 3248–3252.
Harden, N. and Ashburner, M. (1990). *Genetics* **126**, 387–400.
Heidmann, O. and Heidmann, T. (1991). *Cell* **64**, 159–170.
Houck, M. A., Clark, J. B., Peterson, K. R. and Kidwell, M. G. (1991). *Science* **253**, 1125–1129.
Ising, G. and Block, K. (1981). *Cold Spring Harbor Symp. Quant. Biol.* **45**, 527–549.
Ivanov, V. A., Melnikov, A. A., Siunov, A. V., Fodor, I. I. and Ilyin, Y. I. (1991). *EMBO J.* **10**, 2498–2495.

3. Transposable Elements and their Biological Consequences

Jacobson, J. W., Medhora, M. M. and Hartl, D. L. (1986). *Proc. Natl. Acad. Sci. USA* **83**, 8684–8688.
Jakubczak, J. L., Burke, W. D. and Eickbush, T. H. (1991). *Proc. Natl. Acad. Sci. USA* **88**, 3295–3299.
Jakubczak, J. L., Xiong, Y. and Eickbush, T. H. (1990). *J. Mol. Biol.* **212**, 37–52.
Jensen, S. and Heidmann, T. (1991). *EMBO J.* **10**, 1927–1937.
Kaiser, K. and Goodwin, S. F. (1990). *Proc. Natl. Acad. Sci. USA* **87**, 1686–1690.
Kaufman, P. D. and Rio, D. C. (1991). *Proc. Natl. Acad. Sci. USA* **88**, 2613–2617.
Kaufman, P. D., Doll, R. F. and Rio, D. C. (1989). *Cell* **59**, 359–371.
Kidwell, M. G. (1986). In "Drosophila a Practical Approach" (D. B. Roberts, ed.), pp. 59–81. IRL Press, Oxford.
Kidwell, M. G., Frydryk, T. and Novy, J. B. (1983). *Drosophila Information Service* **59**, 63–69.
Kim, H.-Y., Schiefelbein, J. W., Raboy, V., Furtek, D. B. and Nelson, O. E. (1987). *Proc. Natl. Acad. Sci. USA* **84**, 5863–5867.
Laski, F. A., Rio, D. C. and Rubin, G. M. (1986). *Cell* **44**, 7–19.
Lemaitre, B. and Coen, D. (1991). *Proc. Natl. Acad. Sci. USA* **88**, 4419–4423.
Martin, S. M. (1991). *Mol. Cell. Biol.* **11**, 4804–4807.
Mathias, S. L., Scott, A. F., Kazazian, H. H., Boeke, J. D. and Ganriel, A. (1991). *Science* **254**, 1808–1810.
McDonald, J. F., Strand, D. J., Brown, M. R., Pakewitz, S. M., Csink, A. K. and Voss, S. H. (1988). In "Banbury Report 30: Eukaryotic Transposable Elements as Mutagenic Agents" (M. E. Lambert, J. F. McDonald and I. B. Weinstein, eds), pp. 219–234. Cold Spring Harbor Laboratory, Cold Spring Harbor.
McEntee, K. and Bradshaw, V. A. (1988). In "Banbury Report 30: Eukaryotic Transposable Elements as Mutagenic Agents" (M. E. Lambert, J. F. McDonald and I. B. Weinstein, eds), pp. 245–253. Cold Spring Harbor Laboratory, Cold Spring Harbor.
McGinnis, W., Shermoen, A. W. and Beckendorf, S. K. (1983). *Cell* **34**, 75–84.
Miller, D. W. and Miller, L. K. (1982). *Nature* **299**, 562–564.
Miller, L. H., Sakai, R. K., Romans, P., Gwadz, R. W., Kantoff, P. and Coon, H. G. (1987). *Science* **237**, 779–781.
Misra, S. and Rio, D. C. (1990). *Cell* **62**, 269–284.
Moerman, D. G. and Waterston, R. H. (1989). In "Mobile DNA" (D. E. Berg and M. M. Howe, eds), pp. 537–556. American Society for Microbiology, Washington.
Morris, A. C., Eggleston, P. and Crampton, J. M. (1989). *Med. Vet. Entomol.* **3**, 1–7.
Nitasaka, E., Mukai, T. and Yamazaki, T. (1987). *Proc. Natl. Acad. Sci. USA* **84**, 7605–7608.
O'Brochta, D. A. and Handler, A. M. (1988). *Proc. Natl. Acad. Sci. USA* **85**, 6052–6056.
O'Brochta, D. A., Gomez, S. P. and Handler, A. M. (1991). *Mol. Gen. Genet.* **225**, 387–394.
O'Kane, C. J. and Gehring, W. J. (1987). *Proc. Natl. Acad. Sci. USA* **84**, 9123–9127.
Paquin, C. E. and Williamson, V. M. (1988). In "Banbury Report 30: Eukaryotic Transposable Elements as Mutagenic Agents" (M. E. Lambert, J. F. McDonald and I. B. Weinstein, eds), pp. 235–244. Cold Spring Harbor Laboratory, Cold Spring Harbor.
Paro, R., Goldberg, M. L. and Gehring, W. J. (1983). *EMBO J.* **2**, 853–860.
Pélisson, A., Finnegan, D. J. and Bucheton, A. (1991). *Proc. Natl. Acad. Sci. USA* **88**, 4907–4910.
Potter, S. S. (1982). *Nature* **297**, 201–204.
Rio, D. C. (1991). *Trends Genet.* **7**, 282–287.
Rio, D. C. and Rubin, G. M. (1988). *Proc. Natl. Acad. Sci. USA* **85**, 8929–8933.

Robertson, H. M. and Engels, W. R. (1989). *Genetics* **123**, 815–824.
Rogers, J. (1983). *Nature* **306**, 113–114.
Rose, M. R. and Doolittle, W. F. (1983). *Science* **220**, 157–162.
Rubin, G. M. and Spradling, A. C. (1982). *Science* **218**, 348–353.
Scavarda. N. J. and Hartl, D. L. (1984). *Proc. Natl. Acad. Sci. USA* **81**, 7515–7519.
Schwarz-Sommer, Z., Gierl, A., Cuypers, H., Peterson, P. A. and Saedler, H. (1985). *EMBO J.* **4**, 591–597.
Sentry, J. W. and Smyth, D. R. (1989). *Mol. Gen. Genet.* **215**, 349–354.
Strand, D. J. and McDonald, J. F. (1985). *Nucleic Acids Res.* **13**, 4401–4410.
Templeton, N. S. and Potter, S. S. (1989). *EMBO J.* **8**, 1887–1894.
Wang, H. H., Fraser, M. J. and Cary, L. C. (1989). *Gene* **81**, 97–108.
Warren, A. M. and Crampton, J. M. (1991) *Genet. Res., Camb.* **58**, 225–232.
Wilson, C., Pearson, R. K., Bellen, H. J., O'Kane, C. J., Grossniklaus, U. and Gehring, W. J. (1989). *Genes Dev.* **3**, 1301–1313.
Xiong, Y. and Eickbush, T. M. (1988). *Cell* **55**, 235–246.
Xiong, Y. and Eickbush, T. M. (1990). *EMBO J.* **10**, 3353–3362.

Part II.

Genome Mapping, Development and the Immune System

4. Mapping Insect Genomes
   M. ASHBURNER
5. Molecular Aspects of Sex Determination in Insects
   M. BOWNES
6. Gene Control Systems Affecting Insect Development
   M. G. PETITT and M. P. SCOTT
7. Insect Immune Systems
   Y. ENGSTRÖM

4

Mapping Insect Genomes

M. ASHBURNER

I. Introduction	51
II. Genetic Maps	51
III. Polytene Chromosome Maps	56
IV. Microdissection Libraries – "Polytene Chromosomes on a Filter"	59
V. Maps of Restriction Fragment Length Polymorphisms	60
VI. Mini- and Micro-satellite DNA Mapping	62
VII. Whole Genome Molecular Mapping	64
A. Vectors	65
B. YAC Map of the *Drosophila* Genome	66
C. P1 Maps	66
D. Cosmid Maps	67
VIII. The Next Stage	69
References	71

I. INTRODUCTION

A revolution is upon us. For too long *Drosophila melanogaster* has had to hold the standard of insect genetics alone; indeed, so alone that many entomologists do not really think of it as an insect at all. Table I is the catalogue of genetic maps now available, from the scale of a school atlas to one better than any on sale. This chapter is organized, roughly, from the smallest to the largest of these scales.

II. GENETIC MAPS

A genetic map is, to quote a well-known dictionary (Rieger *et al.*, 1976), a "representation of the genetic (relative) distances separating non-allelic

TABLE I. The Ordnance Survey of the genome – map scales.

Map	Kilobases	Scale
1 cM genetic map	600–3000	1 : 600 000[a]–1 : 3 000 000[b]
YAC clones	500	1 : 500 000
P1 clones	100	1 : 100 000
8-cutter restriction enzyme	65	1 : 65 000
polytene chromosomes	20–100	1 : 20 000–1 : 100 000[a]
cosmid clones	40	1 : 40 000
λ clones	15	1 : 15 000
6-cutter restriction enzyme	4	1 : 4 000
4-cutter restriction enzyme	0.256	1 : 256
DNA sequence	0.001	1 : 1

[a] *Drosophila melanogaster*.
[b] *Aedes aegypti*.

loci in a linkage structure". It is, following Sturtevant (1913), constructed from the frequencies of genetic recombination between mutant alleles at different loci. By 1925 *Drosophila* geneticists had constructed a remarkably detailed genetic map of *D. melanogaster*, with 111 different loci placed on the four chromosomes (linkage groups) of this fly (Morgan et al., 1925). This genetic map continues to be refined and improved, both by the classical methods of genetic recombination and by the newer (molecular) techniques introduced in the last few years. Today, the genetic map of *D. melanogaster* has over 3700 loci (Ashburner, 1991a,b; Merriam et al., 1991). By contrast, the next most detailed genetic map for an insect is that of the silkworm moth *Bombyx mori* – this has 207 loci (Doira, 1989; and personal communication to M. Goldsmith). There is only one other insect whose genetic map has more than 100 loci, *Lucilia cuprina* (135 loci) (G. Weller and G. Foster, personal communication), although maps with 10–80 loci have been determined for perhaps ten species (of which several are mosquitoes). This is a deplorable state of affairs, given the great diversity among insects and the economic importance of many species.

The resolution of a genetic map generated by recombination depends on a number of factors – one obvious one is the frequency of meiotic recombination. In *D. melanogaster* the total genetic map is 277 cM. Since each genetic exchange produces 50 cM of map distance this indicates that there are, on average, 5.6 exchanges per meiocyte. Were we to rely on the male of *D. melanogaster* for mapping we would be in bad shape; here the frequency of meiotic exchange is effectively zero, so our mapping would be confined to the determination of syntenic loci, i.e. chromosome

4. Mapping Insect Genomes

assignment, alone. (The absence of crossing-over in males of the higher Diptera (Gethmann, 1988) and in females of the Lepidoptera is, of course, an enormous advantage to mapping, since synteny can be established very easily from a male (Diptera) or female (Lepidoptera) backcross. This is especially useful for multilocus mapping, such as for restriction fragment length polymorphisms (RFLPs).) For *Bombyx mori*, the only other insect for which I have some confidence in the data, the total known genetic map is about 1000 cM. (This figure is almost certainly too low, since the haploid chromosome number of *B. mori* is 28 and one exchange per chromosome would generate a map of 1400 cM.)

Nothing much can be done about the frequency of recombination in one's favourite insect, which leaves hard work as the factor that determines the resolution of the map. *Drosophila* workers are reasonably happy to map to a resolution of 0.1 cM. After all, to ensure the recovery of at least one recombinant between a pair of markers 0.1 cM apart one needs to score only 4600 progeny (with a probability of recovering the recombinant of 99%). (The formula is $N = \log(1-p)/\log(1-f)$ where N is the number of progeny you need to score to detect, with probability p, one recombinant between markers of f cM apart.) Some drosophilists will score 10^6 or even 10^7 zygotes in mapping experiments, but only after devising selective schemes that eliminate as lethals the majority (i.e. non-recombinant) classes. For many interesting organisms experiments on this scale are clearly not practicable. Indeed most of the genetic maps of insects are at a very low resolution. (A practical consequence of this is that, very often, markers being mapped relative to each other will be a relatively long way apart. It must not be forgotten that the linear relationship between recombination frequency and map distance will then not hold, because of the high probability of multiple crossovers. A suitable mapping function must be used; see Ashburner (1989).)

The reasons for the paucity of genetic information in insect species other than *D. melanogaster* and *B. mori* are various. Most obviously, only a minute fraction of the known species of insect have been colonized in laboratories. Even some of these are obligately parthenogenetic, making mapping by genetic recombination an impossibility. However, among sexually reproducing species in culture the chief problem in constructing a map is the availability of genetic markers; without genetic variation, either natural or experimentally produced, genetic mapping by conventional techniques is impossible. Unless the community of researchers (1) is large and (2) has a tradition of free exchange of mutant stocks, constructing a genetic map will be very slow. In practice, for most species of insect either the research community is small and/or those involved have not learnt the lesson of the *Drosophila* community – that is, that the free exchange of experimental materials between laboratories is essential for rapid scientific progress.

To some extent the paucity of conventional (i.e. "visible") genetic markers in many insects was improved by the "isoenzyme" revolution (Hubby and Lewontin, 1966; Harris, 1966; Lewontin, 1991). The discovery that many enzymes could not only be assayed after electrophoresis but also that many showed genetic variation for the character "electrophoretic mobility" led to a minor revolution in genetics. It is interesting that many of the genetic maps of insects rely heavily on isoenzyme markers. For example, the map of *Aedes triseriatus* has 30 markers, of which 21 are isoenzymes (Munstermann, 1990); that of *Anopheles albimanus* has 62 markers, of which 24 are isoenzymes (Narang and Seawright, 1990). Electrophoresis has the advantage that it is a simple method but it suffers from two disadvantages. The first of these is that the number of enzymes for which an assay after electrophoresis is possible is quite small (<2% of the 3000 or so enzymes recognized by the Enzyme Commission). The second is that many enzymes are electrophoretically monomorphic, although more sophisticated methods for detecting variation due to amino acid substitutions that do not affect the net charge of a polypeptide have been developed (MacIntyre and Collier, 1986). I will discuss in some detail, later in this chapter, the practice and consequences of a parallel revolution to that offered by electrophoresis – the discovery of restriction fragment length polymorphisms (RFLPs).

Given the rather sorry state of affairs I have related, the obvious questions are "Does it matter? Do we need detailed genetic maps of many insect species?" I argue that it does and we do. We need them for just the same reason as we require topographical maps if we are hiking in unfamiliar territory – without them we are liable to lose our way. Genetic maps in the broadest sense, including molecular maps, are essential for any genetic analysis, be it of quantitative or "single gene" characters. Let me give two examples.

For those species of insect where quantitative genetic variation is important for exploitation or control, detailed genetic maps are an absolute essential for the manipulation of this variation to advantage. India is now the second-largest producer of raw silk in the world. Sericulture in India employs 5.7 million people and earns India about $100 million in export sales (Central Silk Board, 1989). Most of the Indian silk is produced in tropical regions and the strains are characterized by a low yield and poor silk quality. They are far inferior to the bivoltine strains used for production in Japan and China. However, these superior strains cannot simply be imported into India (political and commercial problems apart) – they are far too susceptible to disease (especially those due to viruses) and are intolerant of the conditions of high temperature and humidity in India. Many of the characteristics that distinguish the bivoltine strains of *B. mori* of China and Japan from the multivoltine strains of South East Asia are quantitative. Indeed directional selection for a number of quantitative traits, e.g. cocoon weight, has been practised in Japan for at least 300 years, and probably for far longer (Yokoyama, 1959). The efficiency of conversion of mulberry leaves into silk – an important factor – is very unlikely to be due to a single genetic difference. Over the years geneticists have developed techniques for handling such polygenic traits. Indeed, given an organism with a "good" genetic map the genes affecting such traits can be located and manipulated through breeding programmes (e.g. Thoday, 1961). However, that is a very difficult, if not impossible, task to do without a detailed genetic map. Now, the ability to construct high-resolution maps using molecular markers has changed the situation beyond all recognition. As I will discuss below, techniques for mapping genes affecting continuously variable characters (quantitative trait loci, or QTLs in the modern jargon) give real hope that a rational improvement, based on genetic models, of insects such as *B. mori* can proceed.

Relatively few insects are of economic importance because of their products. Most impact on human societies because they eat (or infect) either us or the animals and plants upon which we depend. Do we need genetic maps of these? The answer to this question is "yes", if we wish to use genetic (and here I include molecular) techniques for their control. Many genes of interest to insect biologists are those that determine the relative sensitivity of the species to an insecticide, those that affect vectorial capacity or those that cause interspecific hybrid sterility. But why map these genes? The most obvious answer to this question is that only by doing so can one be certain, in many cases, that one is dealing with a single gene character and that similar phenotypes are due to mutations of the same gene. (This last is particularly true of alleles that are dominant to the "wild type".) A dramatic example of the truth of this assertion is seen in the characterization of a genetic variant in *Anopheles gambiae* affecting vectorial competence for *Plasmodium cynomogi*: "Attempts to define the precise genetic basis for refractoriness have been complicated by a lack of genetic markers" (Collins *et al.*, 1986). If these mutations are to be manipulated genetically, for example transferred to other strains, then mapping, and particularly linkage to an obvious "visible" marker, makes this process both quicker and more precise. If the genes are to be cloned, and possibly transformed into other strains (or species), then mapping (especially to a molecular marker or to polytene chromosomes) greatly facilitates the task. Finally, there is great theoretical interest in comparing the genetic maps of related species for this throws considerable light on the important question of the evolution of chromosomes (e.g. Foster *et al.*, 1981).

III. POLYTENE CHROMOSOME MAPS

The discovery (or, rather, rediscovery) of giant polytene chromosomes in larval tissues of *Drosophila* and other Diptera (e.g. Heitz and Bauer, 1933; Painter, 1933) presented the first opportunity to relate a genetic map, an abstraction of the experimental data, to a physical map of the genome, something you can actually see. (This statement is not strictly true, as even earlier Painter and Muller (1929) had been mapping aberration breakpoints on mitotic metaphase chromosomes of *D. melanogaster*.) Painter realized that the banding patterns of the polytene chromosomes were constant features, recognizable from larva to larva, from strain to strain, from year to year. This allowed him to draw a map of these chromosomes, a process taken to its perfection by Calvin Bridges and his son, Philip. The research community is now well served for maps of the polytene chromosomes of *D. melanogaster*. Not only do the Bridges' maps of 50 years ago still form the basis for modern research, but we also have a fine photographic interpretation of these by Lefevre (1976) and a useful series of maps based on electron microscopy of thin sections (Sorsa, 1988b). There are maps of the polytene chromosomes of about 270 other species of drosophilid (Ashburner, 1989) and over 250 other species of Diptera (the majority being culicids and chironomids (Sorsa, 1988a)). Only a fraction (about 20%) of these are of a quality sufficient to make them of any use to the research community at large. But that is a technical problem – more serious for insect geneticists is that only two orders possess polytene chromosomes, the Diptera and the Collembola (Cassagnau, 1966). Even among the Diptera, although polytene chromosomes are known from the most "primitive" tipulid to the most "advanced" muscid, they are not always of sufficient "quality" to make them useful as a research tool. All too often the chromosomes are very poorly synapsed or very difficult to spread (perhaps due to ectopic connections between non-homologous regions) so that their very existence is simply a frustration. However, when "workable" polytene chromosomes are found, they are of enormous utility. So much so, that one should not despair too easily of obtaining workable polytene chromosomes of one's favourite fly – indeed the experience with *Ceratitis capitata* shows that with skill and perseverance polytene chromosomes can be used in the most unpromising of species (Zacharopoulou, 1987; Bedo, 1987).

As Dan Lindsley is fond of pointing out to the younger, more molecularly minded, generation, polytene chromosomes are indeed the "original" physical maps. Polytene chromosomes are important for many different reasons. First, and perhaps foremost these days, they allow the rapid mapping of any cloned DNA sequence, by the simple expedient of *in situ* hybridization. In *D. melanogaster* there are about 110 mb of DNA in the polytenized genome (the fact that repetitive (heterochromatic) sequences may not polytenize must not be forgotten). With of the order of 5000 discrete polytene chromosome bands, this means that the "average" band has some 22 kb of DNA (per haploid equivalent). The range is perhaps from 5 to 150 kb. Resolution of an *in situ* site to a band is not difficult to achieve with non-radioactive detection methods. Hybridization to a region within the largest bands is also not difficult to see, given first-rate preparations. Under optimal conditions fluorescent probes may be resolved to 3 kb (Rykowski *et al.*, 1988).

Another great feature of polytene chromosomes is that they are large enough to be microdissected. When Scalenghe *et al.* (1981) first introduced polytene chromosome microdissection and the subsequent microcloning of the dissected DNA, I was very sceptical that this was a technique of general value. This judgement was wrong, as the number of *Drosophila* genes cloned by this method soon showed. Luckily, nobody listened to my opinions on this matter, for now the polymerase chain reaction (PCR) has obviated the necessity to microclone the microdissected DNA, and has made chromosome microdissection an even more valuable tool in our armoury. I will discuss this method in more detail below.

Perhaps I should add, in passing, that necessity has shown that polytene chromosomes are not required for the viability of a microdissection/PCR strategy – it is possible to microdissect metaphase chromosomes and to amplify the products (Lüdecke *et al.*, 1989). There is no reason why this should not be used for insect chromosomes. I admit I would not like to try it with the chromosomes of many Lepidoptera or Coleoptera – "beetle droppings" is the derogatory term for these small chromosomes, attributed to the late Cyril Darlington. However, metaphase (in particular prometaphase) chromosomes may well be dissectable. Moreover, human cytogeneticists have made fantastic advances in the technique of *in situ* hybridization to metaphase chromosomes, using fluorescent-labelled probes. Not only can single copy probes as short as a few kilobases be detected, but sequences a few megabase pairs apart can be resolved. Even more remarkable is the fact that sequences as close as 100 kb (perhaps even less) can be resolved in interphase nuclei (review: Lawrence, 1990).

There may well be some advantage to such small chromosomes; they could be small enough to separate by pulse field gel electrophoresis. If so, then mapping cloned DNA fragments to chromosomes could be very rapid. In the case of *B. mori* this is unlikely – with 28 chromosomes and a genome size of 530 Mb (Gage, 1974) on average each chromosome will contain 19 Mb, too large a molecule for the current technology to separate. With *Heliothis*, however, at least the smallest chromosomes appear to be separable by pulse field gel electrophoresis (D. Heckel, personal communication).

Despite these advances in molecular techniques, polytene chromosomes retain their original importance to geneticists. The colinearity of the genetic and cytogenetic maps allows the rapid and high-resolution mapping of genes if chromosome aberrations are available. The resolution of this method, in *D. melanogaster*, is to about 20–50 kb, that is to within the size of all but the smallest bands. To achieve such a resolution, however, a large number of aberrations (especially of deletions) covering a reasonably small chromosome interval are required.

The similarity of polytene chromosome maps between closely related species offers a (reasonably) rapid way of inferring phylogenetic relationships. This is because a major class of chromosome rearrangement that distinguishes closely related species is the paracentric inversion. By analysis of the inversions shared between a group of related species it is not too difficult to construct a hypothesis of their phylogenetic relatedness (Sturtevant and Dobzhansky, 1936). Perhaps the best example of such an analysis is that of the endemic Hawaiian *Drosophila* by H. Carson and colleagues. The phylogenetic tree determined from polytene chromosome banding patterns now includes over 100 species (Carson and Yoon, 1982).

There is another fundamental use of polytene chromosome maps that must not go unmentioned. That is their use as diagnostic tools for distinguishing among members of a sibling species complex. Although sibling species with identical polytene chromosome banding patterns are known (homosequential species; they are especially common in the endemic Hawaiian drosophilid fauna) it is usual for sibling species to differ by fixed paracentric inversions. These can then be used as a diagnostic tool for those sibling complexes whose members are morphologically indistinguishable. This technique has been especially useful to distinguish between the different members of the *Anopheles gambiae* species complex, using the nurse cell polytene chromosomes of adult females (e.g. Coluzzi et al., 1979), and between species of simuliid, using larval salivary gland chromosomes (e.g. Vajime and Dunbar, 1975).

IV. MICROSSECTION LIBRARIES – "POLYTENE CHROMOSOMES ON A FILTER"

I have already mentioned the technique of polytene and mitotic chromosome microdissection. Originally, the microdissected DNA was cloned directly in λ-phage vectors (Scalenghe et al., 1981). This is a tricky technique and the yield of phage with inserts can be inconveniently low. Following the lead of Lüdecke et al. (1989) three groups independently circumvented the cloning step by amplifying (with the PCR reaction) microdissected DNA from polytene chromosomes (Saunders et al., 1989; Johnson, 1990; Wesley et al., 1990) (see Johnson (1991) for methods). In D. melanogaster it is now quite feasible to microdissect a single polytene chromosome band, digest the DNA with a restriction enzyme, ligate oligonucleotide linkers to the product and then amplify the DNA in a PCR reaction (Saunders, 1990). This amplified DNA can be labelled and used directly as a probe (e.g. to screen for clones or for in situ hybridization), or it can be cloned into plasmid or phage vectors (the sequences of the oligonucleotide linkers are designed with this in mind) or it can be used as a substrate for probing with other DNA. It is the last of these applications that I wish to discuss here.

The microdissection of polytene chromosomes does not have to be done to the resolution of the single band. For D. melanogaster microdissections of whole numbered polytene chromosome divisions, representing about 1% of the euchromatic genome (say 1.2 Mb), are routine (see below). Enough DNA results from the PCR amplification to spot onto filters. It is then possible to construct a filter with an ordered array of spots, each representing a single division of the polytene chromosome genome. These filters can be hybridized with labelled probes, thus allowing the mapping of the probe to the polytene chromosome without the need for in situ hybridization. It has not, yet, been thought worthwhile to construct such a filter representation of the polytene chromosomes for D. melanogaster, but it most certainly is worth while for species with poor polytenes, or for those for which in situ hybridization is not routine or where the skills to read the polytene chromosomes are scarce. Just such a "filter polytene chromosome" has recently been constructed for Anopheles gambiae sensu strictu. The nurse cell polytene chromosomes were divided into 54 regions by microdissection; each was then amplified and a "filter chromosome" consisting of an array of 54 spots constructed. With a genome content of 240 Mb, each spot represents about 4.5 Mb of sequence information (Zheng et al., 1991). The filters are a reusable mapping resource that could, for example, be distributed to several laboratories working on the same species. They will be of utmost value for mapping both known cloned genes and molecular markers.

V. MAPS OF RESTRICTION FRAGMENT LENGTH POLYMORPHISMS

The base sequence of genomic DNA is simply yet another phenotypic character, yet one that is enormously rich in its information content. Almost as soon as researchers began to use restriction enzymes to cut DNA molecules into specific fragments, polymorphisms for fragment length were discovered. Restriction sites, at least for type II restriction enzymes, are specific base sequences between four and eight base pairs in length. These sequences may be unique, or they may be redundant – that is to say some enzymes will cut a sequence with more than one base at a particular position within its site. At present, there are about 250 different restriction site sequences for which enzymes have been characterized (Roberts, 1991). Intraspecific polymorphism for any restriction site, due either to a simple base substitution or to the insertion (or deletion) of a DNA segment, will lead to an RFLP that can easily be detected if a cloned probe against the DNA fragments is available. Such polymorphisms are inherited as co-dominant Mendelian markers and can be mapped, both with respect to each other and with respect to "classical" visible markers. Although such polymorphisms were used for genetic mapping in adenoviruses (Grodzicker *et al.*, 1974) and yeast (e.g. Petes and Botstein, 1977) it was not until the seminal paper of Botstein *et al.* (1980) that the idea of making an RFLP *map* took off. RFLP mapping has proved to be an extraordinarily powerful technique, not only in organisms in which formal genetic analysis has always been difficult (e.g. humans), but also in organisms which have a long tradition of genetic mapping, e.g. maize (Helentjaris, 1987).

The requirements for constructing an RFLP map are simple: at least two strains of a given species (to provide the necessary genetic polymorphism; if these strains are both inbred but very different then so much the better; they can be different species if at least one backcross is both fertile and shows exchange) and cloned DNA sequences to provide probes (these may be genomic or cDNA). It is surprising, therefore, that insect biologists have made so little use of this method. Although RFLPs have been used as population markers (e.g. between European and African races of the honey bee (Hall, 1990)) I am not aware of any insect for which an RFLP map has been constructed. This should soon change as RFLP maps are now being made for two Lepidoptera, *B. mori* (M. Goldsmith, personal communication) and *Heliothis virescens* (Zraket *et al.*, 1990; D. Heckel, personal communication), the honey bee (Doane, 1990), at least four Diptera, *Anopheles gambiae* (P. Romans, F. Collins, F. C. Kafatos and K. Louis, personal communication), *Mayetiola destructor* (W. C. Black and J. J. Stuart, personal communications), *Ceratitis capitata* (A. Robinson, personal communication), and *Rhagoletis pomonella* (J. Feder, personal communication), and the beetle *Tribolium castaneum* (J. J. Stuart, personal communication; Brown *et al.*, 1990). The only technical problem I can see in constructing an RFLP map is if one's favourite species is exceptionally monomorphic. This appears to be so for some colonizing species, for example the medfly *Ceratitis capitata* (Gasperi *et al.*, 1986). Even then, collections from the original home of the species (in this case East Africa) may well prove to be more polymorphic than those that are derived, presumably as a consequence of a genetic bottleneck. If this fails then interspecific polymorphisms may well be exploitable, if an interspecific hybrid is fertile. Zraket *et al.* (1990) are using hybrids between *Heliothis virescens* and *H. subflexa* to generate an RFLP map of the former species and this method has been used to great advantage by both botanists and mouse geneticists (Copeland and Jenkins, 1991).

One great advantage of an RFLP map over one based on conventional phenotypic markers is, of course, that one is not limited by the availability of spontaneous or induced visible mutations. (I deliberately exclude lethal mutations here. Although lethal mutations are expected to be 5–8 times more frequent than visibles (Muller, 1954) the lack of balancer chromosomes in most insects makes their experimental manipulation tedious.) The work required to make an RFLP map of any given density of markers will largely be a function of genome size and the frequency of meiotic recombination. In organisms with small genomes, it is not difficult to make very high density maps. For example, the weed *Arabidopsis thaliana* has a genome size of 70 Mb and each cM corresponds to only 140 kb of DNA. Here an RFLP map with average density of about two cM has been constructed from only 90 markers (Chang *et al.*, 1988). Even for tomato, with a genome size of 710 Mb (and a map length of about 900 cM), an RFLP map with 300 loci has been made (Paterson *et al.*, 1988). Similarly, the RFLP map of maize has an average resolution of 5 cM, say 15 Mb (given a genome size of 5000 Mb and a map length of 1600 cM). For *Aedes aegypti*, with a genome size of about 780 Mb (Rao and Rai, 1987) and a genetic map length of, say, 220 cM (Munstermann, 1990), each cM corresponds to about 3 Mb. Although I suspect this to be too large a figure (due to an underestimate of the total map length) this is worse than in humans (where $1 \text{ cM} \cong 1 \text{ Mb}$). Nevertheless, an RFLP map of average resolution 5 cM should be quite feasible. For "the mouse" the present RFLP map has a resolution of 3 cM (about 5 Mb) although the objective is a 1-cM (1.5-Mb) map (Copeland and Jenkins, 1991).

Four further points should be made in respect of RFLP mapping in insects. The first is that RFLP mapping is most efficiently done not as two- or three-point crosses but as multilocus crosses. Then data handling and analysis become serious problems. However, human geneticists have solved this with computer programs both to handle the data and make maximum-likelihood estimates of linkage values (e.g. Lathrop *et al.*, 1985; Lander *et al.*, 1987). (For organisms with recombination in only one sex these programs will require some modification.) The second is that many insect genomes are very high in AT content, especially in non-coding regions. For example, *Pseudococcus obscurus* has an overall AT content of nearly 70% (Klein and Eckhardt, 1976). This may affect the choice of restriction enzymes to be used in the search for polymorphism. The third point is that in flies with "workable" polytene chromosomes, mapping probes that detect RFLPs to these by *in situ* hybridization will be a powerful adjunct to mapping and will enormously simplify the formal genetic analysis. The final point may be the most important. Clearly, one of the reasons for so few good genetic maps of insects is economic. To construct, and use, a map with a large number of different markers requires the maintenance of a large number of mutant strains. This is often difficult, and nearly always expensive, in labour if nothing else. RFLP mapping, however, does not demand the maintenance of a large strain collection. In the best circumstances a very few (perhaps only two) strains need to be kept. This has an obvious economic advantage over traditional genetic mapping.

RFLP maps have become of quite exceptional use for those interested in the genetic bases of quantitative traits. Although methods for mapping "polygenes" – i.e. loci contributing to quantitative variation of a character – had been developed for organisms such as *D. melanogaster* (Thoday, 1961) they were impractical for more general use. Lander and Botstein (1989) have developed a method for mapping polygenes, which they call "quantitative trait loci" or QTLs, with respect to RFLP markers. The success of these analytical tools has been demonstrated in plants, most spectacularly in mapping QTLs responsible for several quantitative characters in the tomato (Paterson *et al.*, 1988; Tanksley *et al.*, 1989). It is to be hoped that this technology can be used in insects, for example in improving silk yield and quality in multivoltine races of *B. mori*.

VI. MINI- AND MICRO-SATELLITE DNA MAPPING

The discovery of hypervariable mini-satellite DNA in humans (Jeffreys *et al.*, 1985) has not only revolutionized forensic science but has also provided a very useful tool for genome mapping. For genome fingerprinting, an oligonucleotide probe to a mini-satellite DNA sequence is used to characterize whole genomic DNA. The patterns of hybridization seen are often far too complex to be of much use for mapping. However, if the oligonucleotide probe sequence is first used to isolate a genomic clone that includes not only the hypervariable region, but also flanking unique sequence DNA, then one has a very valuable probe for RFLP analysis (Nakamura et al., 1987). The reason why this is so is that the mini-satellite sequences are typically tandem repeats of a short (10–60-bp) motif. Any given motif may be present at a very large number of chromosomal sites and, moreover, the number of tandem copies at any given site varies from allele to allele. In humans there is at least twice as much genetic polymorphism for these so-called VNTR (variable number of tandem repeat) sites than there is for restriction enzyme sites. Using probes to VNTR sequences in RFLP mapping, therefore, may make it easier to detect polymorphic fragments.

An alternative to RFLP analysis, but one that is probably as, if not more, powerful, is to take advantage of the fact that eukaryotic genomes appear to be full of "simple sequence" DNA (sometimes called "microsatellite" DNA). A simple sequence is just what it says, for example a repeat of the motif $(dG-dT)_n$ or of the triplet $(dC-dA-dG)_n$. The point about such simple sequences is that they are very susceptible to "slippage" during DNA replication (Streisinger and Owen, 1985; Tautz et al., 1986), so that the copy number of the motif, and hence the length of simple sequence DNA at any one chromosomal site, is variable. Using an oligonucleotide of a simple sequence, genomic clones including the simple sequence and adjacent unique DNA can easily be isolated. From these, unique primer oligonucleotides can be made that will amplify, in a PCR reaction, the simple sequence DNA from its site of origin. When this is done from a number of different strains it is usually found that the length of the simple sequence varies between strains (i.e. between alleles). For example Tautz (1989) amplified the simple sequence $(dC-dA-dG)_n \cdot (dG-dT-dC)_n$ known to be located in the *Notch* gene of *D. melanogaster*. Among 11 strains Tautz found four length variants of this sequence, each varying by three nucleotides (i.e. they were 199, 202, 205 and 208 bp long), because it so happens that this simple sequence is within a coding region of *Notch*. Tautz estimates that there will be a length-variable simple sequence about once every 10 kb of eukaryotic DNA – a stupendous reservoir of polymorphism. Indeed, *in situ* hybridization to polytene chromosomes of $(dC-dA)_n \cdot (dG-dT)_n$ probes shows such sequences to be very widespread in *Drosophila* (Pardue et al., 1987; Huijser et al., 1987). It may, at first sight, be surprising that the coding regions of genes should vary in the number of repeats of either a single amino acid or a simple amino acid motif. In fact, however, such variation is common (at least in *Drosophila*) (e.g. Muskavitch and Hogness, 1982; Costa *et al.*, 1991). It is often associated with the so-called "opa" repeat (Wharton *et al.*, 1985; Grabowski *et al.*, 1991) although it may sometimes have a very real functional significance (Costa *et al.*, 1992).

Simple sequence length polymorphisms are already being used for mapping the human genome (review: Weber, 1990) and there is no reason why their power should not be exploited for the mapping of insect genomes. A problem with RAPD markers is that they are not codominent. Moreover, simple sequence length polymorphisms may allow the detection of many alleles at a locus. In fact, PCR-based technologies offer other methods for detecting, and hence mapping, polymorphisms. Two groups have recently shown that oligonucleotides of arbitrary sequence can be used to amplify genomic DNA and to detect length polymorphisms in the amplified products. This technique, dubbed RAPD mapping by one of its inventors, obviates the need for any clone selection or sequencing for mapping polymorphic sites (Williams *et al.*, 1990; Welsh and McClelland, 1990).

VII. WHOLE GENOME MOLECULAR MAPPING

The idea of constructing a complete physical map of an entire eukaryotic genome began when improvements in technology made such a dream realizable. The first eukaryotic organisms for which this has been attempted have two common features, extensive formal genetic analysis and small genome sizes – the yeast *Saccharomyces cerevisiae* (15 Mb) (Olson *et al.*, 1986) and the nematode *Caenorhabditis elegans* (100 Mb) (Coulson *et al.*, 1986). These projects began using λ-phage and cosmids, respectively. Today, two further classes of vector are available, phage P1 and yeast artificial chromosomes (YACs), and I shall begin by considering the relative merits of vectors. The different strategies of genome mapping projects have been characterized as being "top down" or "bottom up" – meaning that the project either starts with the largest DNA inserts possible and works down in size (up in scale) or vice versa (see Evans (1991) for a recent review of strategies).

A. Vectors

The most obvious difference between vectors is the size of DNA they are able to accept as inserts: about 15 kb for λ, 40 kb for cosmids, 75–100 kb for P1 and in excess of 1000 kb (1 Mb) for YACs. The consequence of this is that a physical map made with these vectors differs in its scale, from the largest (λ) to the smallest (YACs). There is, of course, an inverse relationship between scale and the work involved in making the map – it is far harder to construct a cosmid map than one in YACs. YACs (Burke et al., 1987; Schlessinger, 1990; Hieter et al., 1990) differ from all the other vectors in that they are grown in yeast, rather than bacteria. They suffer from a relatively low cloning efficiency and it is difficult to recover large amounts of the cloned DNA from the transformed cells (though this may change (Smith et al., 1990)). There are persistent worries that YAC clones may be unstable (e.g. Garza et al., 1989) and that some YACs include DNA fragments that are non-contiguous in the genome supplying the cloned DNA (Green and Olson, 1990; Bentley and Davies, 1991). Finally, DNA sequences cloned in YACs often have to be subcloned into phage or cosmids for analysis. Against all of these disadvantages, however, YACs are powerful vectors for genome mapping projects, for they allow a comprehensive map to be constructed in the minimum of time.

Bacteriophage P1 has only rather recently been developed as a useful cloning vector (Sternberg, 1990). This phage can accept inserts of between 75 and 100 kb. P1 cloning efficiency is said to be quite high (10^5 clones/μg vector DNA) and the clones can be maintained as single copy plasmids in their host cells. This has an obvious advantage in reducing the probability of clone rearrangement but makes it difficult to obtain a good yield of cloned DNA. This problem is solved for P1 vectors by putting the phage lytic replicon under the control of the *E. coli lac* promoter. P1 libraries have been made for human DNA (Sternberg et al., 1990) and, as described below, for *D. melanogaster*.

Cosmid (and λ-phage) vectors are far more familiar than either YACs or P1s, simply because they have been available for far longer. They have very high cloning efficiencies (up to 10^7 clones/μg vector DNA) and large amounts of cloned DNA can be recovered. They may suffer from problems of instability, especially if they include repetitive sequences and are grown on $recA^+$ hosts. Their small size makes whole genomic mapping a major task.

A consequence of the great size of the inserts in YACs is that on the polytene chromosomes of *D. melanogaster* a single YAC may cover many bands. This means that overlaps between different YACs may often be inferred simply from their *in situ* hybridization sites (but see below). For cosmids and, presumably, P1 clones, this is not possible.

B. YAC Maps of the *Drosophila* Genome

Two groups in St Louis are collaborating on constructing a YAC-based physical map of the genome of *D. melanogaster* (Garza *et al.*, 1989; Hartl *et al.*, 1992; Ajioka *et al.*, 1991). This map is essentially complete. The strategy used was quite straightforward; YACs containing *Drosophila* DNA were mapped to the polytene chromosomes by *in situ* hybridization. Nearly 1000 different YACs have been mapped so far, with an estimated coverage of the euchromatic genome of 1.8 times. Although *in situ* sites can indicate physical overlap between YACs, proof of overlap, and hence of map contiguity, must come from more direct experiments, such as cross-hybridization between YACs, hybridization of other probes to different YACs, or restriction enzyme mapping in from the vector arms (Burke *et al.*, 1987; Riley *et al.*, 1990). The problem facing the YAC projects now is that common to all physical mapping projects in their "end game" period: how to effect closure of the map. In the *ideal* case closure will result in the number of "contigs" being equal to the number of chromosomes (or chromosome arms). Ajioka *et al.* (1991) make a strong case that closure of the YAC map will be achieved by mapping sequence-tagged sites (STSs, see below). Indeed, if the STSs are derived from cDNAs or from known cloned genes then one further important objective of any physical mapping project – tying the physical map to the genetic map – will be achievable. One encouraging feature of the *Drosophila* YACs is that heterochromatic sequences are clonable; for example, YACs derived from Y chromosome DNA have been characterized (Danilevskaya *et al.*, 1991).

C. P1 Maps

There is, so far, little experience with P1 clone libraries of any organism, but the size of P1 inserts (75–100 kb) suggests that they will be useful in filling in the "scale gap" between YACs and cosmids. Two groups are now working on P1 cloning and mapping in *Drosophila*, that of D. Hartl, who has published a description of a P1 library (Smoller *et al.*, 1991), and that of L. Rabinow (personal communication).

D. Cosmid Maps

The largest scale mapping project now underway in an insect is to construct a cosmid-based physical map of *D. melanogaster* by a European consortium (D. M. Glover and R. D. C. Saunders from Dundee, M. Ashburner from Cambridge, F. C. Kafatos, C. Louis, C. Savakis and I. Sidén-Kiamos from Crete and J. Modolell from Madrid). The experimental strategy is very similar to that first used by Coulson and colleagues for the map of *Caenorhabditis elegans* (Coulson and Sulston, 1988) – indeed the mapping method and programs for data analysis and map construction come directly from this group. This strategy is to select clones from a master library of *Drosophila* DNA in cosmids and to "fingerprint" these by restriction enzyme digestion. This does not mean constructing a restriction map of each cosmid; rather, the cosmid DNA is cut into fragments with an enzyme with a 4-bp recognition site that leaves staggered ends (we use *Hin*fI; this actually recognizes a 5-bp site, but the middle base can be anything, GANTC). These ends are then infilled with reserve transcriptase in the presence of a radioactive nucleotide triphosphate. The labelled fragments are separated on a sequencing gel and their sizes automatically read into a computer. Any two cosmids that overlap will possess a certain number of labelled fragments in common. The task of the computer program is to compare the fragment map of any single clone with that of all other clones in its memory and to tell us the clones that overlap (i.e. form a "contig"). This is a statistical exercise; clearly some pairs of clones will share fragments of the same size simply by chance. In the *Drosophila* project we demand that the probability of mistaken similarity be 10^{-5} in the first round of data analysis and 10^{-6} in the final round.

In the *C. elegans* mapping project all clones were considered to come from a single genome. For *D. melanogaster*, however, there are very good theoretical and practical reasons to think of the genome as being composed of 100 different genomes, each corresponding to one numbered division of the polytene chromosome map (Lander and Waterman, 1988). Rather than fingerprint cosmids at random, cosmids are first selected by screening the master cosmid library (which includes about 20 000 clones) with labelled DNA derived by the PCR amplification of DNA microdissected from a single polytene division (see above). In theory, and in practice, this enriches the clones that are derived from that division. The fingerprint databanks for each division are kept separately, making the computational task of detecting contigs easier. There is one major problem with this approach, and that comes from the fact that about 20%

of the genome of *D. melanogaster* is repetitive. Clearly, if the amplified microdissected DNA includes repetitive sequences then clones from anywhere in the genome may be recovered. There is one way in which we minimize (but do not eliminate) this "noise" – that is by taking advantage of the fact that *D. simulans*, an essentially homosequential sibling species to *D. melanogaster*, has far fewer copies of many of the middle-repetitive sequences common in the latter species (see Bingham *et al.*, 1981). By using microamplified *D. simulans* DNA as our probe we reduce the selection of clones from rogue sites. The second check we make is that from each contig selected clones are mapped by *in situ* hybridization. This also means that we can more accurately position the origin of the contigs on the chromosomes.

A detailed analysis of the first fruits of this mapping project, the map of the tip of the X chromosome, has been published (Sidén-Kiamos *et al.*, 1990) and the entire X chromosome has been mapped to a similar density by the end of 1991. For the tip of the X chromosome the theoretical coverage is over 100%; however, it is clear that the map is not closed, i.e. there are gaps between adjacent contigs. Closure will be attempted by several methods – these include cross-hybridization between cosmids (using end probes made from the phage promoters that flank the cloning sites) and by linking contigs with YACs (Coulson *et al.*, 1988, 1991; Kafatos *et al.*, 1991) and, in the future, P1 clones. It is crucial to relate the cosmid physical map both to the genetic map and to other physical maps that are being constructed. This is being done both directly and by mapping clones of known genes and of STSs to the different physical maps.

There is a rather different strategy for the physical mapping of cosmids that is being pursued, for *Drosophila*, by Hoheisel and colleagues (Hoheisel *et al.*, 1991). This is to select overlapping cosmids by filter hybridization with short oligonucleotides of random sequence (Poustka *et al.*, 1986). This strategy has already been used for an organism with a very small genome, Herpes simplex virus type 1 (Craig *et al.*, 1990), and there are theoretical considerations that make it attractive (see Lehrach *et al.*, 1990). A related strategy, in which terminal probes are made from pooled cosmids and used to screen a master cosmid array, has been advocated by Evans and Lewis (1989).

These strategies have been made technically feasible by the development of devices and robots that are able to make filters of clones at very high density (Mackenzie *et al.*, 1989; Nizetic *et al.*, 1991). With the present technology 9216 clones can be ordered onto a single 22×22-cm filter – for cosmids this means that two genomic equivalents of *D. melanogaster* can be contained on one filter. (Such high-density filters are used for screening by the European Consortium.) With even more capable robots promised (C. P. Jones, personal communication) the construction of large ordered libraries will soon be relatively trivial and routine. A great advantage of the robotic technology is that many duplicate copies of high-density filters can be made and distributed to other laboratories for screening. Since the position of the clones is identical on all filter sets, a common "reference library" (Lehrach et al., 1990) can be used by many. If the information derived from these laboratories is gathered centrally on a database, then this is a very powerful strategy.

VIII. THE NEXT STAGE

The physical mapping of an entire genome is not, of course, an end in itself. The physical maps are simply the reference point for future studies. I have already emphasized the necessity of tying any physical map to the classical genetic map. For *Drosophila* this will be done by a variety of techniques, including direct cross-hybridization of the clones that are the actual representation of the map to known (and genetically mapped) genes. There is, however, a strong theoretical and practical case for making any physical map independent of the clones from which it was constructed. Not the least of these is the risk of loss of the master clone library by the failure of a freezer or by fire, earthquake or war. For this reason Olson et al. (1989) have proposed that from each canonical clone on the map a small region (<300 bp) be sequenced and this sequence made publicly available. This sequence would be a "sequence-tagged site" (STS) and could be used not only for establishing the physical relationships among clones ("STS content mapping"; see Palazzolo et al. (1991) for an analysis of strategies using STSs for building maps) but also for selecting the same genomic region from any new clone library. The European cosmid mapping project for *D. melanogaster* is now determining STSs from their cosmids by sequencing from the SP6 and T7 promoters that flank the site of insertion of foreign DNA into the vector.

A different strategy would be to determine sequence motifs from a selected population of sequences – for example, from cDNAs. Sequencing random cDNAs has been proposed as a "short cut" to sequencing an entire genome (Brenner, 1990). As such, it is an attractive strategy, because of the high informational content of cDNAs. One problem is that the frequency of any single cDNA clone in a cDNA library will reflect the abundance of the corresponding mRNA in the population of mRNAs from which the library was constructed. Since the abundances of mRNAs can vary over an enormous range (at least a million-fold) this is a serious obstacle to the strategy. There are signs, however, that techniques are being developed to overcome this problem, and cDNA libraries that are "normalized", with each cDNA species being represented with similar abundance (Ko, 1990; Patanjali *et al.*, 1991; Palazzolo *et al.*, 1989), are being constructed. Partial sequences of cDNAs are not only interesting in their own right (because of the lead they give to the totality of protein coding genes) but also as STSs (or rather ESTs, "expressed sequence tags"). There are several groups now planning to map and sequence cDNAs in *Drosophila* (e.g. M. Palazzolo and colleagues). If the information is as good as that from human cDNAs, where 230 of 435 cDNAs suggest proteins for which nothing similar had been found before (Adams *et al.*, 1991), then exciting times are ahead.

For *Drosophila*, at least, there is a quite different class of sequence that could be used to determine STSs. These are sequences adjacent to genetically characterized and *in situ* mapped P element insertion sites. Very large collections of recessive lethal P element insertions now exist (e.g. Cooley *et al.*, 1988) and a project has begun to recover their insertion sites (by plasmid rescue) for sequencing (G. M. Rubin, A. Spradling and D. L. Hartl, personal communication).

Finally, I must discuss the question of total genome sequencing, with particular reference to *Drosphila*. As yet, nobody to my knowledge has embarked on such a massive project. However, I estimate that about 2% of the euchromatic genome of *D. melanogaster* has been sequenced and of this about 60% (1.4 Mb) is publicly available from the databanks. The longest contiguous sequence in *Drosophila* so far achieved is one of 28.6 kb from the *polyhomeotic* gene (Deatrick *et al.*, 1991). If the rate of increase in *Drosophila* sequence is the same as that for all sequences (i.e. doubling every 24 months, if not less) then the entire sequence of the *Drosophila* genome will have been achieved by 2004. In practice, the rate of growth of sequence data may well slow and, even if not, this sequencing effort is quite unco-ordinated. Not infrequently two or even three different, and often rival, laboratories sequence the same gene and even publish the sequences in the same issue of *Cell*. The question, which is a social and political one rather than scientific, is whether the community should make attempts to co-ordinate what is, after all, an expensive and boring job. One way would be for each of 200 laboratories to sequence one cosmid from a reference library a year, by no means an impossible task (except, perhaps, for the co-ordinators). This would give us the entire euchromatic sequence of *D. melanogaster* in about ten years. The model for such an operation (which is disparaged as a "cottage industry" by some) is the European effort to sequence chromosome III of yeast (Oliver et al., 1992). What is quite certain is that over the next decade the amount of sequence data available from D. melanogaster will increase enormously. This will have a great impact on the experimental analysis of development, evolution and behaviour. It should also have an enormous impact on insect science as a whole.

Acknowledgements

I thank all those who have allowed me to cite their unpublished work. I am very grateful to Dr P. Atkinson, Dr M. Goldsmith, Dr D. Heckel, Dr M. Palazzolo and Dr R. D. C. Saunders for advice. Several colleagues read a manuscript version of this paper; I thank Dr A. T. C. Carpenter, Dr D. M. Glover, Dr M. Goldsmith, Dr D. L. Hartl, Dr D. Heckel, Dr J. Hoheisel, Dr F. C. Kafatos and Dr J. Modolell for their helpful comments. Dr D. L. Hartl, Dr M. Palazzolo and Dr J. Hoheisel sent me preprints of their papers, for which I am most grateful. The cosmid mapping work summarized here is the product of a joint project with Professor D. M. Glover and Dr R. D. C. Saunders (Dundee), Professors F. C. Kafatos, K. Louis and C. Savakis and Dr I. Sidén-Kiamos (Heraklion, Crete) and Dr J. Modolell (Madrid) and is funded as part of the SCIENCE programme of the EEC.

REFERENCES

Adams, M. D., Kelley, J. M., Gocayne, J. D., Dubnick, M., Polymeropoulos, M. H., Xiao, H., Merril, C. R., Wu, A., Olde, B., Moreno, R. F., Kerlavage, A. R., McCombie, W. R. and Venter, J. C. (1991). *Science* **252**, 1651–1656.
Ajioka, J. W., Smoller, D. A., Jones, R. W., Carulli, J. P., Vellek, A. E. C., Garza, D., Link, A. J., Duncan, I. W. and Hartl, D. L. (1991). *Chromosoma* **100**, 495–509.
Ashburner, M. (1989). *"Drosophila*: A Laboratory Handbook". Cold Spring Harbor Press, New York.
Ashburner, M. (1991a). *Drosophila Information Service* **69**, 1–399.
Ashburner, M. (1991b). FlyBase – A *Drosophila* genetic database. Electronic readable files available from the EMBL fileserver, SEQNET and FTP.BIO.INDIANA.EDU.
Bedo, D. G. (1987). *Genome* **29**, 598–611.
Bentley, D. R. and Davies, K. E. (1991). *Genome News* **7**, 17–19.
Bingham, P. M., Levis, R. and Rubin, G. M. (1981). *Cell* **25**, 693–704.
Botstein, D., White, R. L., Skolnick, M. and Davis, R. W. (1980). *Am. J. Human Genet.* **32**, 314–331.
Brenner, S. (1990). *In* "Human Genetic Information: Science, Law and Ethics", *Ciba Foundation Symp.* **149**, 6–17.
Brown, S. J., Henry, J. K., Black, W. C. and Denell, R. E. (1990). *Insect Biochem.* **20**, 185–193.
Burke, D. T., Carle, G. F. and Olson, M. V. (1987). *Science* **236**, 806–812.
Carson, H. L. and Yoon, J. S. (1982). *In* "The Genetics and Biology of Drosophila", Vol. 3b (M. Ashburner, H. L. Carson and J. N. Thomson, eds), pp. 296–344. Academic Press, London.
Cassagnau, P. (1966). *C.R. Acad. Sci. Paris* **262D**, 168–170.
Central Silk Board (1989). "Silkman's Companion". Central Silk Board, Bangalore.

Chang, C., Bowman, J. L., DeJohn, A. W., Lander, E. S. and Meyerowitz, E. M. (1988). *Proc. Natl. Acad. Sci. USA* **85**, 6856–6860.
Collins, F. H., Sakai, R. K., Vernick, K. D., Paskewitz, S., Seeley, D. C., Miller, L. H., Collins, W. E., Campbell, C. C. and Gwadz, R. W. (1986). *Science* **234**, 607–610.
Coluzzi, M., Sabatini, A., Petrarca, V. and Di Deco, M. A. (1979). *Trans. R. Soc. Trop. Med. Hyg.* **73**, 483–497.
Cooley, L., Kelley, R. and Spradling, A. (1988). *Science* **239**, 1121–1128.
Copeland, N. G. and Jenkins, N. A. (1991). *Trends Genet.* **7**, 113–118.
Costa, R., Peixoto, A. P., Thackeray, J. R., Dalgleish, R. and Kyriacou, C. P. (1991). *J. Mol. Evol.* **32**, 238–246.
Costa, R., Peixoto, A. P., Barbujani, G. and Kyriacou, C. P. (1992). *J. Mol. Evol.* (in press).
Coulson, A. and Sulston, J. (1988). In "Genome Analysis: A Practical Approach" (K. E. Davies, ed.), pp. 19–39. IRL Press, Oxford.
Coulson, A., Sulston, J., Brenner, S. and Karn, J. (1986). *Proc. Natl. Acad. Sci. USA* **83**, 7821–7825.
Coulson, A., Waterston, R., Kiff, J., Sulston, J. and Kohara, Y. (1988). *Nature* **335**, 184–186.
Coulson, A., Kozono, Y., Lutterbach, B., Shownkeen, R., Sulston, J. and Waterston, R. (1991). *BioEssays* **13**, 413–417.
Craig, A. G., Nizetic, D., Hoheisel, J. D., Zehetner, G. and Lehrach, H. (1990). *Nucleic Acids Res.* **18**, 2653–2660.
Danilevskaya, O. N., Kurenova, E. V., Pavlova, M. N., Bebehov, D. V., Link, A. J., Koga, A., Vellek, A. and Hartl, D. L. (1991). *Chromosoma* **100**, 118–124.
Deatrick, J., Daly, M., Randsholt, N. B. and Brock, H. W. (1991). *Gene.* **105**, 185–195.
Doane, W. W. (1990). *Insect Molecular Genetics Newsletter* **5**, 1–2.
Doira, H. (1989). "Proceedings of the 6th International Congress SABRAO", pp. 961–964.
Evans, G. A. (1991). *BioEssays* **13**, 39–44.
Evans, G. A. and Lewis, K. A. (1989). *Proc. Natl. Acad. Sci. USA* **86**, 5030–5034.
Foster, G. G., Whitten, M. J., Konovalov, C., Arnold, J. T. A. and Maffi, G. (1981). *Genet. Res.* **37**, 55–69.
Gage, L. P. (1974). *Chromosoma* **45**, 27–42.
Garza, D., Ajioka, J. W., Burke, D. T. and Hartl, D. L. (1989). *Science* **246**, 641–646.
Gasperi, G., Malacrida, A. R. and Milani, R. (1986). In "Fruitflies. Proceedings of the 2nd International Symposium" (A. P. Economopoulos, ed.), pp. 149–157. Elsevier, Amsterdam.
Gethmann, R. C. (1988). *J. Hered.* **79**, 344–350.
Grabowski, D. T., Carney, J. P. and Kelley, M. R. (1991). *Nucleic Acids Res.* **19**, 1709.
Green, E. D. and Olson, M. V. (1990). *Science* **250**, 94–98.
Grodzicker, T., Williams, J., Sharp, P. and Sambrook, J. (1974). *Cold Spring Harbor Symp. Quant. Biol.* **39**, 439–446.
Hall, H. G. (1990). *Genetics* **125**, 611–621.
Harris, H. (1966). *Proc. R. Soc. Lond.* **164B**, 298–310.
Hartl, D. L., Duncan, I. W., Ajioka, J. W., Cai, H., Garza, D., Jones, R. W., Kieffel, P., Lackey, M. A., Link, A. J., Lohe, A. R., Lozovskaya, E. R., Martin, C. H., Palazzolo, M. J. Petrov, D., Sawyer, S. A., Smoller, D. A., Vellek, A. E. C. and Yee, J. (1992). *Trends Genet.* **8**, 70–75.
Heitz, E. and Bauer, H. (1933). *Mikrosk. Anat.* **17**, 67–82.
Helentjaris, T. (1987). *Trends Genet.* **3**, 217–221.
Hieter, P., Connelly, C., Shero, J., McCormick, M. K., Antonarakis, S., Pavan, W. and

Reeves, R. (1990). *In* "Genetic and Physical Mapping" (K. E. Davies and S. M. Tilghman, eds), pp. 83–120. Cold Spring Harbor Press, New York.
Hoheisel, J. D., Lennon, G. G., Zehetner, G. and Lehrach, H. (1991). *J. Mol. Biol.* **221**, 903–914.
Hubby, J. L. and Lewontin, R. C. (1966). *Genetics* **54**, 577–594.
Huijser, P., Hennig, W. and Dijkhof, R. (1987). *Chromosoma* **95**, 209–215.
Jeffreys, A. J., Wilson, V. and Thein, S. L. (1985). *Nature* **314**, 67–73.
Johnson, D. H. (1990). *Genomics* **6**, 243–251.
Johnson, D. H. (1991). *In* "PCR A Practical Approach" (M. J. McPherson, P. Quirke and G. R. Taylor, eds), pp. 121–155. IRL Press, Oxford.
Kafatos, F. C., Louis, C., Savakis, C., Glover, D. M., Ashburner, M., Link, A. J., Sidén-Kiamos, I. and Saunders, R. D. C. (1991). *Trends Genet.* **7**, 155–161.
Klein, A. S. and Eckhardt, R. A. (1976). *Chromosoma* **57**, 333–340.
Ko, M. S. H. (1990). *Nucleic Acids Res.* **18**, 5705–5711.
Lander, E. S. and Botstein, D. (1989). *Genetics* **121**, 185–199.
Lander, E. S. and Waterman, R. (1988). *Genomics* **2**, 231–239.
Lander, E. S., Green, P., Abrahamson, J., Barlow, A., Daly, M. J., Lincoln, S. E. and Newburg, L. (1987). *Genomics* **1**, 174–181.
Lathrop, G. M., Lalouel, J. M., Julier, C. and Ott, J. (1985). *Am. J. Human Genet.* **37**, 482–498.
Lawrence, J. B. (1990). *In* "Genetic and Physical Mapping", Vol. 1 (K. E. Davies and S. M. Tilghman, eds), pp. 1–38. Cold Spring Harbor Press, New York.
Lefevre, G. (1976). *In* "The Genetics and Biology of Drosophila", Vol. 1a (M. Ashburner and E. Novitski, eds), pp. 31–66. Academic Press, London.
Lehrach, H., Drmanac, R., Hoheisel, J., Larin, Z., Lennon, G., Monaco, A. P., Nizetic, D., Zehetner, G. and Poustka, A. (1990). *In* "Genetic and Physical Mapping", Vol. 1 (K. E. Davies and S. M. Tilghman, eds), pp. 39–81. Cold Spring Harbor Press, New York.
Lewontin, R. C. (1991). *Genetics* **128**, 657–662.
Lüdecke, H.-J., Senger, G., Claussen, U. and Horsthemke, B. (1989). *Nature* **338**, 348–350.
MacIntyre, R. J. and Collier, G. E. (1986). *In* "The Genetics and Biology of Drosophila", Vol. 3e (M. Ashburner, H. L. Carson and J. N. Thompson, eds), pp. 39–146. Academic Press, Orlando.
Mackenzie, C., Stewart, B. and Kaiser, K. (1989). *Technique* **1**, 49–52.
Merriam, J. M., Ashburner, M., Hartl, D. L. and Kafatos, F. C. (1991). *Science* **254**, 221–225.
Morgan, T. H., Bridges, C. B. and Sturtevant, A. H. (1925). *Bibliog. Genet.* **II**, 1–262.
Muller, H. J. (1954). *In* "Radiation Biology, High Energy Radiation", Vol. 1, part 1 (A. Hollaender, ed.), pp. 351–473. McGraw-Hill, New York.
Munstermann, L. E. (1990). *In* "Genetic Maps", 5th edn (S. J. O'Brien, ed.), pp. 3.179–3.183. Cold Spring Harbor Press, New York.
Muskavitch, M. A. T. and Hogness, D. S. (1982) *Cell.* **29**, 1041–1051.
Nakamura, Y., Leppert, M., O'Connell, P., Wolff, R., Holm, T., Culver, M., Martin, C., Fujimoto, E., Hoff, M., Kumlin, E. and White, R. (1987). *Science* **235**, 1616–1622.
Narang, S. K. and Seawright, J. A. (1990). *In* "Genetic Maps", 5th edn (S. J. O'Brien, ed.), pp. 3.190–3.193. Cold Spring Harbor Press, New York.
Nizetic, D., Zehetner, G., Monaco, A. P., Gellen, L., Young, B. D. and Lehrach, H. (1991). *Proc. Natl. Acad. Sci. USA* **88**, 3233–3237.
Oliver, S. G., van der Aart, Q. J. M., *et al.* (1992) *Nature* **357**, 38–46.
Olson, M. V., Dutchik, J. E., Graham, M. Y., Brodeur, G. M., Helms, C., Frank, M., MacCollin, M., Scheinman, R. and Frank, T. (1986). *Proc. Natl. Acad. Sci. USA* **83**, 7826–7830.

Olson, M. V., Hood, L., Cantor, C. and Botstein, D. (1989). *Science* **245**, 1434–1435.
Painter, T. S. (1933) *Science* **78**, 585–586.
Painter, T. S. and Muller, H. J. (1929). *J. Hered.* **20**, 287–298.
Palazzolo, M. J., Hyde, D. R., VijayRaghavan, K., Mecklenburg, K., Benzer, S. and Meyerowitz, E. (1989). *Neuron* **3**, 527–539.
Palazzolo, M. J., Sawyer, S. A., Martin, C. H., Smoller, D. A. and Hartl, D. L. (1991). *Proc. Natl. Acad. Sci. USA* **88**, 8034–8038.
Pardue, M. L., Lowenhaupt, K., Rich, A. and Nordheim, A. (1987). *EMBO J.* **6**, 1781–1789.
Patanjali, S. R., Parimoo, S. and Weissman, S. M. (1991). *Proc. Natl. Acad. Sci. USA* **88**, 1943–1947.
Paterson, A. H., Lander, E. S., Hewitt, J. D., Peterson, S., Lincoln, S. E. and Tanksley, S. D. (1988). *Nature* **335**, 721–726.
Petes, T. D. and Botstein, D. (1977). *Proc. Natl. Acad. Sci. USA* **74**, 5091–5095.
Poustka, A., Pohl, T., Barlow, D. P., Zehetner, G., Craig, A., Michiels, F., Ehrlich, E., Frischauf, A.-M. and Lehrach, H. (1986). *Cold Spring Harbor Symp. Quant. Biol.* **51**, 131–139.
Rao, P. N. and Rai, K. S. (1987). *Heredity* **59**, 253–281.
Rieger, R., Michaelis, A. and Green, M. M. (1976). "Glossary of Genetics and Cytogenetics". Gustav Fischer-Verlag, Jena.
Riley, J., Butler, R., Ogilvie, D., Finniear, R., Jenner, D., Powell, S., Anand, R., Smith, J. C. and Markham, A. F. (1990). *Nucleic Acids Res.* **18**, 2887–2890.
Roberts, R. J. (1991). REBASE, Release 9108. Electronically readable files available from GenBank and EMBL Fileservers.
Rykowski, M. C., Parmelee, S. J., Agard, D. A. and Sedat, J. W. (1988). *Cell* **54**, 461–472.
Saunders, R. D. C. (1990). *BioEssays* **12**, 245–248.
Saunders, R. D. C., Glover, D. M., Ashburner, M., Sidén-Kiamos, I., Louis, C., Monastirioti, M., Savakis, C. and Kafatos, F. C. (1989). *Nucleic Acids Res.* **17**, 9027–9037.
Scalenghe, F., Turco, E., Edström, J. E., Pirrotta, V. and Melli, M. (1981). *Chromosoma* **82**, 205–216.
Schlessinger, D. (1990). *Trends Genet.* **6**, 248–258.
Sidén-Kiamos, I., Saunders, R. D. C., Spanos, L., Majerus, T., Trenear, J., Savakis, C., Louis, C., Glover, D. M., Ashburner, M. and Kafatos, F. C. (1990). *Nucleic Acids Res.* **18**, 6261–6270.
Smith, D. R., Smyth, A. P. and Moir, D. T. (1990). *Proc. Natl. Acad. Sci. USA* **87**, 8242–8246.
Smoller, D. A., Petrov, D. and Hartl, D. L. (1991). *Chromosoma* **100**, 487–494.
Sorsa, V. (1988a). "Polytene Chromosomes in Genetic Research". Ellis Horwood Ltd, Chichester.
Sorsa, V. (1988b). "Chromosome maps of *Drosophila*". 2 Vols. CRC Press, Boca Raton, Florida.
Sternberg, N. (1990). *Proc. Natl. Acad. Sci. USA* **87**, 103–107.
Sternberg, N., Reuther, J. and de Riel, K. (1990). *New Biology* **2**, 151–162.
Streisinger, G. and Owen, J. E. (1985). *Genetics* **109**, 633–659.
Sturtevant, A. H. (1913). *J. Exp. Zool.* **14**, 43–59.
Sturtevant, A. H. and Dobzhansky, T. (1936). *Proc. Natl. Acad. Sci. USA* **22**, 448–450.
Tanksley, S. D., Young, N. D., Paterson, A. H. and Bonierbale, M. W. (1989). *Biotechnology* **7**, 257–264.
Tautz, D. (1989). *Nucleic Acids Res.* **17**, 6463–6471.

Tautz, D., Trick, M. and Dover, G. A. (1986). *Nature* **322**, 652–656.
Thoday, J. M. (1961). *Nature* **191**, 368–370.
Vajime, C. G. and Dunbar, R. W. (1975). *Tropenmed. Parasitol.* **26**, 111–138.
Weber, J. L. (1990). *In* "Genetic and Physical Mapping", Vol. 1 (K. E. Davies and S. M. Tilghman, eds), pp. 159–181. Cold Spring Harbor Press, New York.
Welsh, J. and McClelland, M. (1990). *Nucleic Acids Res.* **18**, 7213–7218.
Wesley, C. S., Ben, M., Kreitman, M., Hagag, N. and Eanes, W. F. (1990). *Nucleic Acids Res.* **18**, 599–603.
Wharton, K. A., Yedvobnick, B., Finnerty, V. G. and Artavanis-Tsakonas, S. (1985). *Cell* **40**, 55–62.
Williams, J. G. K., Kubelik, A. R., Livak, K. J., Rafalski, J. A. and Tingey, S. V. (1990). *Nucleic Acids Res.* **18**, 6531–6535.
Yokoyama, T. (1959). "Silkworm Genetics Illustrated". Japan Society for the Promotion of Science, Tokyo.
Zacharopoulou, A. (1987). *Genome* **29**, 67–71.
Zheng, L., Saunders, R. D. C., Fortini, D., Coluzzi, M., Glover, D. M. and Kafatos, F. C. (1991) *Proc. Natl. Acad. Sci. USA* **88**, 11187–11191.
Zraket, C. A., Barth, J. L., Heckel, D. G. and Abbott, A. G. (1990). In "Molecular Insect Science" (H. H. Hagedorn *et al.*, eds), pp. 13–20. Plenum Press, New York.

5

Molecular Aspects of Sex Determination in Insects

M. BOWNES

I. Introduction	76
II. The Genetics of Sex Determination in *Drosophila*	77
III. The Genetics of Sex Determination in Diptera	80
IV. The Genetics of Sex Determination in Haplodiploid Hymenopterans	84
V. Is a Single Model for Sex Determination in Insects Possible?	85
VI. The Molecular Basis of Sex Determination in *Drosophila*	89
VII. Molecular Analysis of Sex Determination in Other Insects	95
References	99

I. INTRODUCTION

Understanding sex determination in insects would be of great value for controlling insect populations. The classical genetic approach of using mutations which perturb normal sexual development in males or females, often transforming the sexual phenotype of the adult fly, has been extremely valuable in *Drosophila* for unravelling the hierarchy of genes which governs sex determination. Given this genetic background and the advances in modern molecular techniques it has now been possible to dissect at the molecular level the genes involved and largely to understand how sex-specific gene expression is achieved in *Drosophila*. Some genetic information has been collected on sex determination in other insects, but the sophisticated genetic approach used in *Drosophila* would certainly not be feasible in other insects. My aim here will be to look at what is known about sex determination in other insects and ask if it is sufficiently similar to *Drosophila* that we may be able to unravel sex determination in other more environmentally important insects by molecular techniques. I will confine my discussions to the control of sex determination in the somatic cells, since different genes seem to be involved in the determination of sex in the germ line, at least in *Drosophila*.

The genetic basis of sex determination in insects looks extremely diverse, since in some insects sex is determined by the Y chromosome, and in some by the ratio of X chromosomes to autosomes; other insects have no sex chromosomes at all but have male-determining factors located on the autosomes. Yet others use a haplodiploid system with diploids developing into females and unfertilized eggs into males. Nothiger and Steinmann-Zwicky (1985) have presented a single unifying principle for sex determination in all insects despite this variation. Is it possible that this common underlying genetic mechanism could share any homologous genes at the molecular level? I will argue that it may be the case but that we should proceed with caution and not assume that a similar regulatory cascade will necessarily be achieved by the same molecular mechanism.

II. THE GENETICS OF SEX DETERMINATION IN *DROSOPHILA*

The primary sex determination signal in *Drosophila* is the ratio of X chromosomes to autosomes. Males have one X and one Y chromosome (the Y chromosome is required for fertility but not for sex determination) and females have two X chromosomes. Males thus have an X : A ratio of 0.5, whilst females have an X : A ratio of 1.0. The X : A ratio is translated into sexual differentiation via a hierarchical series of regulatory genes. These genes were identified because of the sexual transformations observed when there are mutations at these loci. The key genes involved are *Sex-lethal* (*Sxl*), *transformer* (*tra*), *transformer-2* (*tra-2*), *doublesex* (*dsx*) and *intersex* (*ix*). There are also a number of important genes required to ensure that this system is set in motion. These include the maternally expressed gene *daughterless* (*da*), and various X-linked numerator elements (including *sisterless-a* and *sisterless-b*, *sis-a* and *sis-b*) which are involved in assessing the X : A ratio. At the end of the pathway are the sex differentiation genes which actually confer the sexual phenotype of the fly. This hierarchy has been recently reviewed by Slee and Bownes (1990), Steinmann-Zwicky *et al.* (1990) and Cline (1989).

Mutations in these genes cause transformations in the sex of the fly as shown in Table I. The state of activity of *Sxl* determines alternative pathways in males and females. It is inactive in males, and thus recessive mutations have no effect on males, but cause the transformation of XX

TABLE I. Genes involved in sex determination in *Drosophila melanogaster*

Gene	Mutation(s)	Phenotype	Function	Chromosomal location
daughterless (*da*)	Recessive: maternally acting	Female embryonic lethal	Necessary for activation of *Sxl* in female embryos	2
sisterless a and *b* (*sis-a* and *sis-b*)	Recessive	Female embryonic lethal	X chromosome numerator elements involved in *Sxl* activation in females	X
Sex-lethal (*Sxl*)	Recessive	XX Cells → male phenotype lethal to XX flies	Active in females to determine somatic differentiation by activation of *tra*	X
	Dominant	X0 cells → female phenotype lethal to XY flies	Inactive in males	
transformer (*tra*)	Recessive	XX flies → pseudomales	Active in conjunction with *tra-2* in regulation of *dsx* in females	3
transformer-2 (*tra-2*)	Recessive	XX flies → pseudomales	Active in females to induce female-specific *dsx* expression and repress male-specific *dsx* expression. Inactive in male sex determination	2
doublesex (*dsx*)	Recessive 3 alleles	XX flies → intersex or XY flies → intersex or XX and XY flies → intersex	Active in males to repress female differentiation functions	3
	Dominant	XX flies → intersex	Active in females to repress male differentiation functions	
intersex (*ix*)	Recessive	XX flies → intersex	Active in females in conjunction with *dsx* product to repress male differentiation functions	2

5. Molecular Aspects of Sex Determination in Insects

cells into a male phenotype, and are lethal to XX female flies. Conversely, dominant mutations cause XO cells to develop a female phenotype and are lethal to XY male flies, but have no effect on females. The reason why the mutations at *Sxl* are lethal to whole flies and the sexual transformations must be observed in clones of cells is that *Sxl* is also the key gene for controlling dosage compensation in *Drosophila*. This is achieved by hyperactivity of the one X chromosome in males and unless this is accurately controlled lethality results. Recessive alleles of *tra* and *tra-2* transform XX flies into pseudomales. These flies look morphologically like normal males, but they are sterile. Recessive alleles of *dsx* can transform both XX and XY flies into intersexual flies. These show both male and female characteristics, but are not mosaics. Dominant mutations at *dsx* and recessive mutations at *ix* transform XX flies into intersexes. Detailed genetic analysis of the interactions of these genes with each other and the phenotypes of the various mutations enabled a picture of the sex determination pathway to be built up (e.g. Cline, 1978, 1983; Baker and Ridge, 1980).

Mutations in the genes which regulate the expression of *Sxl* have phenotypes affecting just one sex. Duplication of *sis-a* and *sis-b* is lethal to chromosomal males, whereas deletion of one copy of both *sis-a* and *sis-b* is lethal to chromosomal females. There are mutant alleles of *da* where at specific temperatures all daughters of homozygous mothers die whilst their sons survive. Mutations in these genes also cause lethality as they are upstream of *Sxl*; thus it fails to function in dosage compensation as well as in sex determination.

As I describe the basics of how the hierarchy works in genetic terms I will use "on" for the production of an active gene product and "off" for an inactive product. Although it is easiest conceptually to think of the genes as being transcribed or not transcribed we will see that this is not always the case at the molecular level. I should also point out that sex determination is a cell-autonomous event, with each cell responding individually to the genetic cascade. This can be best visualized by the existence of gynandromorph flies which are part male and part female in phenotype, due to some regions of the fly having two X chromosomes and some regions having only one X chromosome.

The gene at the head of the hierarchy is *Sxl*; it is "on" in females with an X : A ratio of 1.0 and "off" in males with a ratio of 0.5. The product of the *da* gene and the X-linked numerator elements *sis-a* and *sis-b* are required for the activation of *Sxl*. The product of the *da* gene is present in both male and female embryos, as are the products of *sis-a* and *sis-b*; however, since the latter two genes are present in two copies in females and one copy in males there will be more gene product in females prior to the activation of *Sxl* and the onset of dosage compensation. The result of *Sxl* being "on" is that *tra* is also "on" in females, whereas it remains "off" in males where *Sxl* is "off". The final regulatory gene is different in that it is "on" in both males and females but expressed in two different forms, a male form and a female form. When *tra* is "on", *dsx* in the female mode is "on", and this leads to the repression of the male differentiation genes in females, thus promoting female differentiation. When *tra* is "off", *dsx* is "on" in the male mode, repressing the female differentiation genes and permitting male differentiation. The *tra-2* gene seems to be similarly expressed in male and female somatic cells and is on a side branch of the hierarchy. The precise role of *ix* is not yet clear, but it probably interacts with the female product of the *dsx* gene to regulate the expression of the sex differentiation genes. This scheme is outlined in Fig. 1, and should be seen as a simplified version of the *Drosophila* sex determination hierarchy.

Since the pivotal point is whether *Sxl* is "on" or "off", it is crucial to understand the method of assessment of the X : A ratio. One theory to account for this is that the autosomes encode a repressor of *Sxl* in equal amounts in males and females. Although Sex-lethal would have low-affinity binding sites for this repressor there would be high-affinity sites on the X chromosome. These would bind all of the repressor in females but in males some would remain bound to *Sxl*, thus repressing it (Chandra, 1985). Other models assume a balance between these putative negative regulators and the identified positive regulators such as *sis-a* and *sis-b*. There would thus be both repressors and activators of *Sxl* being produced in both males and females but the overall balance in their activity would lead to inactivation of *Sxl* in males and activation in females.

III. THE GENETICS OF SEX DETERMINATION IN DIPTERA

Although *Musca* has X and Y chromosomes, they are completely heterochromatic. The Y chromosome of some strains carries a male sex-determining factor *M*. Thus XX are females and XY are males. In other strains both males and females are XX and have *M* located on an autosome but the particular autosome varies between strains (Franco *et al*., 1982). *M* is usually found in a heterochromatic region and is behaving as if it is located on a transposable element (Denholm *et al*., 1986). Sex is determined by *M*, with *M/m* being males and *m/m* being female (Franco *et al*., 1982). Yet more strains were found where both males and females were homozygous for *M*. They carry a female-determining factor F^D at a

Maternal Genes

da +

♀ zygote (X:A=1)   ♂ zygote (X:A=0.5)

2 copies   1 copy

sis-a +		sis-a +	
sis-b +	Sxl +	sis-b +	Sxl +
	active		inactive

Zygotic Genes

| tra + | | tra + |
| active | | inactive |

| tra-2 + | | tra-2 + |
| active | dsx + | inactive | dsx + |

| ix + | ♀ form | ix + | ♂ form |
| active | | inactive | |

♀ differentiation genes ON   ♂ differentiation genes ON
♂ differentiation genes OFF   ♀ differentiation genes OFF

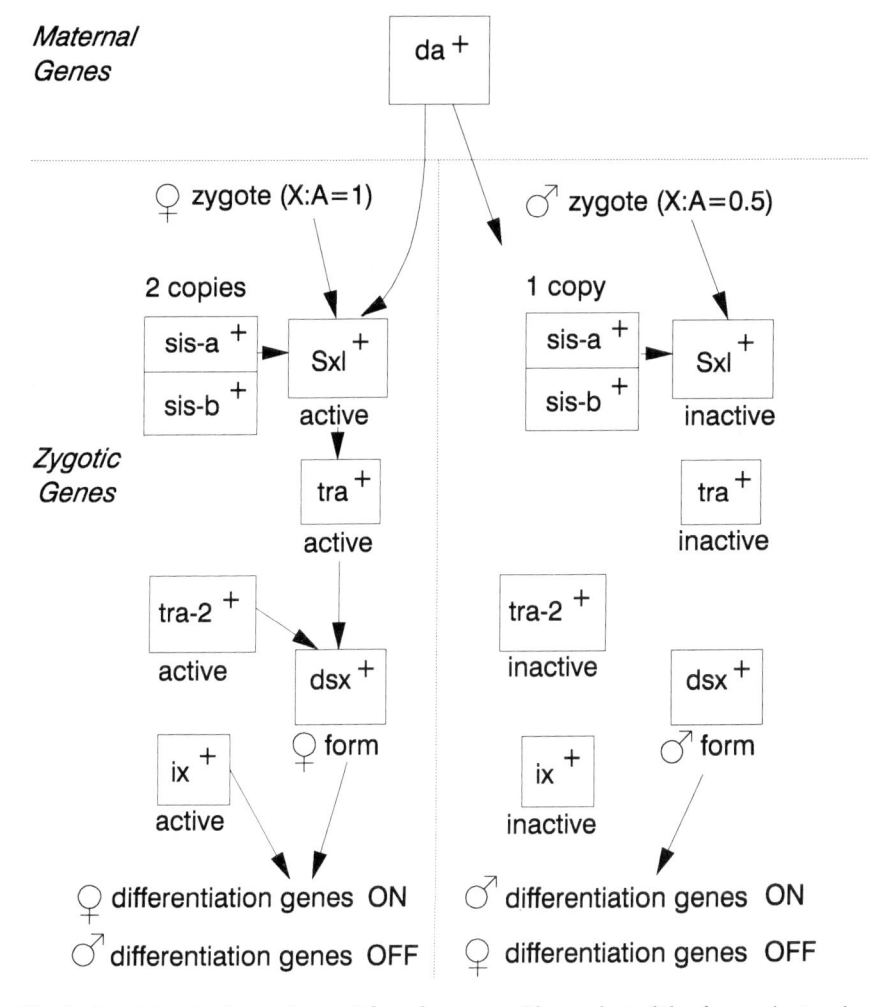

Fig. 1. Sex determination in *Drosophila melanogaster*. The product of the *da* gene is stored in the oocyte and is present in the zygote of both sexes. In female eggs the X : A ratio is 1. This information is used to activate *Sxl* in females and requires the products of the *sis-a* and *sis-b* genes along with the product of *da*. Since *sis-a* and *sis-b* are located on the X chromosome they are present in two copies in females. Thus *Sxl* is active in female zygotes. This leads to *tra* being active and the product of *tra* along with that of *tra-2* leads to the bifunctional *dsx* gene being active in the female form. This female (*dsx*) product along with the product of the *ix* gene represses the male differentiation genes in females, leading to the expression of the female differentiation genes. In the male zygote the X : A ratio is 0.5. Although the products of *da*, *sis-a* and *sis-b* are present, *Sxl* is not activated as insufficient *sis-a* and *sis-b* products are made from the single X chromosome present in male cells. Thus *Sxl* is inactive. In turn *tra* is inactive and *tra-2* has no function in sex determination in males. As a result *dsx* is expressed in the male mode and represses the female differentiation genes, leading to the expression of the male differentiation genes in males. This is a cell-autonomous process as this cascade operates independently in each cell.

fixed location on an autosome 4 which is dominant and epistatic to *M* and thus produces a female fly (Franco *et al.*, 1982; McDonald *et al.*, 1978). There are also genes called *Arrhenogenic* (*Ag*) and *transformer* (*tra*) which encode maternal effect female-determining factors. *Ag*/+ and *tra*/*tra* mothers produce fertile pseudomales, or intersexual offspring (Inoue and Hiroyoshi, 1986). The model proposed by Inoue and Hiroyoshi (1986) for sex determination in *Musca domestica* is shown in Fig. 2.

Calliphora also have an XY pair which is largely heterochromatic and which has no sex-determining function. The male-determining function is autosomally located (Lewis and John, 1968).

Like some strains of *Musca* the Chironomids have no morphologically different sex chromosomes, and sex is determined by a dominant male-determining factor. In some species this is linked to a specific chromosomal inversion, but in others its location is variable, thus behaving as if it is associated with a transposable element (Hagele, 1985; Martin and Lee, 1984). A female-determining factor has also been identified in some strains but the male-determining factor was shown to be epistatic to this (Thompson and Bowen, 1972). Thus sex determination in *Chironomus* looks essentially similar to that in *Musca* except that in *Musca* the female-determining factor is epistatic to the male-determining factor.

Sex determination amongst the mosquitoes is quite variable. *Anopheles gambiae* and *Anopheles culicifacies* show classical sex linkage, with the male being XY and the female being XX. In *Aedes* sex is determined by a male-determining factor which is dominant (McClelland, 1962; review: Clements, 1991). Gynandromorphs of male and female tissues have been described in *Culex* where sex is determined by a single gene on an autosome rather than a sex chromosome (Gilchrist and Haldane, 1946), suggesting that sex determination is cell autonomous as in *Drosophila*. Furthermore intersex flies similar in phenotype to the *ix*, *dsx* and *tra* mutants of *Drosophila* have been identified in *Aedes aegypti* and *Culex pipiens*. The recessive *ix* mutation of *Aedes aegypti* is on chromosome II, and transforms males into intersexes, but has no effect on females. This allele is temperature sensitive; at 27°C male larvae develop into male adults, at 30°C they develop into intersexes, and at 37°C they develop into females (Craig, 1965). This is similar to the *tra-2^{ts}* allele of *Drosophila*; however, *tra-2^{ts}* transforms chromosomal females into intersexes and at the higher temperatures into pseudomales. Certain alleles of *dsx* operate on chromosomal males, transforming them to a female developmental pathway, and constitutive *Sxl* mutations can also have this effect in single cells, though they are lethal to flies. Temperature-sensitive

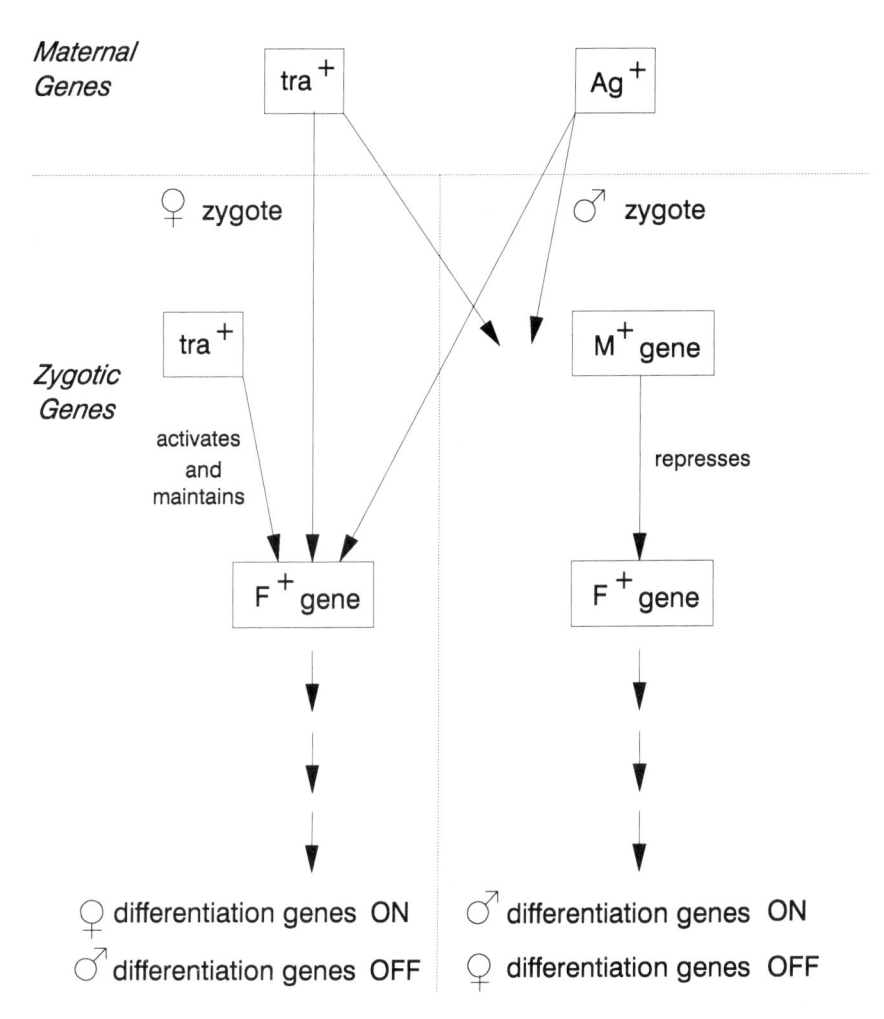

5. Molecular Aspects of Sex Determination in Insects

Fig. 2. Sex determination in *Musca domestica*. The maternal gene products *tra* and *Ag* are stored in the oocyte. In the female zygote these activate the *F* gene. The zygotic expression of *tra* is also needed to maintain *F* function. This ultimately leads to expression of the female differentiation genes but the other genes involved have not been identified. In male zygotes the *M* gene product is present which represses the *F* gene function. This ultimately leads to repression of the female differentiation genes and activation of the male differentiation genes. Again the steps involved after *F* are unknown.

alleles of *dsx* and *Sxl* have not been found in *Drosophila*. Since *ix* is epistatic to *M* in the mosquito, it seems to be downstream of *M* in the sex determination pathway. The temperature-sensitive period for *ix* is during larval development, as for *tra-2^{ts}* of *Drosophila* (Craig and Hickey, 1967), suggesting that perhaps it acts further down the regulatory cascade than a gene equivalent to *Sxl*.

The sex of some northern strains of *Aedes* depends upon the temperature at which they are reared, suggesting that in these species the native *ix* allele is temperature sensitive (Horsfall, 1974). As with the *ix* mutation described above, males are transformed into intersexes at the higher temperatures.

In *Culex pipiens* the autosomal mutation *zwitterfaktor* (*zwi*) transforms genotypic males into intersexes. *Zwitterfaktor* translates as gynandromorph factor but this is not an accurate description of the phenotype, which is intersexual. Another sex-linked gene *cercus* (*c*) causes the development of intersexes from chromosomal females, and has no effect on males (Barr, 1975). The resulting flies are sterile and do not take blood meals. Thus the *c* gene is similar to *tra* and *ix* of *Drosophila*, though as so few mutations have been studied in *Culex* that transform sex it could also be similar to *dsx*, but alleles which transform genetic males have not been found yet. Clearly in mosquitoes there are a number of genes involved in sex determination, suggesting a regulatory cascade similar to that of *Drosophila*; however, further genetic experiments are needed to place these genes into a hierarchical sequence to determine which ones might be functionally equivalent to specific *Drosophila* genes.

IV. THE GENETICS OF SEX DETERMINATION IN HAPLODIPLOID HYMENOPTERANS

These species have a most unusual sex determination mechanism, with fertilized eggs developing into diploid females and unfertilized eggs into haploid males. The occurrence of some diploid males in brother–sister matings led Whiting (1943) to propose that in *Bracon hebetor* sex is determined by the existence of a number of alleles at a single sex-determining locus, and that heterozygotes for these alleles become females whereas homozygotes and hemizygotes become males. Similar mechanisms have been suggested for several Hymenoptera, including the ant, *Solenopsis invicta*, and the honey bee, *Apis mellifera* and *Apis cerana indica* (Wyoke, 1979). Others have proposed for the parasitoid wasps that there are multiple sex-determining loci and that heterozygotes for any of the alleles are females, and that males are hemizygous or must be homozygous at all of the loci (Crozier, 1971). An alternative model suggesting that sex in the Hymenoptera is determined by a balance between non-additive male-determining genes and additive female-determining genes scattered throughout the genome has also been proposed (e.g. Kerr and Nielson, 1967); this was largely based on the observation that triploids were always females.

Recently this debate has been tested experimentally in the saw fly *Athalia rosae ruficornis* and the production of triploid males by repeated brother–sister matings provided clear evidence that in this species there is a single locus with multiple alleles determining sex (Naito and Suzuki, 1991). The triploid males were morphologically normal but were larger than the diploid and haploid males. It seems likely then that in some species there is a single locus and in others several loci which must be heterozygous to select the female pathway of sex determination and differentiation (Bull, 1981). Thus although this is referred to as haplodiploid sex determination it is not whether the flies are haploid or diploid *per se* which is critical, but the state of heterozygosity of specific alleles which is essential for initiating female development. A scheme for sex determination in this system is shown in Fig. 3.

V. IS A SINGLE MODEL FOR SEX DETERMINATION IN INSECTS POSSIBLE?

With such a diversity of mechanisms governing sex and the variability seen even between closely related species, and even between strains, it may not seem possible to bring all these processes together in one model. Nöthiger and Steinmann-Zwicky (1985), however, succeeded in doing this, and it seems that this is still the best interpretation of the data available. Their idea is that in all insects there is a gene equivalent to *Sxl*, a repressor which inactivates *Sxl*, a gene which activates *Sxl* and a gene at the end of the pathway which is equivalent to *dsx*, and which is expressed in two alternative forms to interact with one or other of the two sets of male and female downstream differentiation genes.

Since sex determination seems to be essentially similar in all of the dipteran insects, I will make the comparisons with *Musca* as most genetic data exist for this insect. The male-determining factor *M* described for *Musca* and many other dipterans would correspond to the repressors of *Sxl* which are involved in assessing the X : A ratio in *Drosophila*. These have been postulated to exist but have not yet been found in *Drosophila*. The recessive gene *m* would not encode a functional product so *Sxl* would be active. In *Drosophila* there are also maternal genes such as *da* which

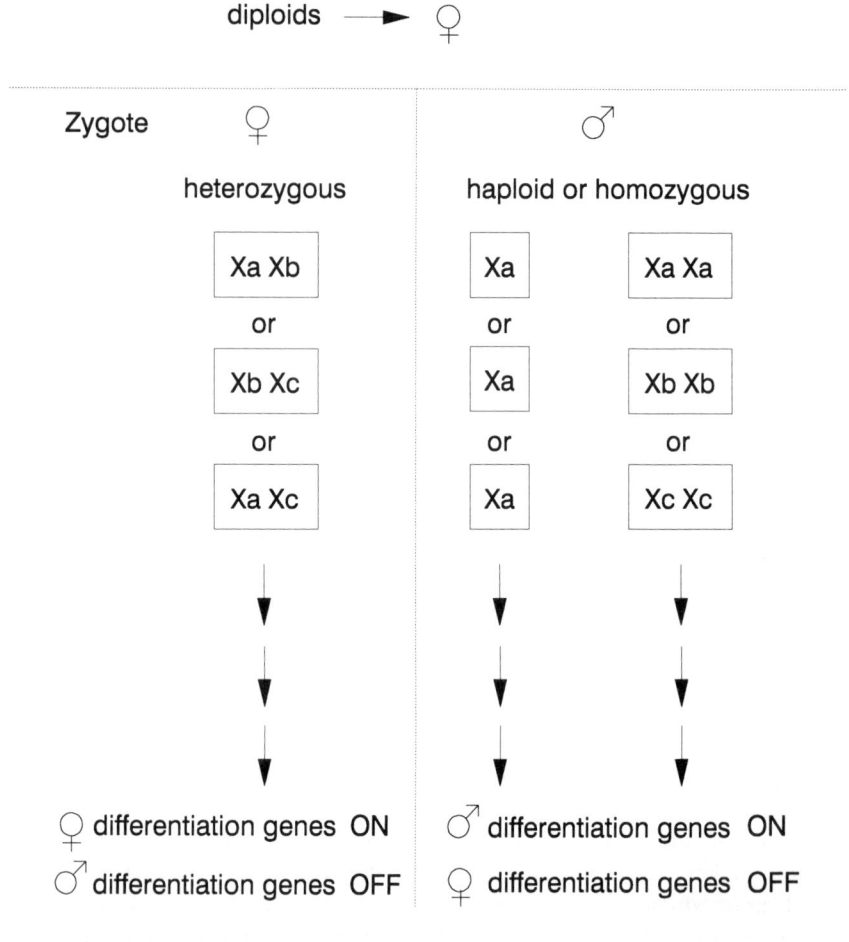

haploids ⟶ ♂
diploids ⟶ ♀

Zygote	♀ heterozygous	♂ haploid or homozygous	
	Xa Xb	Xa	Xa Xa
	or	or	or
	Xb Xc	Xa	Xb Xb
	or	or	or
	Xa Xc	Xa	Xc Xc

♀ differentiation genes ON
♂ differentiation genes OFF

♂ differentiation genes ON
♀ differentiation genes OFF

Fig. 3. Sex determination in *Athalia rosae*. Under normal circumstances haploids develop into males and diploids into females. This depends upon a locus X being heterozygous in the females. Any zygotes homozygous for an allele at X develop into males also. How such a system where multiple alleles for a given locus can promote male or female development dependent upon whether it is homozygous, hemizygous or heterozygous could operate is unknown.

are essential to activate *Sxl*. The equivalent genes in *Musca* would be *tra* and *Ag*. The *F* gene of *Musca* would be equivalent to *Sxl*. The phenotypes of mutations in these genes are not quite the same between the species. I assume that this is because of the fact that in *Drosophila* there is a need for dosage compensation because there are many genes located on the X chromosome which must be expressed to equivalent levels in males and females. This is achieved by higher levels of expression of the single X chromosome in males (Lucchesi et al. 1981) and its regulation is intricately linked with sex determination in *Drosophila* and is controlled by *Sxl* (Cline, 1983). This mechanism is not needed in insects with heterochromatic sex chromosomes or no sex chromosomes; thus mutations in the genes equivalent to or upstream of *Sxl*, namely *tra*, *Ag* and *F* of *Musca*, cause sex transformations, whereas in *Drosophila* they are lethal to one sex in whole flies and the sexual transformations can only be observed in cells by producing genetic mosaics.

No gene equivalent to any of the genes downstream of *Sxl* has been definitively identified in any other insect, although some of the mutations observed in mosquitoes may be in this category. A common scheme for sex determination in insects is shown in Fig. 4.

The haplodiploid mechanism of sex determination described for hymenopterans was also brought in line with the common sex determination scheme proposed by Nothiger and Steinmann-Zwicky (1985). The multiple alleles which must be heterozygous to confer female development are proposed to be mutations at *Sxl* which produce an inactive gene product and therefore lead to male development when they are homozygous or hemizygous. For this model to work it has to be proposed that the different alleles are in some way able to complement each other so that when two different mutations are present in a diploid animal then they are able to make a functional product. That this gene is equivalent to *Sxl* is a little more difficult to justify as it seems just as likely to me that products of this locus could have to interact with an *Sxl*-like gene and therefore be upstream of it. Perhaps as heterodimers they can activate a gene equivalent to *Sxl* but as homodimers they cannot. It is possible then that these genes are more similar to *sis-a* and *sis-b* of *Drosophila*. This could work equally well, imagining them to be repressors which only function as homodimers.

So it is possible to envisage genes being formally equivalent to each other in terms of their genetic function and even insects with environmentally determined sex can be accommodated. In the paedogenetic gall midge *Heteropeza* sex is determined by the nutritional status of the mother. This could reflect conditional mutations in maternal factors affecting the expression of *Sxl*; such conditional mutations have been

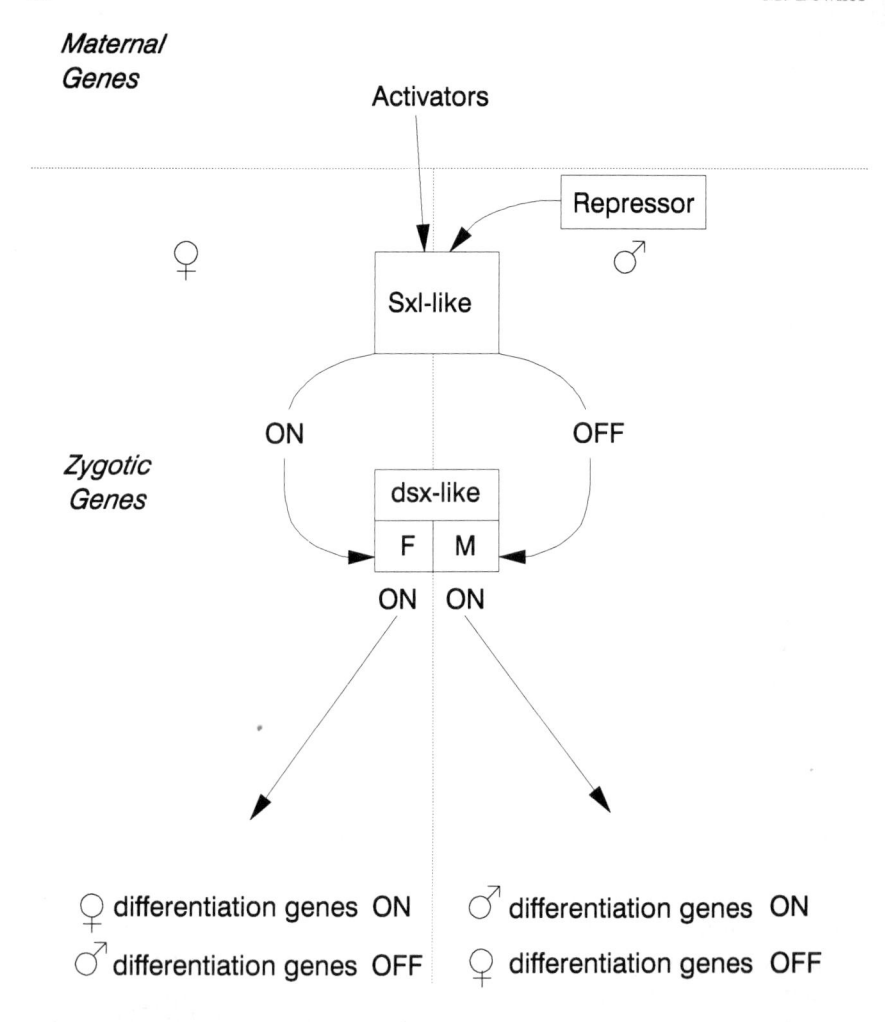

Fig. 4. A common scheme for insect sex determination. The common genetic cascade involves the maternal production of activators of an *Sxl*-like gene. There may also be other activators produced zygotically. The male zygotes produce a repressor of the *Sxl*-like gene. Ultimately this leads to a functional *Sxl* product in females but not males. This would result in the expression of a gene at the end of the cascade equivalent to *dsx* either in the female or male mode. The products of this locus would ultimately select the appropriate male or female differentiation genes for activation or repression. Only genes which function at the top of the hierarchy have been identified in *Musca* and a number of insects other than *Drosophila*, but some downstream genes may have been identified in *Culex* and *Aedes*.

identified in zygotic components of the pathway such as the temperature-sensitive *tra-2* allele of *Drosophila* which alters the sex of the organism according to the temperature at which it is reared. However, what I want to ask in this chapter is what is the likelihood that any of these genes are homologous at the molecular level and which genes are they likely to be. I will therefore describe the molecular basis of sex determination in *Drosophila*.

VI. THE MOLECULAR BASIS OF SEX DETERMINATION IN *DROSOPHILA*

The initial step in the process of sex determination for the embryo is the activation of *Sxl* in females and its repression in males. This requires a number of gene products which are maternally supplied, but which are present in both males and females. The best characterized of these is *da*, which encodes a putative regulator of transcription within the helix-loop-helix class (Cronmiller and Cline, 1987). The predicted *da* product is part of a group of proteins which generally act as heterodimers to bind to enhancers in the DNA (Murre *et al.*, 1989). Thus the product of *da* probably interacts with another protein, e.g. *sis-b*, to bring about its regulation of *Sxl*, and its function as a heterodimer is consistent with this gene having other roles in development in addition to *Sxl* activation.

Sxl activation also requires the zygotically derived products of the numerator elements which are located on the X chromosome and would thus yield higher amounts of gene products in female than in male zygotes. These include *sis-a* and *sis-b*, the latter being located within the achaete–scute complex which is important in neurogenic development. The *sis-b* function corresponds to the T4 protein of the achaete–scute complex and like *da* is a member of the helix-loop-helix regulatory protein class (Torres and Sanchez, 1989; Erickson and Cline, 1991). Thus the products of the genes activating *Sxl* have a number of different functions in development; for example, *da* is involved in neurogenesis, and this is probably mediated by them interacting in different combinations in different tissues at different stages of development. In the case of sex determination it is probably the products of *da* and *sis-b* which interact to activate *Sxl*.

There may also be repressors of *Sxl*, perhaps autosomally encoded helix-loop-helix proteins, which could act by sequestering *da* or *sis-b* products, but these have not yet been identified. The balance of these regulatory molecules leads to the expression of *Sxl* in females, but not in males. There are numerous roles for the *Sxl* gene products, including autoregulation of the *Sxl* gene itself. The action of *Sxl* does not depend upon a simple on/off switch. There are multiple transcripts produced from this gene and these include female-specific forms and male-specific forms. The *Sxl* product itself is required to generate female-specific transcripts. The difference between the male and female transcripts depends upon alternative processing of a primary transcript, such that different exons are included in the transcripts unique to each sex. This sex-specific splicing to generate sex-specific products is a theme we will meet again and again in the somatic control of sex determination in *Drosophila*. When *Sxl* is spliced in the female mode it encodes a protein of 354 amino acids, but when it is spliced in the male pattern it encodes a protein of only 48 amino acids which is unlikely to be functional. This is shown in Fig. 5, and has been described in detail by Bell *et al.* (1988) and Maine *et al.* (1986). Quite how this female-specific splicing pattern is initiated in females remains unclear, but there are some early and transient transcripts from the *Sxl* gene which may have a role in this process. Certainly these early transcripts are initiated at a different site to the later ones (Salz *et al.*, 1989), and may be confined to the female embryo. Thus it may be that the initial activation of the gene is by transcriptional control involving the various numerator elements, maternal factors and repressors, and that once this early female-specific transcript has been made the later transcript is initiated, and by this time there would be enough of the early female-specific *Sxl* product to initiate the female-specific splicing pattern of the late transcripts in females. As one might expect from its function, once *Sxl* was sequenced and analysed it was found to encode an RNA-binding protein.

Genetic evidence indicates that *Sxl* regulates *tra*. Given that *Sxl* autoregulates by means of differential splicing it might be expected to regulate *tra* by a similar mechanism, and indeed this is what happens (see Fig. 5). When the *tra* gene was characterized it was shown to produce two transcripts. One transcript is found in both males and females, and one transcript is found only in females. The female-specific and non-sex-specific transcripts differ in their splicing pattern, and the latter has no long open reading frame whilst the former encodes a protein of 211 amino acids. The characteristics of this protein mean that it is likely to bind to nucleic acids (Boggs *et al.*, 1987). Using *tra* cDNAs and expressing them in *Drosophila* by P element transformation it was shown that the non-sex-specific transcript has no sex-transforming function but the female cDNA-directed female development in chromosomal males (McKeown *et al.*, 1988). The function of *tra* is in the regulation of *dsx*.

Thus there are in the transcript of the *tra* gene two alternative choices for splicing, one of which is the preferred site seen in the non-sex-specific

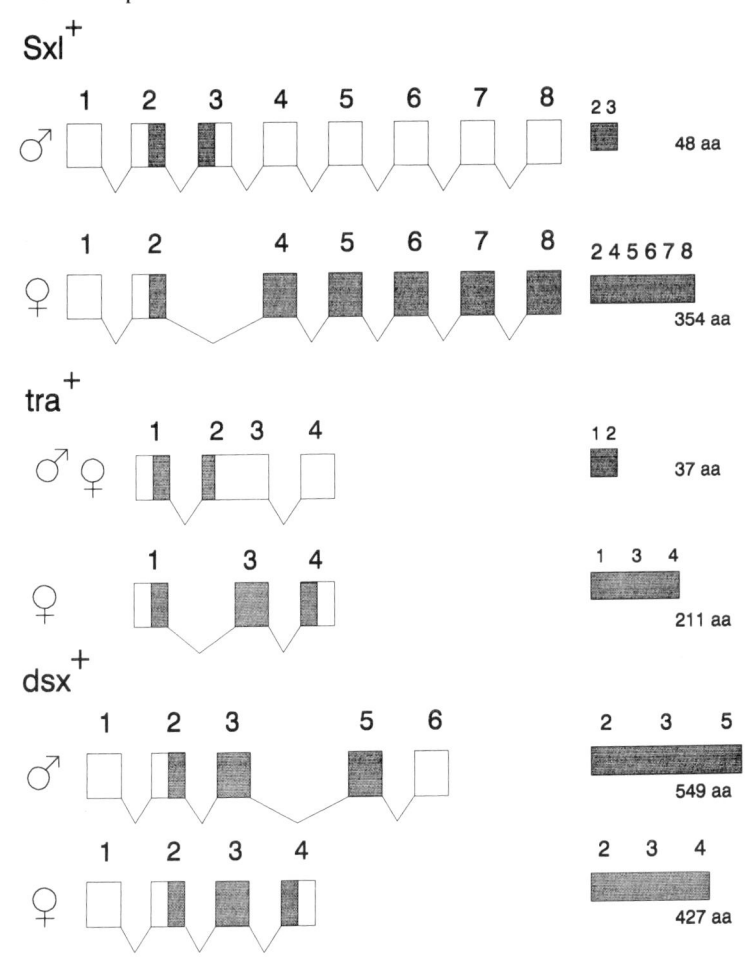

Fig. 5. Sex-specific RNA processing in *Drosphila*. □, exons (not drawn to scale); ■, coding regions (not drawn to scale); ∨, introns (not drawn to scale). The RNA splicing is shown to the left and the final protein product to the right. The size of the protein product in terms of number of amino acids is shown as are the exons which contribute to the protein. In *Sxl*, exon 3 is included in the final product of males, but contains a stop codon, thus producing a small non-functional protein. In females, however, exon 3 is omitted and exons 2 and 4 are spliced together. This results in a long open reading frame. In *tra* the non-sex-specific transcript includes exon 2 which contains a stop codon and produces a non-functional truncated protein. The female splicing omits this exon and splices within the 2/3 exon to include exon 3 which encodes a function protein. In *dsx* there are two alternative splicing patterns. In males exon 4 is omitted and a polyA tail is added to exon 6. In females, exon 4 is included and a polyA tail is added here, thus omitting exons 5 and 6. Both transcripts encode functional proteins.

product and the other only being used in the presence of the product of the *Sxl* gene. Thus somehow *Sxl* must guide the choice of this splice acceptor site in females, and one could imagine this could be by repressing the nonspecific acceptor site or promoting the use of the female-specific acceptor site. This has been tested in Kc cells by co-transfection of specific *Sxl* cDNAs and test constructs carrying genes that will generate unspliced *tra* transcripts. Only female-specific *Sxl* cDNA can direct the female-specific splicing of *tra*, and male cDNAs and female cDNAs carrying a frame shift mutation are unable to do this. By using mutant constructs which deleted either the nonspecific or the female-specific acceptor site (Inoue *et al.*, 1990), it was shown that the *Sxl* product binds to the non-sex-specific acceptor site and represses its use. *Sxl* protein produced in *E. coli* binds to *tra* RNA which contains this site but not to RNA in which this site has been deleted (Inoue *et al.*, 1990). These splice choice decisions are described in Fig. 6.

The *tra-2* gene is rather different and is similarly transcribed in both males and females. Thus it is not regulated by *Sxl* and is on a side branch of the hierarchy (see Fig. 1). Nonetheless it is required along with the *tra* gene to bring about female-specific splicing of the *dsx* transcript, and is critical for the correct expression of the downstream differentiation genes. The cDNA corresponding to the major transcript has been sequenced and encodes a protein of 264 amino acids which has homology to ribonucleoproteins and is thus likely to have a direct function in the correct splicing of *dsx* transcripts (Amrein *et al.*, 1988).

The products of both *tra* and *tra-2* are required for the correct splicing of the *dsx* transcript (Hoshijima *et al.*, 1991). The genetic evidence suggested that there were both male and female products from the *dsx* gene which were important for the regulation of expression of the downstream differentiation genes. This was shown to be the case by the molecular analysis of the gene. *dsx* encodes a number of transcripts, including some non-sex-specific and some sex-specific transcripts. When sex-specific cDNAs were sequenced they were found to come from a common precursor, which is differentially spliced and has alternative polyadenylation sites in the male and female. Both transcripts, however, give rise to a functional protein that is 549 amino acids in the male and 427 amino acids in the female. As can be seen in Fig. 5 these have a common N-terminus but have different C-termini (Burtis and Baker, 1989). The predicted protein products of this gene have little in common with other proteins in database searches, but they do have long stretches of repeated amino acids which have been detected in other regulatory proteins. That these cDNAs perform the predicted genetic function has been shown by transformation experiments. The expression of the female

5. Molecular Aspects of Sex Determination in Insects

Regulation of tra$^+$ splicing by Sxl products

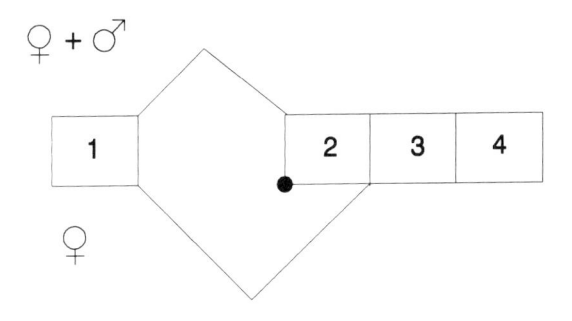

Regulation of dsx$^+$ splicing by tra$^+$ and tra-2$^+$ products

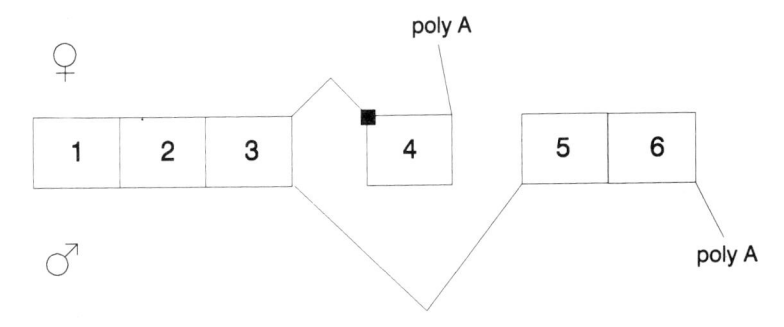

Fig. 6. Splicing choices in *Drosophila*. ●, repressor (product of *Sxl* gene); ■, activator (products of *tra* and *tra-2* genes). Only the splice junctions where there is a choice in which exons are ligated are shown in these diagrams. When *tra* is spliced in the female form the *Sxl* product acts as a repressor, binding to the splice acceptor site at exon 2 and forcing the use of another site at the start of exon 3. When *dsx* is spliced in females the products of *tra-2* and *tra* bind to the acceptor site in exon 4, promoting its use. If no activator is bound here the splice acceptor site at the start of exon 5 is used instead; this is the default, male splicing pattern. The positions of the polyadenylation sites are marked for the two transcripts.

cDNA brings about female development in both chromosomal males and females (Burtis and Baker, 1989).

The regulation of splicing of *dsx* does not seem to involve a negative mechanism as in the case of *tra*, but rather seems to be an active promotion of the use of the splicing of exon 4 to exon 3 in the female-specific transcript by both *tra* and *tra-2* products, which are therefore acting as activators of a specific splice site choice. If these products are not present then *dsx* is spliced in the male mode; thus the male product is the default state and the female product must be actively promoted, and there are sequences within exon 4 which are critical for this splice site to be used (Nagoshi and Baker, 1990). Co-transfection experiments expressing both *tra* and *tra-2* in Kc cells lead to the female-specific splicing of the *dsx* transcript (Hoshijima *et al.*, 1991). Deletions of various parts of the transcript showed that this was an activation which involved the use of sequences within exon 4. This requires both *tra* and *tra-2* products. The binding of *tra-2* protein produced in *E. coli* is critical for the female-specific splicing and the female-specific polyadenylation (Hedley and Maniatis, 1991). This is summarized in Fig. 6.

The role of the sex-specific *dsx* proteins is to bring about the appropriate expression of the downstream differentiation genes. Whilst a number of sex-specific genes have been identified which might come into this category, most of them are expressed in the reproductive tissues, such as the chorion genes in females (Mariani *et al.*, 1988) and several male-specific transcripts expressed in the testis and accessory glands (Monsma and Wolfner, 1988; Schäfer, 1986). These do not seem to be directly regulated by *dsx* and it seems likely that *dsx* is involved in the early decisions as to the sex these tissues will develop into; thereafter they are under tissue-specific regulatory mechanisms (Bownes *et al.*, 1990). To date the only genes shown to be directly downstream of the sex determination hierarchy in somatic cells are the female-specific yolk protein genes (*yp1*, *yp2*, *yp3*). Mutations in any of the genes in the pathway interfere with the correct expression of the *yps* (Bownes and Nöthiger, 1981). *Cis*-regulatory regions upstream of the *yp* genes which confer their sex-specific fat body expression have been described (Garabedian *et al.*, 1986; Liddell and Bownes, 1991). The *yp1* and *yp2* genes which are divergently transcribed and separated by 1.2 kb have been recently further analysed in our laboratory and we have identified other regions which also confer sex-specific fat body expression, suggesting that there is redundancy within this regulatory region (Abrahamsen *et al.*, 1992). Burtis *et al.* (1991) have shown that *in vitro* the products of the *dsx* gene produced in *E. coli* can bind to the *cis*-regulatory element shown to give sex-specific expression in transformation experiments. Somewhat surprisingly, however, both the male and female products bound to the same DNA sequence. Since the *yp* genes are only expressed in females, presumably some other factors are involved *in vivo* to ensure that only the appropriate gene expression is repressed by *dsx*. One of the genes involved may be the *ix* gene, which has not yet been analysed at the molecular level, but from genetic evidence is thought to act at the end of the pathway and to interact with *dsx* in the repression of the downstream differentiation genes. Ultimately one should be able to analyse the DNA sequences involved and then determine the combination of proteins needed to confer the appropriate sex and tissue specificity on the downstream genes.

VII. MOLECULAR ANALYSIS OF SEX DETERMINATION IN OTHER INSECTS

There have been very few attempts to analyse any of the genes involved in sex determination in any insects other than *Drosophila*, and nothing has been published in this area to my knowledge. However, some progress is now being made. As described earlier, sex in *Chironomus* is determined by a male sex-determining factor *M*, which in *Chironomus thummi* is located on chromosome III. This is present in the genome of males and not females and is associated with a transposable element. Repetitive DNA present in the male-determining region shown in Fig. 7(a) has been cloned and hybridizes with a family of highly tandem repetitive DNA sequences which is present in both males and females at any chromosomal locations. This repeat DNA was used as a probe to screen a male genomic library and clones were isolated which flank the repetitive element and come from the male-determining region. This single copy DNA corresponds to an additional single restriction fragment in Southern blots of male but not female *Chironomus thummi* DNA (Fig. 7(b)), and therefore comes from the site of integration of the male-specific insertion. Clearly there is a male-specific integration of a transposable element in the sex-determining region of *Chironomus*, though as the element is present many times in the genome it is unlikely to determine sex itself. This flanking single copy region is now being analysed for transcription units which will hopefully identify the gene(s) involved in sex determination. This work has been done by Kraemer and Schmidt in Mainz (personal communication), and means that information on the molecular nature of *M* in this species should soon be forthcoming. It will be of great interest to see the sequences of these genes and to ascertain whether or not they have any homology to known *Drosophila* genes, though it should be pointed out that if they are equivalent to a repressor of a *Sxl*-like gene then this is the one class of gene which has not been identified in *Drosophila*. Furthermore it seems possible now that the molecular analysis of *da* and *sis-b* has been undertaken that these

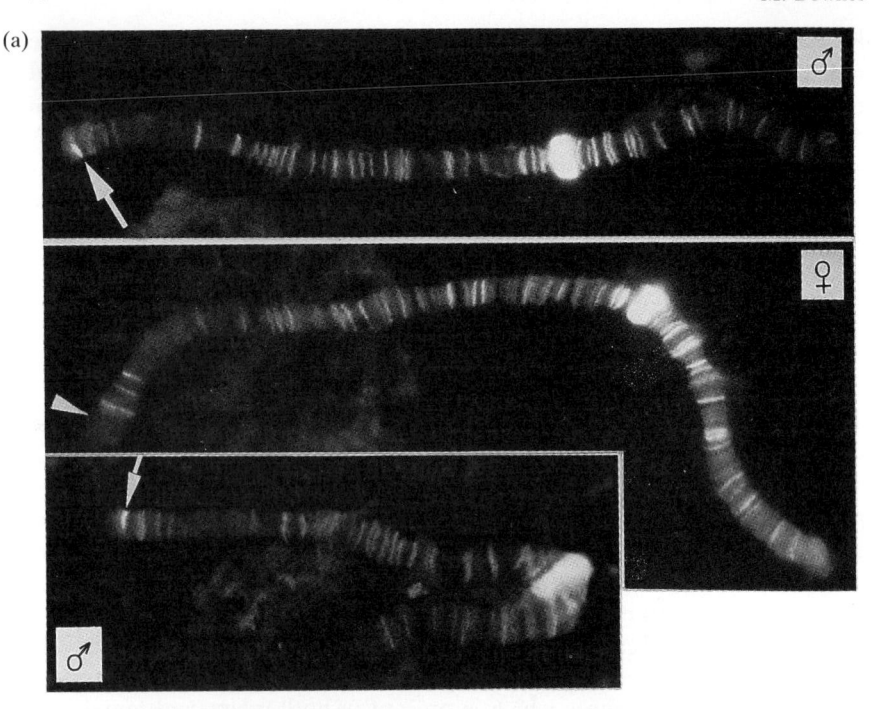

Fig. 7. (a) The salivary gland chromosome III of *Chironomus thummi thummi* (2 males and 1 female) after an *in situ* hybridization with biotin-labelled Cla elements (clustered and interspersed tandem repetitive DNA family). The hybridized Cla elements were made visible by fluorescent antibodies. Near the left end of the male chromosomes there is a hemizygous hybridization signal (arrows) clearly visible, which is completely absent in the homologous region of the female chromosome.

products could achieve the activation of *Sxl* without the need to postulate the existence of repressors as well.

Nothing is known about genes which may lie between the genes which initiate sex and the differentiation genes at the end of the pathway in any other insect, yet there must be a gene which can distinguish between and lead to the expression of two alternative sets of genes in males and females. One is tempted to speculate that a gene genetically equivalent to *dsx* should exist in other insects. One way of establishing if this were the case would be to see if the downstream differentiation genes of other insects could be correctly expressed in *Drosophila*. If this were the case then the molecular interactions leading to the sex-specific expression would be conserved and we could assume that *dsx* exists in the species

5. Molecular Aspects of Sex Determination in Insects

(b)

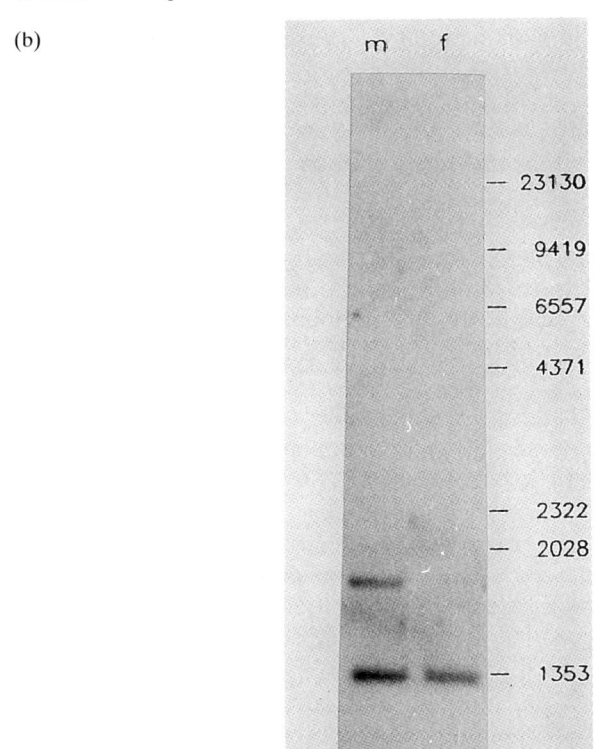

Fig. 7. (b) Southern hybridization of male (m) and female (f) DNA of *Chironomus thummi thummi* with a single copy probe flanking the male-specific transposable element. In females one homozygous band appears whereas in males the signal reflects the hemizygous/ heterozygous state. The difference between the molecular weights of both signals corresponds exactly to the length of the male-specific transposable element (i.e. 400 bp).

tested. Towards this end we have sequenced two of the female-specific downstream genes: the *yolk protein* (*yp*) genes of *Calliphora* (Martinez and Bownes, unpublished), which had been cloned and identified by Rabacha *et al.* (1988). They are surprisingly well conserved when compared to the *Drosophila yp* genes. Furthermore, when we analysed the expression of these genes in *Calliphora* using a combination of northern blots and whole mount *in situ* hybridization techniques we find that the distribution of transcripts is similar to that of *Drosophila*. The *yp* transcripts are found in the vitellogenic female but not the male fat body of adults and in the follicle cells of the same stages of oogenesis as they are seen in *Drosophila*. It seems possible then that similar *trans*-acting proteins will be used to achieve a similar pattern of regulation. To test this we will fuse the upstream region of the *Calliphora yp* gene to a *lacZ* reporter gene, and transform *Drosophila* using P element transformation. If the reporter gene is expressed in the female but not the male fat body then it is likely that a protein which can substitute for the *dsx* protein will exist in *Calliphora* and it will be likely to be encoded by a homologous gene.

It is very difficult to assess whether there should be any genes homologous to other genes in the sex determination pathway. As can be seen there are genes which have a similar function to *Sxl* in *Musca*, namely *F*, and there are also maternal genes similar in function to *da*, namely *Ag* and *tra*. It is tempting to assume that these genes will be homologous in molecular terms, but we should not forget that *Sxl* in *Drosophila* has functions in dosage compensation which are not needed for many other insects as they have heterochromatic sex chromosomes and have no need for such a mechanism. Also *da* and *sis-b* are members of classes of genes which function to regulate transcription in particular tissues at particular times in development by acting together in various combinations. It seems that many such molecules could be used to achieve the same end and that one should not assume that they will be closely related. The genetic similarity between *Sxl* and *F* and the need for a gene to head the pathway and set into motion the difference between male and female development again makes it tempting to speculate that an *Sxl* homologue will be found in other insects; however, I think that some degree of caution should be exercised, since one can envisage a scheme which in genetic terms is equivalent in other organisms such as *Caenorhabditis elegans*, where some of the genes bear the same names as in *Drosophila*, such as *tra*, yet there is no similarity in the underlying molecular mechanism between the two species and alternative splicing is not involved in sex determination in *Caenorhabditis* (Hodgkin, 1990). For there to be a gene homologous to *Sxl* the mechanism of sex determination will need to be by alternate splicing since this gene encodes an RNA-binding protein.

The way forward then seems to be to proceed with the cloning and analysis of the genes which can be identified genetically in other insects such as *M*, and to look for the function of downstream genes in *Drosophila* as described. It is also worth looking for homologous genes in other species using heterologous probes. The divergence of the genomes often means that finding such genes by simply looking for cross-hybridization fails; however, by using selected oligonucleotides and using PCR with the DNA from the species of interest it may be easier to see if a similar gene exists. This is slightly hampered in that at present only the sequences of the *Drosophila melanogaster* genes are generally known so it is more difficult to select the oligonucleotides from what are likely to be conserved regions of the molecule. Some progress is being made with molecular analysis of some of these genes from other *Drosophila* species, which should help with this approach. Another possibility is to look for related proteins using antibodies to the *Drosophila* proteins. At present there are monoclonal antibodies to *Sxl*, but polyclonal antibodies would be preferable for these experiments.

In summary there is little molecular analysis of sex determination in other insects, but the technology is now at the point where this should be feasible and will be a very exciting area of future research.

Acknowledgements

I would like to thank colleagues for allowing me to include their unpublished work as referred to in the text, Roger Slee for comments on the manuscript and Mike Greaves for drawing the figures.

REFERENCES

Abrahamsen, N., Martinez, A., Kjaer, T., Sondergaard, L. and Bownes, M. (1992). Submitted.
Amrein, H., Gorham, M. and Nöthiger, R. (1988). *Cell* **55**, 1025–1035.
Baker, B. and Ridge, K. (1980). *Genetics* **94**, 383–423.
Barr, A. R. (1975). *J. Med. Ent.* **12**, 567–570.
Bell, L. R., Maine, E. M., Schedl, P. and Cline, T. W. (1988). *Cell* **55**, 1037–1046.
Boggs, R. T., Gregor, P., Idriss, S., Belote, J. M. and McKeown, M. (1987). *Cell* **50**, 739–747.
Bownes, M. and Nöthiger, R. (1981). *Mol. Gen. Genet.* **182**, 222–228.
Bownes, M., Steinmann Zwicky, N. and Nothiger, R. (1990). *EMBO J.* **9**, 3975–3980.
Bull, J. J. (1981). *Evolution* **35**, 568–580.
Burtis, K. C. and Baker, B. S. (1989). *Cell* **56**, 997–1010.
Burtis, K. C., Coshigano, K. T., Baker, B. S. and Wensink, P. C. (1991) *EMBO J.* **10**, 2577–2582.
Chandra, H. S. (1985). *Proc. Natl. Acad. Sci. USA* **82**, 1165–1169.
Clements, A. N. (1991). "Biology of Mosquitos", Vol. 1. Chapman & Hall, London.
Cline, T. W. (1978). *Genetics* **90**, 683–698.
Cline, T. W. (1983). *Dev. Biol.* **95**, 260–274.
Cline, T. W. (1989). *Cell* **59**, 231–234.

Craig, G. B. (1965). "Proceedings of the 12th International Congress Ent." p. 263.
Craig, G. B. and Hickey, W. A. (1967). In "Genetics of Insect Vectors of Disease" (Wright and Pal, eds), pp. 105-108. Elsevier, Amsterdam.
Cronmiller, C. and Cline, T. W. (1987). Cell **48**, 231-234.
Crozier, R. H. (1971) Am. Nat. **105**, 399-412.
Denholm, I., Franco, M. G., Rubini, P. G. and Vecchi, M. (1986). Genet Res. **47**, 19-27.
Erickson, J. W. and Cline, T. W. (1991). Science **251**, 1071-1074.
Franco, M. G., Rubini, P. G. and Vecchi, M. (1982). Genet. Res. **40**, 279-293.
Garabedian, M. J., Shepherd, B. M. and Wensink, P. C. (1986). Cell **45**, 859-867.
Gilchrist, B. M. and Haldane, J. B. S. (1946). Hereditas **3**, 175-190.
Hagele, K. (1985). Chromosoma **91**, 167-171.
Hedley, M. L. and Maniatis, P. (1991). Cell **65**, 579-586.
Hodgkin, J. (1990). Nature **344**, 721-728.
Horsfall, W. R. (1974). Ann. Zool. Fennici **11**, 224-236.
Hoshijima, K., Inoue, K., Higuchi, I., Sakamoto, H. and Shimura, Y. (1991). Science **252**, 833-836.
Inoue, H. and Hiroyoshi, T. (1986). Genetics **112**, 469-482.
Inoue, K., Hoshijima, K., Sakamoto, H. and Shimura, Y. (1990). Nature **344**, 461-463.
Kerr, W. E. and Nielson, R. A. (1967). J. Apicult Res. **6**, 3-9.
Lewis, K. R. and John, B. J. (1968). Int. Rev. Cytol. **23**, 277-378.
Liddell, S. and Bownes, M. (1991). Mol. Gen. Genet. **230**, 219-224.
Lucchesi, J. C., Skripsky, T. and Tax, F. E. (1981). Chromosoma **82**, 217-227.
Maine, E. M., Salz, H. K., Schedl, F. and Cline, T. W. (1986). Cold Spring Harbor Symp. Quant. Biol. **50**, 595-604.
Mariani, B. D., Lingappa, J. R. and Kafatos, F. C. (1988). Proc. Natl. Acad. Sci. USA **85**, 3029-3033.
Martin, J. and Lee, B. T. O. (1984). Chromosome **90**, 190-197.
McClelland, G. A. H. (1962). Trans. R. Soc. Trop. Med. Hyg. **56**, 4.
McDonald, I. C., Evenson, P., Nickel, C. A. and Johnson, O. A. (1978). Ann. Entomolog. Soc. Am. **71**, 692-694.
McKeown, M., Belote, J. M. and Boggs, R. T. (1988). Cell **53**, 887-895.
Monsma, S. A. and Wolfner, M. F. (1988). Genes Dev. **2**, 1063-1073.
Murre, C., McCaw, P. S., Vossin, H., Caudy, M., Jan, L. Y., Jan, Y. N., Cabrera, C. V., Buskin, J. N., Hauschka, S. D., Lassar, A. B., Weintraub, H. and Baltimore, D. (1989). Cell **58**, 537-544.
Nagoshi, R. N. and Baker, B. S. (1990). Genes Dev. **4**, 89-97.
Naito, T. and Suzuki, H. (1991) J. Hered. **82**, 101-104.
Nöthiger, R. and Steinmann-Zwicky, M. (1985). Cold Spring Harbor Symp. Quant. Biol. **50**, 615-621.
Rabacha, A., Tucker, M. A., De Valoir, T., Berlikoff, E. J. and Beckingham, K. (1988). Dev. Biol. **129**, 449-463.
Salz, H. K., Maine, E. M., Keyes, L. N., Samuels, M. E., Cline, T. W. and Schedl, P. (1989). Genes Dev. **3**, 708-719.
Schäfer, U. (1986). Mol. Gen. Genet. **202**, 219-225.
Slee, R. and Bownes, M. (1990). O. Rev. Biol. **65**, 175-204.
Steinmann-Zwicky, M., Amrein, H. and Nöthiger, R. (1990). Adv. Genet. **27**, 189-237.
Thompson, P. E. and Bowen, J. S. (1972). Genetics **70**, 491.
Torres, M. and Sanchez, L. (1989). EMBO J. **8**, 3079-3086.
Whiting, P. W. (1943). Genetics **28**, 365-382.
Wyoke, J. (1979). J. Apicult. Res. **18**, 122-127.

6

Gene Control Systems Affecting Insect Development

M. G. PETITT AND M. P. SCOTT

I. Introduction 101
II. Pattern Formation in the *Drosophila* Embryo 102
III. Current Issues and Strategies 110
   A. Control of Complex Transcription Patterns 113
   B. Cell–Cell Communication Mediated by Segment Polarity Gene Products 116
   C. The Search for the Regulatory Targets of Homeotic Genes 117
   D. Structure–Function Relationships Among Homeodomain Proteins 120
IV. Summary and Outlook 121
   References 122

I. INTRODUCTION

The genetic control of development necessarily involves the regulation of basic cellular activities, such as growth, division, migration, adhesion, shape changes, and the synthesis of products that define a cell's specialized role in the biology of the organism. Somehow the genome orchestrates these cellular activities reproducibly from generation to generation to create the highly ordered multicellular organism characteristic of its species. How do genes accomplish this evidently elaborate and precise control of cell behaviour, accurately co-ordinating the activities of thousands to billions of cells, so that each cell's developmental programme matches its position in the embryo? These are questions currently being addressed by developmental biologists working on a variety of organisms, but nowhere has gene function been connected more directly to the creation of pattern than in *Drosophila*.

One of the key concepts to emerge from developmental genetics is that the behaviour of cells in development is guided by multiple regulatory inputs. Master regulatory genes, which co-ordinate and control cellular differentiation, act in complex hierarchies. Regulatory pathways diverge, so that a single regulatory gene may control multiple pathways. Regulatory pathways also come together, so that multiple pathways will control the activity of a single gene or cell. For example, *Drosophila*'s sex combs, which are comb-like structures that only develop on the front legs of adult males, form as a result of regulatory inputs from both the sex determination gene hierarchy and the positional information gene hierarchy. Such convergence or combinatorial action of regulatory pathways helps to explain both how complexity is generated and how cell fates are precisely specified during development.

Although combinatorial action of regulators may be a general theme in pattern formation, the molecular mechanisms of regulation are diverse. Some of the examples in *Drosophila* include regulated RNA splicing in sex determination, hormonal activation of transcriptional regulators in moulting, intercellular signalling through receptor tyrosine kinases in the specification of photoreceptor cell fates in the eye, and spatial control of translation in some of the early steps of segmentation. Our laboratory has focused primarily on the question of how genes control the creation of the segmented body, a process that involves both the spatial control of transcription and intercellular signalling.

The goal of geneticists and molecular biologists studying development is to understand the link between genes and development. What are the genetic and molecular components of a developmental process? Can the developmental process be explained in terms of the genetic and molecular functions we have identified? We will outline our current understanding of how regulatory genes control pattern formation in the *Drosophila* embryo and describe problems of current interest in the field with a view to how they can be addressed experimentally.

II. PATTERN FORMATION IN THE *DROSOPHILA* EMBRYO

One of the most well-understood examples of pattern formation is the segmentation of the larval cuticle of *Drosophila*. The segmental organization of the cuticle is reflected in the repeating pattern of bristles, called *denticles*, on the ventral surface (Fig. 1). There is one so-called *denticle belt* per segment. Most segments can be clearly identified by the

6. Gene Control Systems Affecting Insect Development

morphology of the denticle belt, so that each segment has a unique identity. The readily visible segmental markers on the larval cuticle have facilitated a remarkably systematic study of how genes control the creation of pattern.

Nüsslein-Volhard, Wieschaus, and Jürgens performed a now famous set of screens for all mutations that affect the larval cuticle pattern (Nüsslein-Volhard and Wieschaus, 1980; Jürgens et al., 1984; Nüsslein-Volhard et al., 1984; Wieschaus et al., 1984). The screens identified nearly all the zygotically active genes that control segmentation. Subsequent screens identified a large number of maternally active genes involved in some of the same developmental events (Anderson and Nüsslein-Volhard, 1984; Schüpbach and Wieschaus, 1986, 1989; Nüsslein-Volhard et al., 1987). With respect to these maternal genes, the genotype of the mother determines the phenotype of the embryo. The products of the maternal genes are deposited in the oocyte and participate in the earliest steps in the development of the zygote. Some genes that control segmentation probably were not identified in the screens because they act *both* maternally and zygotically. In these cases, homozygous mutant mothers do not survive and therefore cannot be tested for maternal effects, and homozygous mutant embryos may not show a phenotype because the wild type maternal product is present.

Among the zygotic lethal mutations generated in the screen, about 3% caused gross defects in the cuticle pattern. The patterning mutations displayed a variety of phenotypes, but some of the phenotypes identified genes controlling the number, orientation, and identity of segments. On the basis of the results, it was estimated that between 1% and 2% of the 5000–10 000 genes in *Drosophila* are involved in the regulation of segmentation. The remainder of the mutations affecting cuticle development identified genes involved in several other processes, including the dorsoventral organization of the cuticle, head development, and the specification of neural precursors, which arise from the same population of cells that make part of the cuticle.

The surprising result of the genetic studies was how revealing they were about the organization of the genetic functions controlling segmentation. There are two general classes that organize the cuticle pattern along the anterior–posterior axis. Mutations in the more than 20 known *segmentation* genes cause the deletion of pattern elements, sometimes accompanied by mirror-image duplications of remaining pattern elements. Mutations in the eight *homeotic* genes, on the other hand, cause segments improperly to differentiate structures normally only found in other segments. In other words, homeotic mutations cause segmental transformations from one segmental identity to another. During the 80 or so previous years of study in *Drosophila*, the homeotic genes had been identified serendipitously by dominant and recessive mutations that cause transformations in adult flies. One kind of mutation in the homeotic gene *Antennapedia* (*Antp*), for example, causes legs to grow in place of antennae (Fig. 2).

The segmentation genes can be further categorized into three subclasses, according to their mutant phenotypes. *Gap* gene mutations cause large deletions of several contiguous segments. *Pair-rule* gene mutations cause the deletion of pattern elements at two-segment intervals, giving embryos with half the usual number of segments. *Segment polarity* gene mutations cause the deletion of pattern elements in every segment, accompanied by the mirror-image duplication of some remaining pattern elements. The cuticle phenotypes of representative segmentation mutants are illustrated in Fig. 3.

In outline, the segmentation genes sequentially subdivide the embryo into finer and finer repeating units. The gap genes are active first, defining broad subdivisions of the embryo, whereas the segment polarity genes, acting last, control intrasegmental patterning. The homeotic genes modify the differentiation of individual units of the repeating pattern. The maternal genes, which we will not consider in detail, act the earliest of all, organizing the egg into anterior and posterior domains, as well as dorsal and ventral domains.

We now know that the segmentation and homeotic genes act in a regulatory hierarchy (Akam, 1987; Scott and Carroll, 1987; Ingham, 1988). For the most part, the genes are transcribed in precise spatial patterns that coincide with where they function (Figs 4 and 5). The correct spatial expression of a later-acting gene depends on the activities of several of the earlier-acting genes, although the relationships are not simple. For example, the correct spatial expression of the pair-rule gene *fushi tarazu* (*ftz*) depends on the activities of the gap genes and three of the other pair-rule genes (Carroll and Scott, 1986). The *ftz* protein product normally accumulates in seven transverse stripes in the early embryo (Fig. 4). A mutation in one of the regulators of *ftz* transcription

---

Fig. 1. The larval cuticle. Top and centre: a darkfield image (top) and phase contrast image (centre) of the ventral surface of the larval cuticle (anterior is to the left). The ventral surface of the larval cuticle contains a repeating pattern of *denticle belts*, one per segment. The denticle belts of the eight abdominal segments are relatively broad and visible, whereas the thoracic denticle belts (near the anterior end) are less obvious. The larval head is entirely internal and not easily visible at this stage. Bottom: three denticle belts corresponding to the third thoracic segment (T3) and the first two abdominal segments (A1 and A2). The denticle belts in many of the larval segments have distinct morphologies that give segments their unique identities. Note especially the differences between T3 and A1.

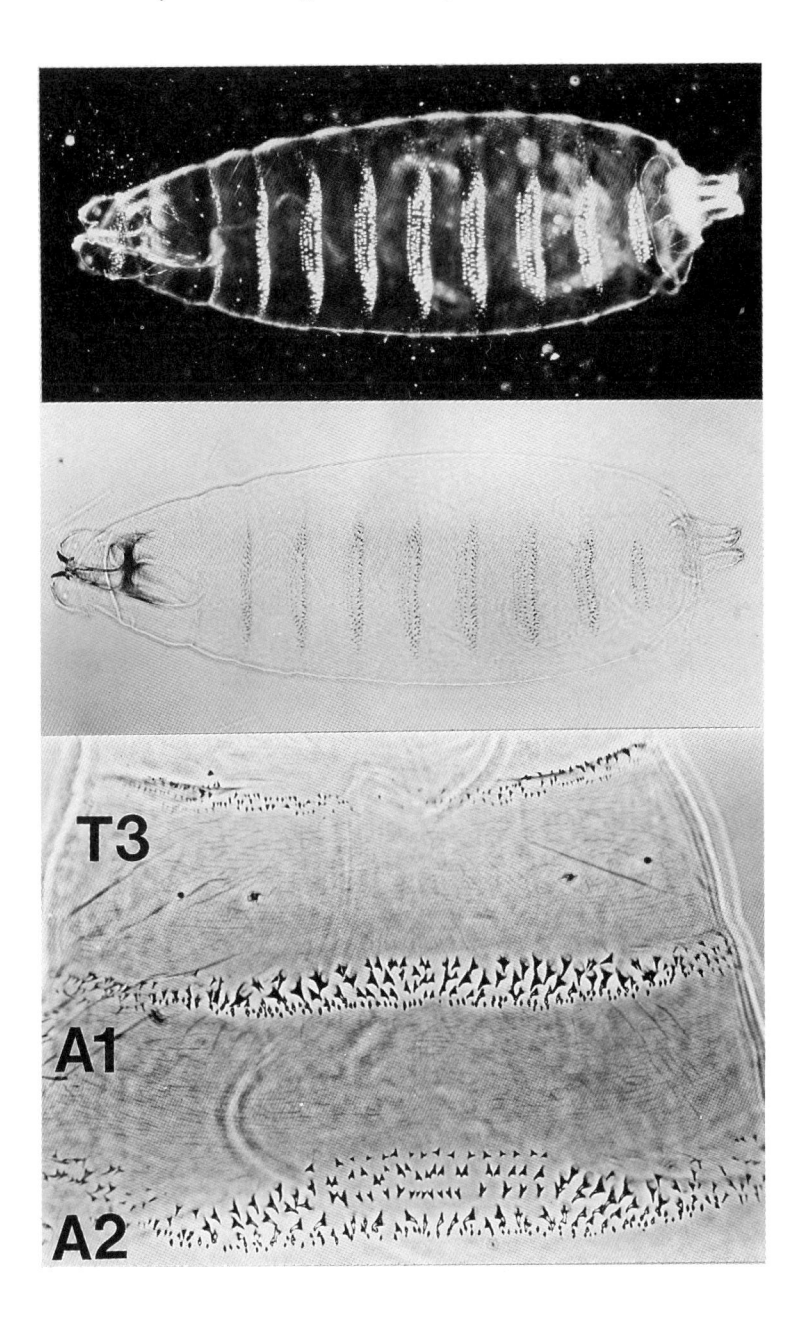

6. Gene Control Systems Affecting Insect Development   105

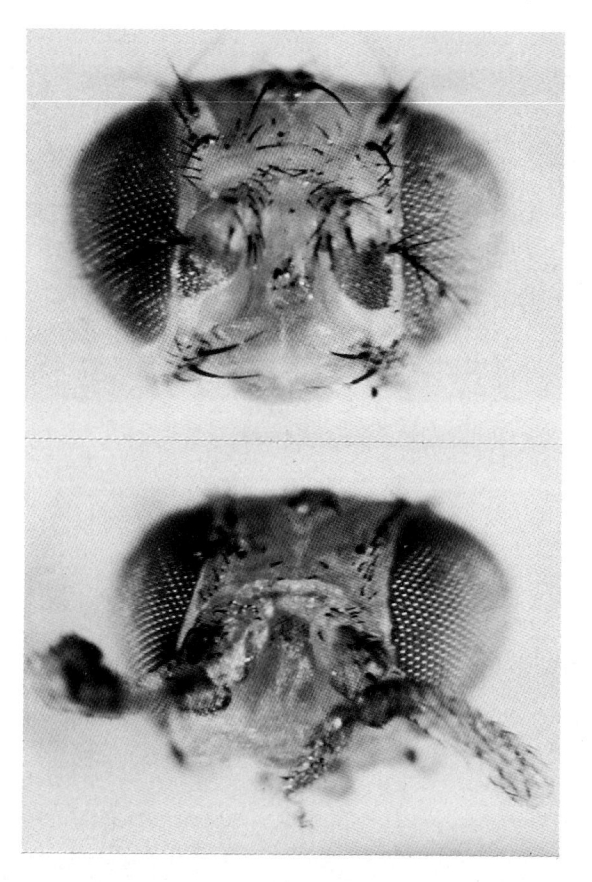

Fig. 2. The dominant *Antennapedia* phenotype. Top: a wild type fly with normal antennae. Bottom: a fly with legs in place of antennae caused by a mutation in the homeotic gene *Antennapedia* (*Antp*). The mutation leads to the abnormal expression of the *Antp* gene in the antennal primordium.

Fig. 3. Defects in the larval pattern caused by mutations in segmentation genes. Darkfield images of the lateral surface of larval cuticles from wild type (top), pair-rule mutant (centre) and segment polarity mutant (bottom) larvae (anterior is to the left). The wild type cuticle has eight easily visible abdominal denticle belts (top). The pair-rule mutant has about four (centre). The segment polarity mutant (bottom) has deleted the region of naked cuticle between denticle belts in every segment and duplicated the remaining cuticle in reverse orientation. The net effect of replacing naked cuticle with extra denticles is the "lawn" of denticles shown.

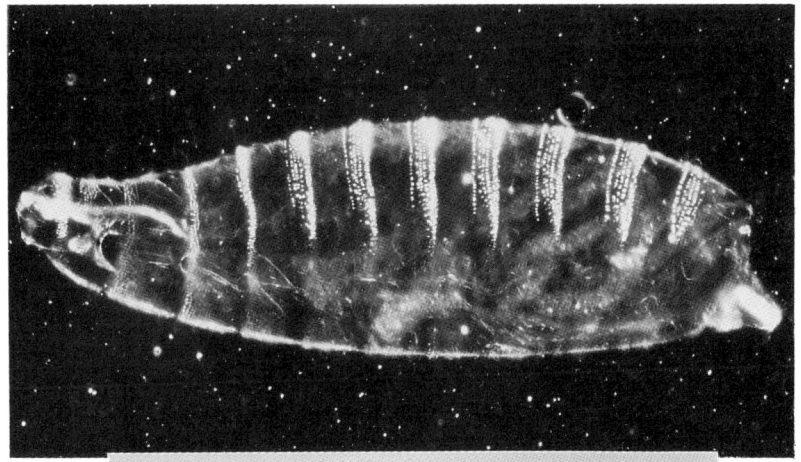

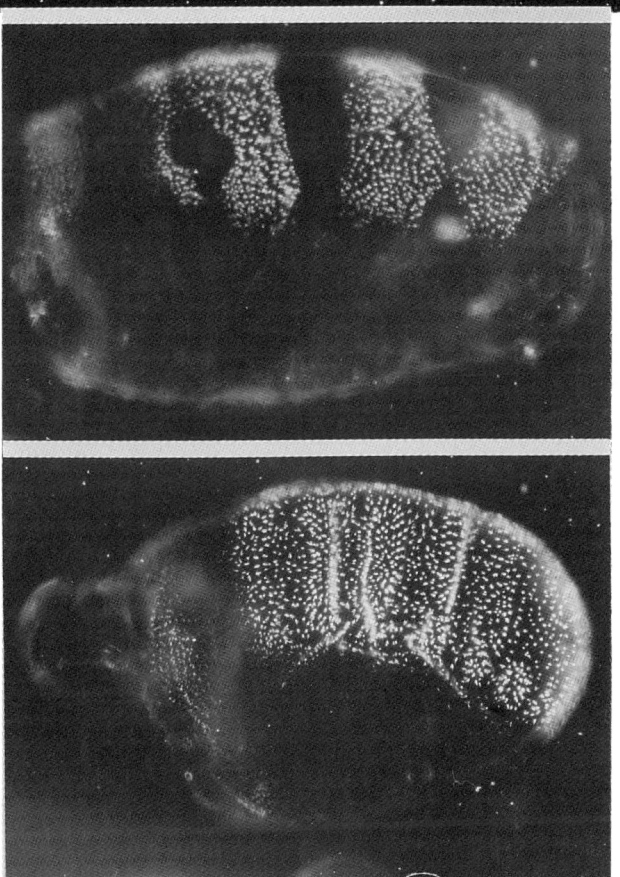

6. Gene Control Systems Affecting Insect Development   107

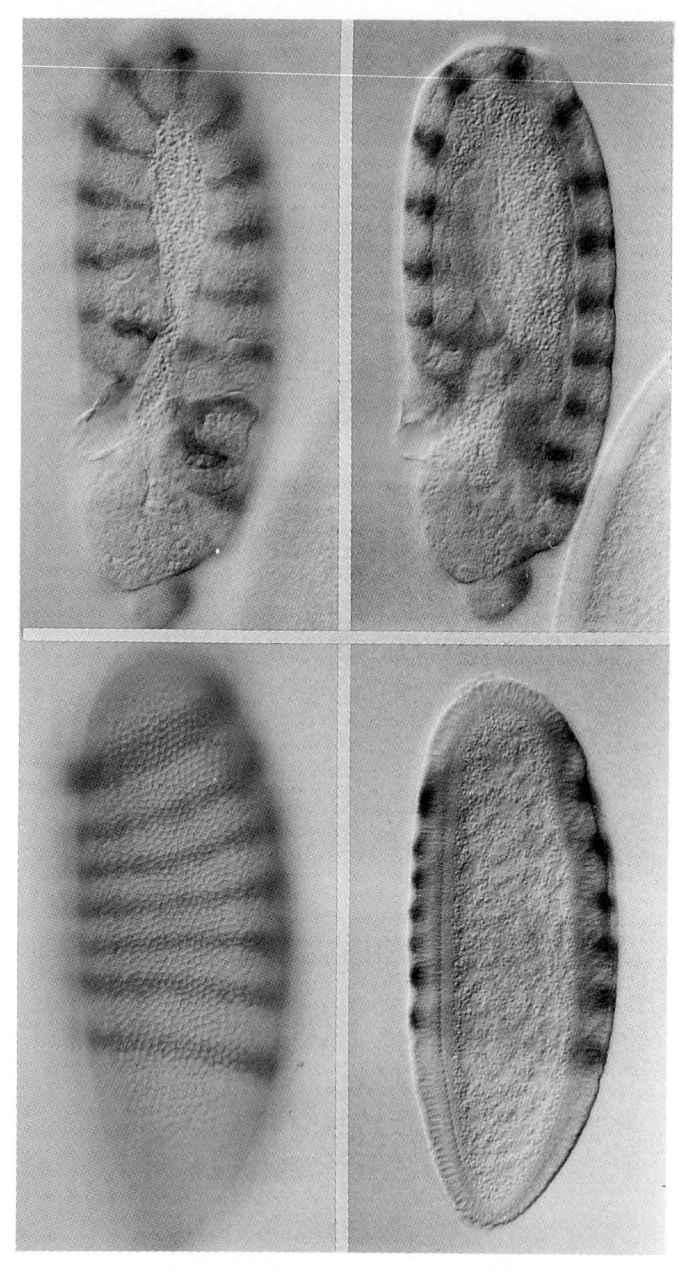

Fig. 4. Segmentation gene expression. Top left: expression of the protein product of the pair-rule gene *fushi tarazu* (*ftz*) in seven stripes in an early embryo. The protein has been detected by standard immunolabelling techniques. In focus is the lateral surface of the embryo, where the nuclei are visible as small bumps. Bottom left: the same embryo in a different plane of focus. Cells are visible on the perimeter. The central part of the embryo is yolk. Top right: expression of the protein product of the segment polarity gene *engrailed* (*en*) in 14 stripes, focusing on the lateral surface of the embryo. Bottom right: the same embryo in a different plane of focus. In all panels anterior is to the left and ventral is downwards.

6. Gene Control Systems Affecting Insect Development

does not cause the simple loss or gain of part of the expression pattern. Instead, the mutation causes a reduction in the number of *ftz* stripes, as well as changes in the positioning and width of the remaining stripes. The complex effect is probably due to the fact that the regulators interact with *each other*, as well as with *ftz*. The regulatory hierarchy is summarized in Fig. 6.

The exquisite control over the spatial expression of segmentation and homeotic genes is critical for normal development. When they are expressed outside their normal domains of expression, either by mutations affecting their regulation (Frischer *et al.*, 1986; Schneuwly *et al.*, 1987) or by the use of inducible promoters driving the ubiquitous expression of their cDNAs, the organization of the cuticle is altered (e.g. Struhl, 1985; Ish-Horowicz and Pinchin, 1987; Gibson and Gehring, 1988). Apparently, the master regulators control cell behaviour in all or most cells where they are expressed. It is only by restricting their expression to precise spatial domains that the organism develops normally. In order to understand the logic of pattern formation, we need to know how these complex transcription patterns are generated.

The inferred protein sequences of many segmentation and homeotic genes provide the key to understanding the molecular mechanisms that underlie the regulatory interactions. The gap and pair-rule segmentation genes, and the homeotic genes, encode proteins that regulate transcription. The protein products of the eight homeotic genes and several of the pair-rule genes each contain a homeodomain, a DNA-binding motif first identified in *Drosophila* (McGinnis *et al.*, 1984; Scott and Weiner, 1984; Laughon and Scott, 1984; Shepherd *et al.*, 1984; Akam, 1987; Scott and Carroll, 1987; Ingham, 1988). Each of the gap genes encodes a protein containing a zinc finger DNA-binding domain (Rosenberg *et al.*, 1986; Tautz *et al.*, 1987; Nauber *et al.*, 1988). The homeodomain and zinc finger proteins look like DNA-binding proteins and have in several cases been shown to act as transcription factors (Jaynes and O'Farrell, 1988; Krasnow *et al.*, 1989; Winslow *et al.*, 1989; Zuo *et al.*, 1991). Some of the maternal genes, which control zygotic gene activity, also encode transcription factors (Driever and Nüsslein-Volhard, 1989).

Evidently, the creation and diversification of segments depends upon a cascade of transcriptional regulatory interactions during early development (Fig. 6). The proteins encoded by earlier-acting genes probably regulate the spatial expression of later-acting members of the hierarchy by binding to *cis*-regulatory sequences at the target gene promoter, although such direct interactions have been demonstrated in only a few cases. Some of the regulators must also control downstream genes, such as those encoding cytoskeletal components and cell surface proteins, that directly affect cell behaviour. Through these interactions, cell fates, controlled by master regulatory genes, are co-ordinated with the generation of pattern by the segmentation and homeotic gene hierarchy.

Unlike other members of the hierarchy, most of the segment polarity genes that have been characterized do not encode proteins with DNA-binding motifs. Instead, the inferred protein products show homologies to proteins known to act as membrane components, kinases, or diffusible signalling molecules (Rijsewijk *et al.*, 1987; Hooper and Scott, 1989; Bourois *et al.*, 1990; Peifer and Wieschaus, 1990; Préat *et al.*, 1990). The implication of the protein structures is that segment polarity gene products mediate intercellular signalling, either as extracellular signalling molecules, or as components of signal transduction pathways. Since a cell's polarity is only defined with respect to its environment, it makes intuitive sense that cells should learn polarity through direct communication with their neighbours.

Segment polarity gene products are still present in the developing embryo long after the products of the gap and pair-rule genes have disappeared, suggesting that segment polarity genes play a role in maintaining cell fates once they are specified. The communication between cells after the initial specification by transcription factors provides a way for cell fates to adjust to perturbations or minor mistakes in development. Such flexibility would help to explain the apparent precision of developmental events.

III. CURRENT ISSUES AND STRATEGIES

Despite the remarkable progress in understanding early *Drosophila* development, several critical issues remain unresolved. Segmentation can be explained conceptually, but in many cases, the molecular details are

Fig. 5. Homeotic gene expression. Top: expression of the protein product of the homeotic gene *Antennapedia* (*Antp*) during the first half of embryogenesis. *Antp* is primarily expressed in the thoracic segments, the site of heavy dark staining. At this stage, the staining corresponds to developing epidermal and muscle cells. The protein can also be detected in a small subset of cells in most abdominal segments. These cells probably correspond to neural precursors. Centre: *Antp* protein is found in the developing midgut, where it is required for the formation of a constriction. Note the dark staining at the site of a constriction separating two compartments of the gut (arrowheads). Out of focus to the left is staining in the developing central nervous system. Bottom: *Antp* protein is found in the developing central nervous system, which consists of a longitudinal cord of densely packed neurones and support cells just inside the ventral surface of the embryo. The darkest staining corresponds to the thoracic part of the CNS. In all panels, anterior is to the left.

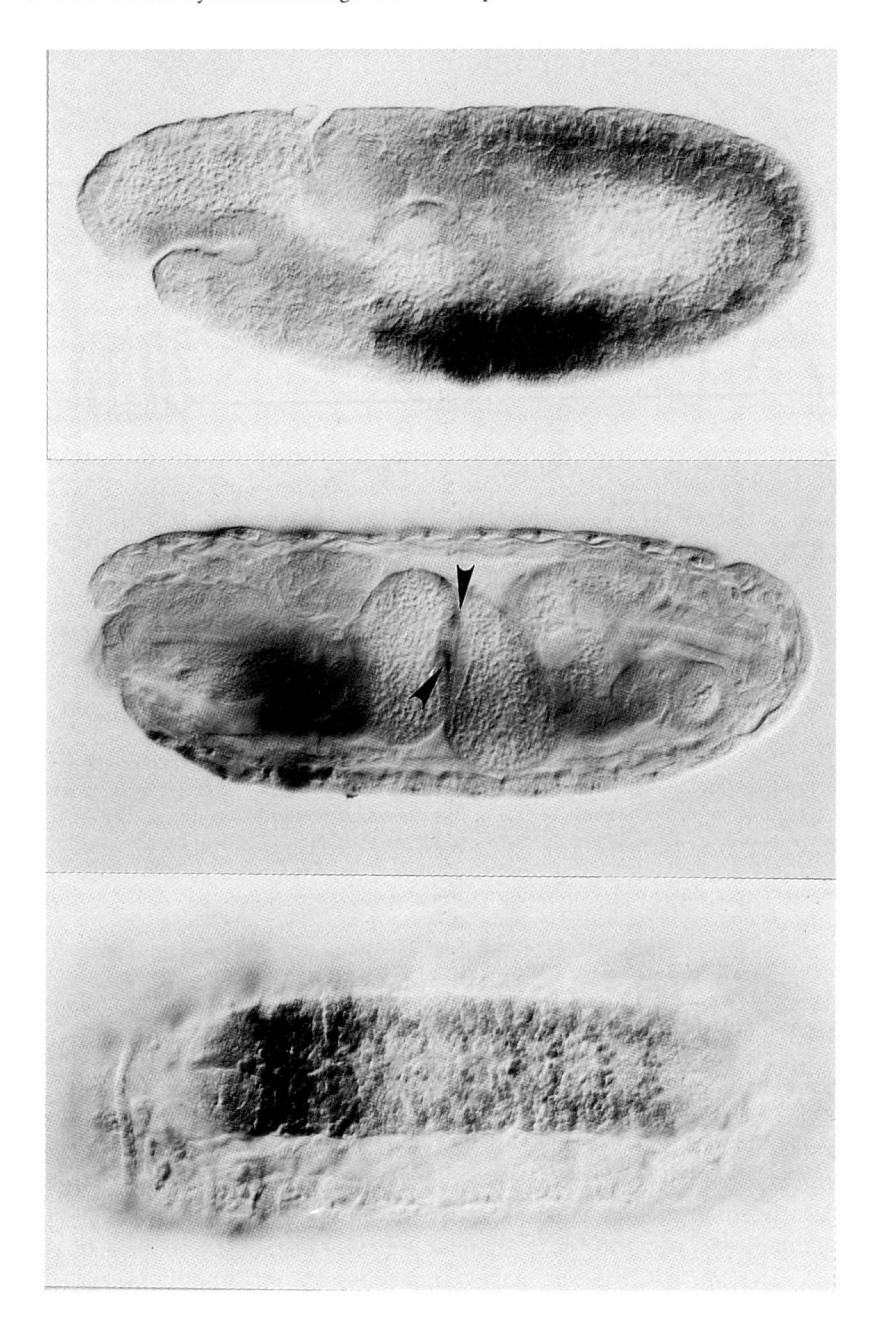

6. Gene Control Systems Affecting Insect Development   111

Maternal
↓
Gap
↓
Pair-rule
↓
Segment polarity
↓
Homeotic

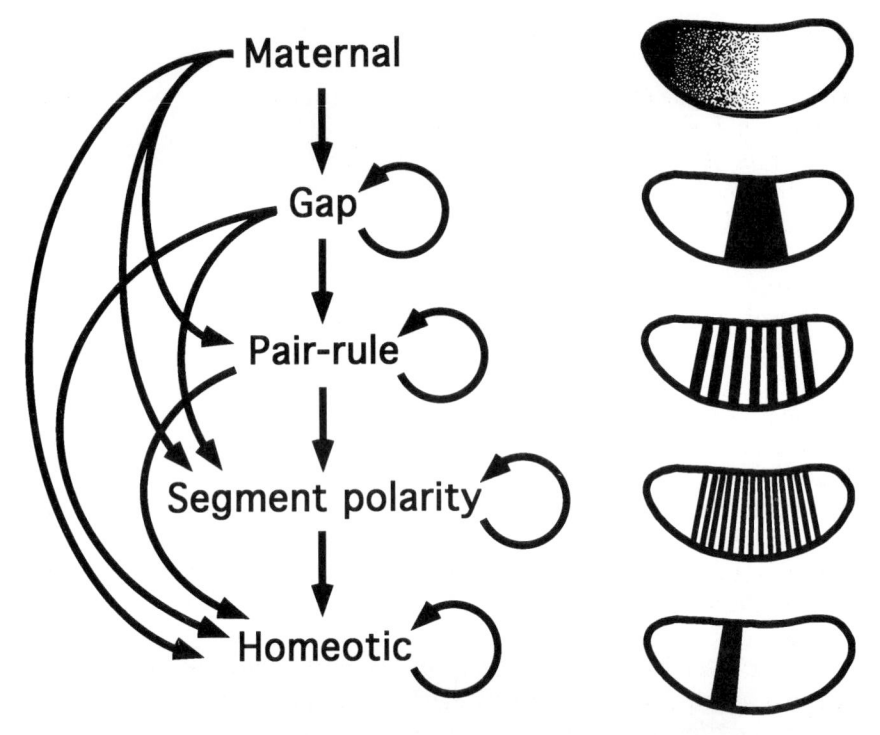

Fig. 6. The gene regulatory hierarchy controlling segmentation. Left: a simplified schematic representation of the gene classes that participate in creating the larval cuticle pattern and the regulatory relationships between them. Earlier-acting genes regulate the spatial expression of later-acting genes. In many cases, genes within a given class regulate other members of the class (depicted as arrows beginning and ending at the same class). Many other genes not shown in the diagram may also play a role in segmentation. Not shown are the putative targets of regulation by homeotic and segmentation genes. Right: representative expression patterns of each gene class are shown on the outline of an early embryo. The larval cuticle pattern is determined by where each of the master regulatory genes is expressed. Some maternal gene products are present in concentration gradients and regulate downstream genes in a concentration-dependent manner.

unknown. We would like to understand better, for example, how pattern is generated through transcriptional regulatory interactions. Another major unanswered question is how transcriptional regulation by the segmentation and homeotic proteins leads to specific cell behaviours. Similarly, how do the segment polarity gene products mediate cell–cell signalling, and how do these signalling events lead to the choice of cell fate? Here we will present current questions in our laboratory and the field as a whole, and outline strategies that are being used to answer those questions.

A. Control of Complex Transcription Patterns

The chain of cause and effect in the early embryo can be explored by examining the regulation of specific segmentation and homeotic gene promoters. Segmentation and homeotic genes are transcribed in complex patterns during development. In asking how the segmentation and homeotic gene promoters are controlled, there are two general goals. One is to identify the actual spatial cues or transcription factors that regulate the promoter. Some of these will be the products of previously identified members of the regulatory hierarchy. Others may be the products of genes not previously known to be involved in segmentation, perhaps because their mutant phenotypes do not specifically include segmentation defects. The second goal is to elucidate the general principles of complex regulation. How are multiple regulatory inputs integrated? Since homeotic gene promoter activity serves as an indicator of pattern formation, understanding the organization of discrete regulatory sequences in the promoter reveals how multiple regulatory pathways converge to generate pattern. Studies in our laboratory provide examples of how the study of spatial regulation has led to new insights in pattern formation.

One of the most well-studied examples of regulation in the early embryo is the control of the *ftz* gene promoter. Mutations in *ftz* cause embryos to develop with about half the normal number of segments. *ftz* transcripts are normally first detected just prior to blastoderm formation, when they form a continuous block of expression in the central two thirds of the embryo. Slightly later, at the blastoderm stage, the *ftz* transcripts are detected in seven transverse stripes, in the positions where *ftz* is required for the development of seven of the 14 metameres. At still later stages, *ftz* is transcribed in the developing nervous system, where it controls neural cell fates. The repeated use of a master regulatory gene for many different processes is a general theme in development.

The regulation of *ftz* has been examined by fusing *ftz* promoter sequences to a so-called reporter gene, which encodes a product that is easily detected in embryos, and introducing the fusion gene into the *Drosophila* germ line genome as part of a P transposable element. The method allows promoter regulation to be examined *in vivo*. Using this approach, Hiromi *et al.* (1985) defined sequences sufficient to confer the normal pattern of *ftz* regulation *in vivo*, as well as a 600-bp subsequence (the "zebra" element) capable of directing expression in the characteristic seven-stripe pattern in the mesoderm of early embryos. Our laboratory sought to define sequences important for the regulation of the striped pattern by identifying sequences within the zebra element conserved in evolution (S. Sonoda and M. P. Scott, in preparation). In a three-way comparison among *Drosophila melanogaster*, *D. virilis* and *D. pseudoobscura*, we found many short, conserved sequences interspersed with non-conserved sequence. Because the three species diverged from one another tens of millions of years ago, the average rate of genetic change predicts that non-functional sequences will have diverged completely.

In order to test the hypothesis that sequence conservation implies regulatory function, we mutated a highly conserved sequence in an otherwise normal zebra element and found that the mutation altered the expression pattern in an unusual way. Instead of the normal seven stripes, the altered zebra element caused expression in a 14-stripe pattern, a result that would not have been easily predicted based upon current understanding.

The examination of two other conserved sequences found in the zebra element has revealed additional regulatory complexity in the early embryo. One sequence coincides with a binding site for a zinc finger protein found in early embryos (Brown *et al.*, 1991). When the binding site was altered slightly to prevent binding of the protein *in vitro*, the effect *in vivo* was to allow expression from the *ftz* promoter in very early embryos, during the first few cleavage divisions. Only a very few genes are normally active during the early cleavage divisions of *Drosophila*. The zinc finger protein, encoded by the gene *tramtrak*, appears to prevent the premature expression of master regulators like *ftz* in the early cleavage stage embryo. *ftz* and other regulators would probably interfere with normal development if activated prematurely. This hypothesis can be tested by expressing an *ftz* cDNA under the control of the mutant promoter and asking whether it supports normal development.

The other conserved sequence coincides with the binding site for the protein FTZ-F1, which is expressed during embryogenesis. Mutations in the sequence that prevent binding by FTZ-F1 *in vitro* also reduce *ftz* transcription *in vivo* and show differential effects on the different stripes of *ftz* expression (Ueda *et al.*, 1990). FTZ-F1 and the *tramtrak* protein are involved in the process of segmentation, but neither was identified in genetic screens for segmentation genes, perhaps because they regulate many other genes as well.

We have also examined the spatial regulation of the segment polarity gene *patched* (*ptc*) (Y. Higashi and M. P. Scott, unpublished data). Mutant *ptc* embryos have pattern deletions in every segment accompanied by mirror-image duplications of some of the remaining pattern elements. Like many other segment polarity genes, *ptc* is transcribed in an evolving pattern that at one point consists of 14 transverse stripes, one for each segment primordium (e.g. Fig. 4). Eventually, the central part of each of the 14 original stripes disappears, and the pattern resolves into 28 very narrow stripes.

Two features of segment polarity genes suggest that they respond to regulatory inputs other than merely the earlier-acting gap and pair-rule gene products. One is that segment polarity genes continue to be expressed in precise spatial patterns long after the gap and pair-rule proteins have disappeared. Because most of the segment polarity genes do not encode transcriptional regulators, they do not simply maintain their own transcription through autoactivation. What controls the maintenance of the pattern? The other intriguing feature is that segment polarity gene promoters respond to signals sent from other cells. The pathways that lead from extracellular signals to nuclear gene regulation are not understood and can be discovered in part through studies of *ptc* regulation. Using promoter–reporter gene fusions introduced into flies with P elements, we are currently exploring the complex regulation of the *ptc* promoter.

The homeotic genes present a still different set of regulatory questions. Because each of the homeotic genes is expressed in unique, rather than repeating, domain in the embryo, some of the rules that define segmentation gene expression will not apply to homeotic genes. Homeotic genes are called upon repeatedly during development for fate decisions in many tissues (e.g. Fig. 5). The complexity of the task of regulating homeotic gene promoters is reflected in the enormity of their *cis*-regulatory domains: functionally important regulatory sequences span a region of more than 100 kb surrounding the *Ultrabithorax* (*Ubx*) promoter (Bender *et al.*, 1983). In contrast, typical genes in *Drosophila* are less than 5 kb in size.

We are interested in the regulation of the homeotic genes for several reasons. First, because homeotic genes sit at the end of the regulatory hierarchy as it is now understood, their expression is sensitive to nearly the entire hierarchy (Fig. 6). By studying homeotic gene regulation, the entire system of spatial information in the early embryo can be explored. Second, in many tissues where homeotic genes are expressed, little or nothing is known about how spatial information is generated. The homeotic gene promoters serve as probes for spatial regulators in those tissues. Third, the eight homeotic genes are organized in clusters that are conserved in evolution between flies and mammals, indicating that they are part of an ancient system for creating pattern (Akam, 1989). Like the *Drosophila* homeotic genes, the vertebrate homologues control development in restricted spatial domains (e.g. Chisaka and Capecchi, 1991). The conserved gene organization suggests that clustering is functionally important. Perhaps the clusters represent structural domains in chromatin that allow co-ordinate regulation of the homeotic genes. Finally, the *cis*-regulatory functions defined in our studies will allow us to ask questions about homeotic protein function because it will become possible to express homeotic cDNAs in specific tissues and patterns.

We have examined the *cis*-regulation of the two *Antp* promoters, P1 and P2, using promoter–reporter gene fusions. We have shown that 16 kb of sequence flanking P1 and 10 kb of sequence flanking P2 is sufficient to give nearly normal P1 and P2 expression patterns in embryos, including those tissues for which the basis of spatial regulation is unknown (M. G. Petitt and M. P. Scott, unpublished data). By examining nested deletions of these fragments, sequences required for both tissue specificity and correct spatial patterning within tissues have been detected. When subfragments of the P1 regulatory domain are tested independently for their ability to regulate a heterologous promoter, they show complex patterns that are not simple components of the normal P1 pattern. We believe this reflects a requirement for combinatorial action among many widely dispersed sequences. We are now beginning to ask how upstream regulators interact with the *Antp* promoters in specific tissues.

B. Cell–Cell Communication Mediated by Segment Polarity Gene Products

Although the maintenance and refinement of the segmental pattern is achieved in part through cell–cell signalling mediated by the products of segment polarity genes, the signalling pathways are poorly understood. There is no clear understanding of the chain of events that constitutes a signal sent from one cell to another. Not all of the components of the process have been identified, and for those that have been identified, there is little understanding of how they function. A current goal of studies on segmentation is to identify all the components of the signalling pathways and understand how they interact with one another.

One example of a segment polarity gene involved in cell–cell communication is the gene *ptc*. In addition to serving as a model of complex transcriptional regulation, the study of *ptc* will provide insights into how the segment polarity genes control cell behaviour through intercellular signalling. The *ptc* protein appears to be an integral membrane protein with seven membrane-spanning segments (Hooper and Scott, 1989; Nakano *et al.*, 1989), suggesting that it may act as a receptor. If it is a receptor, what is its ligand? Mutations in *ptc* affect the transcription of other segment polarity genes, indicating that the *ptc* protein, although acting in the cell membrane, controls events in the nucleus. How do the intracellular parts of the *ptc* protein communicate with other parts of the cell? The answers to these questions will reveal the way cells control fate decisions in their neighbours and how they learn their polarity.

C. The Search for the Regulatory Targets of Homeotic Genes

The transcription factors encoded by the homeotic genes and some of the segmentation genes appear to regulate arrays of downstream "target" genes, yet we know of few, if any, *bona fide* targets. Why have targets not been identified in screens for mutations that affect the larval cuticle pattern? Apparently, mutations in target genes do not give cuticle phenotypes that reveal their roles in segmentation. The target genes may play many roles in many cell types, and the loss of target gene function may lead to multiple developmental defects that mask their individual functions. Alternatively, target genes may play relatively subtle roles individually, so that the loss of a single target gene function has only a weak effect on the cuticle. Another type of genetic screen has identified a number of genes that can enhance or suppress homeotic phenotypes in a dosage-dependent manner, suggesting that they may functionally interact with homeotic genes (Kennison and Tamkun, 1988; Tamkun *et al.*, 1992). Some of the modifiers of homeotic gene function are likely to act downstream of the homeotic genes. However, many of the modifiers characterized so far appear to act upstream, as regulators of homeotic gene expression.

Two genes now thought likely to be targets of homeotic proteins in the developing midgut were originally identified by their functions in processes unrelated to midgut development. Only subsequently was it discovered that they were expressed in the visceral mesoderm of the midgut and that the visceral mesoderm expression depended on homeotic genes. One of the putative target genes, *decapentaplegic* (*dpp*), encodes a secreted protein related to mammalian growth factors. *dpp* expression in a restricted part of the midgut visceral mesoderm requires the homeotic gene *Ubx* (Padgett *et al.*, 1987; Immerglück *et al.*, 1990; Reuter *et al.*, 1990). Both *dpp* and *Ubx* are required for the formation of a constriction that separates two compartments of the midgut. In addition, *dpp* is required for the expression of the homeotic gene *labial* in adjacent endoderm cells. The other putative target, the segment polarity gene *wingless* (*wg*), is the *Drosophila* homologue of the mammalian protooncogene *wnt-1* (Rijsewijk *et al.*, 1987). Like *dpp*, *wg* and *wnt-1* encode secreted proteins thought to act as extracellular signalling molecules. The *wg* protein has been detected on the surface of neighbouring endoderm cells, although the potential function of *wg* in signalling to the endoderm is not understood (van den Heuvel *et al.*, 1989). *wg* expression in the developing midgut depends on the homeotic gene *abdominal-A*, which is required for another of the gut constrictions to form (Immerglück *et al.*, 1990; Reuter *et al.*, 1990). However, *wg* itself is not required for the constriction.

The fortuitous discovery of two putative targets of the homeotic genes has provided a narrow but instructive look at what lies downstream of the homeotic genes. The two examples indicate that one way homeotic genes control differentiation is by controlling the expression of intercellular signalling molecules, although there may be a variety of other ways. The expression of target genes need not simply follow the expression of a homeotic gene. There is no simple relationship between the expression patterns of the two targets and the expression patterns of the homeotic genes that regulate them. The example of *dpp* suggests that some target genes might be identified genetically, since both *dpp* and *Ubx* mutations give similar midgut phenotypes. Perhaps genetic screens that identify genes controlling a discrete morphogenetic event will be successful in identifying tissue-specific targets, such as *dpp* and *wg*.

Given the paucity of genetically identified target genes, what non-genetic properties could we expect among target genes that would aid in their identification? The potential relationship between homeotic genes and the *dpp* and *wg* genes was initially inferred on the basis of the overlap in their expression patterns. Perhaps predictions about where target genes are expressed would be useful. Because homeotic gene activity controls the diversification of segments, target genes are expected to be differentially transcribed between segments. This criterion has been used as the basis for a new approach to the identification of target genes that will be described below.

In order to identify the regulatory targets of homeotic genes, our laboratory and others have begun to use two non-genetic methods that make use of unique attributes of *Drosophila*. One is the ability to map positions in the genome cytologically using the giant polytene chromosomes of larval salivary glands (Fig. 7). Potential targets of homeotic proteins are identified by asking where homeotic proteins accumulate on the polytene chromosomes. Such sites may correspond to genes regulated by the homeotic proteins. Using immunolabelling techniques, we have found that homeotic proteins bind to many sites on the salivary gland chromosomes (D. Andrew, M. Horner, V. Yorke, A. McCormick, and M. P. Scott, unpublished data). By using chromosomal deficiencies and other breakpoint mutations, it is possible to map the binding sites more precisely, in some cases to a single gene.

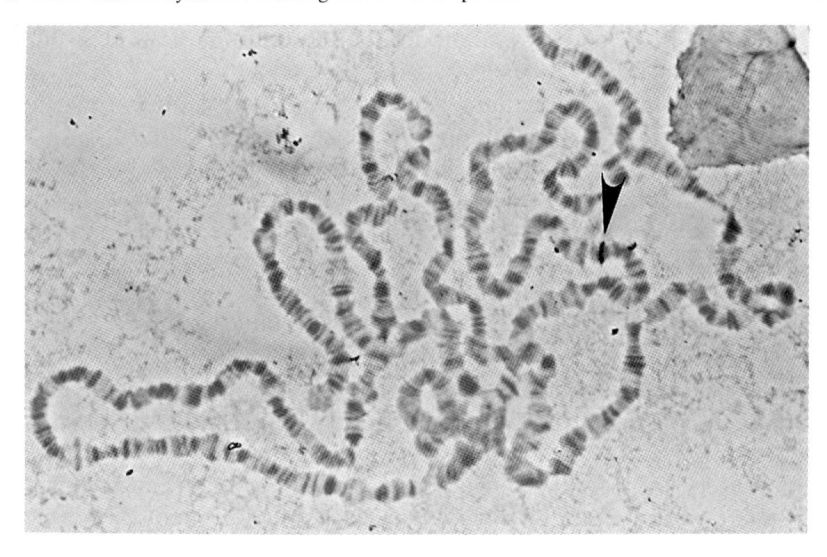

6. Gene Control Systems Affecting Insect Development

Fig. 7. Polytene chromosomes. The polytene chromosomes come from larval salivary glands that have been flattened for viewing by phase contrast microscopy. Immunostaining to detect proteins associated with the chromosomes, or *in situ* hybridization to detect specific DNA sequences, produce bands of staining such as the one indicated by the arrowhead. The sites can be mapped with respect to the chromosomal bands.

A critical question in the search for targets by immunostaining polytene chromosomes is whether the binding sites truly reflect sites of regulation, or whether the proteins merely bind to the chromosomes without regulating nearby genes. One important criterion that helps distinguish regulatory from non-regulatory interactions is whether the expression of putative target genes changes in homeotic mutant backgrounds. For those potential target loci that have been cloned, it is possible to ask whether the target depends on homeotic genes for its expression and whether homeotic proteins bind to the gene *in vitro*. Subsequent *in vivo* analysis using P-element-mediated germ line transformation may tell us whether such binding sites are functionally important. We are currently studying the regulation of two putative target genes identified in this way; both appear to be regulated by homeotic genes.

A second approach used in our laboratory to find targets utilizes an important new general technique in *Drosophila* for identifying the genetic components of a process. The method uses transposable P elements containing a naïve promoter fused to a reporter gene (*lacZ*) to detect enhancers near the chromosomal site of P element insertion (O'Kane and

Gehring, 1987). A transposase-producing chromosome is used to jump the P element to new locations. Frequently, the reporter genes are expressed in patterns that mimic the expression patterns of nearby genes. These "enhancer traps" identify genes on the basis of their expression patterns, which provide strong clues as to their roles in developmental processes. For example, one might expect a gene that specifically controls cell fate decisions in the nervous system to be expressed in early neuroblasts.

We have used enhancer traps to identify genes involved in several developmental processes (e.g. Doe et al., 1991). Several of our enhancer trap insertions identify genes whose expression patterns suggest they are regulated by the homeotic genes *Sex combs reduced* (*Scr*) or *Antp* (D. Andrew, A. McCormick and M. P. Scott, unpublished). To determine whether expression of the nearby gene in fact depends on *Scr* or *Antp*, the enhancer trap chromosome is crossed into homeotic mutant backgrounds. Conveniently, the enhancer trap constructs are equipped with a bacterial replication origin and antibiotic drug resistance gene so that sequences flanking the site of enhancer trap insertion can be cloned directly from genomic DNA. Genomic DNA is merely digested and circularized. Only those fragments containing the enhancer trap sequences will transform bacteria. The enhancer trap approach should allow for the efficient isolation of many potential target genes (e.g. Wagner-Bernholz et al., 1991).

D. Structure–Function Relationships Among Homeodomain Proteins

One of the intriguing questions in the field of homeotic genes is how the different homeotic proteins direct the development of different structures. There are only minor differences in the ANTP and UBX homeodomains, for example, and *in vitro* DNA-binding studies suggest that they recognize similar DNA sequences (Beachy et al., 1988; Laughon et al., 1988; Müller et al., 1988; Affolter et al., 1990; Hayashi and Scott, 1990; S. Hayashi, unpublished data). If, in fact, they bind to the same genes *in vivo*, how are their different functions accounted for? One way to answer these questions is to examine the relationship between the structure of the homeodomain protein and its *in vivo* function. Does the functional specificity of the homeotic proteins reside in the DNA-binding domain, or does it reside in some other part of the protein?

We and others have begun to define domains of the *Antp*, *Ubx* and *Scr* proteins required for their normal functions by testing deleted proteins and hybrid proteins in an *in vivo* assay. Assays for testing homeotic gene function *in vivo* rely on the fact that homeotic genes cause homeotic phenotypes when misexpressed. When an *Antp* cDNA is expressed ubiquitously under the control of a heat-inducible promoter during the third larval instar, one of the phenotypic consequences is the transformation of antennae to legs. Similarly, when *Antp* is expressed ubiquitously during embryogenesis, instead of in its normal, restricted pattern, the first thoracic segment and head segments develop with characteristics of the second thoracic segment. Altered cDNAs can be tested in the same assay. Surprisingly, much of the *Antp* protein sequence is dispensable for causing both the larval and adult cuticle transformations (Gibson *et al.*, 1990).

In order to understand the basis for the functional specificity of homeotic proteins, our laboratory has examined the function of several hybrid cDNAs that encode parts of both *Antp* and *Ubx* or both *Antp* and *Scr* (W. Zeng, M. Horner, and M. P. Scott, unpublished). Some of the hybrid cDNAs exhibit activities of both of the component cDNAs. In other words, the functional specificity may not reside in a single domain. However, most of the functional specificity appears to reside in and around the homeodomain, or DNA-binding portion of the protein (Kuziora and McGinnis, 1989; Gibson *et al.*, 1990; Mann and Hogness, 1990). It is now crucial to learn whether different homeodomains in fact bind different sequences *in vivo* and how the other parts of the homeotic proteins contribute to function.

Since a substantially deleted *Antp* protein behaves like the full-length *Antp* protein in the ubiquitous expression assay, the assay may not be sensitive enough to detect all the functions of homeotic proteins. Altered proteins have been tested only for their effects on cuticular derivatives, but *Antp* is expressed and functions in several other tissues. How would altered cDNAs function if expressed in the normal *Antp* pattern? In order to ask that question we will use the results of analysing *cis*-regulatory sequences of homeotic genes to express homeotic cDNAs in specific patterns.

IV. SUMMARY AND OUTLOOK

Drosophila has served as one of the most powerful systems for the study of animal development, particularly with respect to the question of how genes control developmental patterning and cell fates. Although our attention in this chapter has been focused on the early *Drosophila* embryo, combined genetic and molecular analysis has been successfully applied to many other problems in *Drosophila* development, such as the specification of cell fates in the compound eye, neural pathfinding, and sex determination. A general finding from the study of many different developmental problems in *Drosophila*, as well as other organisms, is that the specification of cell fates is accomplished through a variety of molecular strategies. Nonetheless, many of these basic strategies are used repeatedly in development, and principles that are learned in the study of *Drosophila* will clearly apply to other organisms, including vertebrates. The structure and expression patterns of all classes of segmentation genes in *Drosophila* are conserved among arthropods, and, to a lesser degree, among annelids and chordates (Patel *et al.*, 1989; Sommer and Tautz, 1991). Already much of the current research on mammalian development, where genetic analysis has been limited, focuses on understanding the function of genes first identified as the homologues of master regulatory genes in *Drosophila* (Balling *et al.*, 1989; Wolgemuth *et al.*, 1989; Zimmer and Gruss, 1989; Chisaka and Capecchi, 1991).

Because the basic conceptual understanding of pattern formation is coherent, even if incomplete, attention in the field has turned more and more to the question of how regulatory gene hierarchies are linked to cell biology. It seems likely that the pathways leading from homeotic gene function to the cellular functions that must participate in differentiation of, for example, the leg, will be many and complex. However, as with pattern formation, there may be organizing principles that simplify understanding.

Acknowledgements

We thank Deborah Andrew, Shigeo Hayashi, Yujiro Higashi, Michael Horner, Alison McCormick, Sandra Sonoda, Victoria Yorke, and Wenlin Zeng for conveying results prior to publication; Laura Mathies, Kim Schuske, and Wenlin Zeng for contributions to figures; and Deborah Andrew for comments on the manuscript. M. G. Petitt is a graduate student in the Department of Molecular, Cellular, and Developmental Biology, at the University of Colorado, Boulder. Research in our laboratory is supported by grants from the National Institutes of Health (#18163) and the American Cancer Society.

REFERENCES

Affolter, M., Percival-Smith, A., Müller, M., Leupin, W. and Gehring, W. J. (1990). *Proc. Natl. Acad. Sci. USA* **87**, 4093–4097.
Akam, M. (1987). *Development* **101**, 1–22.
Akam, M. (1989). *Cell* **57**, 347–349.
Anderson, K. V. and Nüsslein-Volhard, C. (1984). *Nature* **311**, 223–227.
Balling, R., Mutter, G., Gruss, P. and Kessel, M. (1989). *Cell* **58**, 337–347.

6. Gene Control Systems Affecting Insect Development

Beachy, P. A., Krasnow, M. A., Gavis, E. R. and Hogness, D. S. (1988). *Cell* **55**, 1069–1081.
Bender, W., Akam, M., Karch, F., Beachy, P. A., Peifer, M., Spierer, P., Lewis, E. B. and Hogness, D. S. (1983). *Science* **221**, 23–29.
Bourois, M., Moore, P., Ruel, L., Grau, Y., Heitzler, P. and Simpson, P. (1990). *EMBO J.* **9**, 2877–2884.
Brown, J. L., Sonoda, S., Ueda, H., Scott, M. P. and Wu, C. (1991). *EMBO J.* **10**, 665–674.
Carroll, S. B. and Scott, M. P. (1986). *Cell* **45**, 113–126.
Chisaka, O. and Capecchi, M. R. (1991). *Nature* **350**, 473–479.
Doe, C. Q., Chu, L. Q., Wright, D. M. and Scott, M. P. (1991). *Cell* **65**, 451–464.
Driever, W. and Nüsslein-Volhard, C. (1989). *Nature* **337**, 138–143.
Frischer, L. E., Hagen, F. S. and Garber, R. L. (1986). *Cell* **47**, 1017–1023.
Gibson, G. and Gehring, W. J. (1988). *Development* **102**, 657–675.
Gibson, G., Schier, A., LeMotte, P. and Gehring, W. J. (1990). *Cell* **62**, 1087–1103.
Hayashi, S. and Scott, M. P. (1990). *Cell* **63**, 883–894.
Hiromi, Y., Kuroiwa, A. and Gehring, W. J. (1985). *Cell* **43**, 603–613.
Hooper, J. E. and Scott, M. P. (1989). *Cell* **59**, 751–765.
Immerglück, K., Lawrence, P. A. and Bienz, M. (1990). *Cell* **62**, 261–268.
Ingham, P. W. (1988). *Nature* **335**, 25–34.
Ish-Horowicz, D. and Pinchin, S. M. (1987). *Cell* **51**, 405–415.
Jaynes, J. B. and O'Farrell, P. H. (1988). *Nature* **336**, 744–749.
Jürgens, G., Wieschaus, E., Nüsslein-Volhard, C. and Kluding, H. (1984). *Wilhelm Roux Arch. Dev. Biol.* **196**, 141–157.
Kennison, J. A. and Tamkun, J. W. (1988). *Proc. Natl. Acad. Sci. USA* **85**, 8136–8140.
Krasnow, M. A., Saffman, E. E., Kornfeld, K. and Hogness, D. S. (1989). *Cell* **57**, 1031–1043.
Kuziora, M. A. and McGinnis, W. (1989). *Cell* **59**, 563–571.
Laughon, A. and Scott, M. P. (1984). *Nature* **310**, 25–31.
Laughon, A., Howell, W. and Scott, M. P. (1988). *Development* (Suppl.) **104**, 85–93.
Mann, R. S. and Hogness, D. S. (1990). *Cell* **60**, 597–610.
McGinnis, W., Levine, M. S., Hafen, E., Kuroiwa, A. and Gehring, W. J. (1984). *Nature* **308**, 428–433.
Müller, M., Affolter, M., Leupin, W., Otting, G., Wüthrich, K. and Gehring, W. J. (1988). *EMBO J.* **7**, 4299–4304.
Nakano, Y., Guerrero, I., Hidalgo, A., Taylor, A., Whittle, J. R. S. and Ingham, P. W. (1989). *Nature* **341**, 508–513.
Nauber, U., Pankratz, M. J., Kienlin, A., Seifert, E., Klemm, U. and Jackle, H. (1988). *Nature* **336**, 489–492.
Nüsslein-Volhard, C. and Wieschaus, E. (1980). *Nature* **287**, 795–801.
Nüsslein-Volhard, C., Wieschaus, E. and Kluding, H. (1984). *Wilhelm Roux Arch. Dev. Biol.* **193**, 267–282.
Nüsslein-Volhard, C., Frohnhofer, H. G. and Lehmann, R. (1987). *Science* **238**, 1675–1681.
O'Kane, C. J. and Gehring, W. J. (1987). *Proc. Natl. Acad. Sci. USA* **84**, 9123–9127.
Padgett, R. W., St, J. R. D. and Gelbart, W. M. (1987). *Nature* **325**, 81–84.
Patel, N. H., Martin-Blanco, E., Coleman, K. G., Poole, S. J., Ellis, M. C., Kornberg, T. B. and Goodman, C. S. (1989). *Cell* **58**, 955–968.
Peifer, M. and Wieschaus, E. (1990). *Cell* **63**, 1167–1178.
Préat, T., Thérond, P., Lamour, I. C., Limbourg, B. B., Tricoire, H., Erk, I., Mariol, M. C. and Busson, D. (1990). *Nature* **347**, 87–89.

Reuter, R., Panganiban, G. E. F., Hoffmann, F. M. and Scott, M. P. (1990). *Development* **110**, 1031–1040.
Rijsewijk, F., Schuermann, M., Wagenaar, E., Parren, P., Weigel, D. and Nusse, R. (1987). *Cell* **50**, 649–657.
Rosenberg, U. B., Schroder, C., Preiss, A., Kienlin, A., Kôte, S., Riede, I. and Jäckle, H. (1986). *Nature* **319**, 336–339.
Schneuwly, S., Kuroiwa, A. and Gehring, W. J. (1987). *EMBO J.* **6**, 201–206.
Schüpbach, T. and Wieschaus, E. (1986). *Roux's Arch. Dev. Biol.* **195**, 302–317.
Schüpbach, T. and Wieschaus, E. (1989). *Genetics* **121**, 101–117.
Scott, M. P. and Carroll, S. B. (1987). *Cell* **51**, 689–698.
Scott, M. P. and Weiner, A. J. (1984). *Proc. Natl. Acad. Sci. USA* **81**, 4115–4119.
Shepherd, J. C., McGinnis, W., Carasco, A. E., De, R. E. M. and Gehring, W. J. (1984). *Nature* **310**, 70–71.
Sommer, R. and Tautz, D. (1991). *Development* **113**, 419–430.
Struhl, G. (1985). *Nature* **318**, 677–680.
Tamkun, J. W., Deuring, R., Scott, M. P., Kissinger, M., Pattatucci, A., Kaufman, T. C. and Kennison, J. A. (1992). *Cell* **68**, 561–572.
Tautz, D., Lehmann, R., Schnurch, H., Schuh, R., Seifert, E., Kienlin, K. and Jäckle, H. (1987). *Nature* **327**, 383–389.
Ueda, H., Sonoda, S., Brown, J. L., Scott, M. P. and Wu, C. (1990). *Genes Dev.* **4**, 624–635.
van den Heuvel, M., Nusse, R., Johnston, P. and Lawrence, P. (1989). *Cell* **59**, 739–749.
Wagner-Bernholz, J. T., Wilson, C., Gibson, G., Schuh, R. and Gehring, W. J. (1991). *Genes Dev.* **5**, 2467–2480.
Wieschaus, E., Nüsslein-Volhard, C. and Jürgens, G. (1984). *Wilhelm Roux Arch. Dev. Biol.* **193**, 267–282.
Winslow, G. M., Hayashi, S., Krasnow, M., Hogness, D. S. and Scott, M. P. (1989). *Cell* **57**, 1017–1030.
Wolgemuth, D. J., Behringer, R. R., Mostoller, M. P., Brinster, R. L. and Palmiter, R. D. (1989). *Nature* **337**, 464–467.
Zimmer, A. and Gruss, P. (1989). *Nature* **338**, 150–153.
Zuo, P., Stanojevic, D., Colgan, J., Han, K., Levine, M. and Manley, J. L. (1991). *Genes Dev.* **5**, 254–264.

7

Insect Immune Systems

Y. ENGSTRÖM

   I. Introduction 125
  II. Background 126
      A. Cellular Reactions 126
      B. Humoral Reactions 127
 III. Insect Immune Proteins 128
      A. Antibacterial Proteins 128
      B. Recognition Molecules 130
  IV. *Drosophila* as an Insect Immune Model System 131
      A. *Drosophila* Immune Response 132
      References 135

I. INTRODUCTION

Insects are living in an environment which is occupied by a variety of microorganisms like bacteria, fungi, protozoa, as well as parasitizing insects. Therefore, insects must be equipped with an effective immune system that has the power to eliminate the invaders, and to ensure the survival of the individual and the species. The insect immune system involves both cellular and humoral reactions that in a close interplay defend the organism against attacking microbes and parasites.

This chapter is not intended to cover all the different areas of past and present work on insect immunity, but rather to give an introduction into a few topics of intense research. Several reviews in recent years have carefully covered both cellular and humoral immunity in insects (Boman *et al.*, 1991; Boman and Hultmark, 1987; Faye and Hultmark, 1992; Götz and Boman, 1985; Kanost *et al.*, 1990; Lackie, 1988; Ratcliffe *et al.*, 1985). I will concentrate on the structure and function of some of the humoral factors isolated from Lepidoptera and Diptera, as well as the structure, organization and regulation of genes encoding antibacterial factors in *Drosophila melanogaster*.

The ability of insects to escape infection was already a matter of interest at the beginning of this century, as the health of silkworm moths was an important economic issue. Several groups of investigators have, since then, been involved in studying and characterizing the properties of the insect immune systems, as well as the different mechanisms used by pathogenic organisms and parasitoids to counteract the immune system of the host (reviews: Götz and Boman, 1985; Vinson, 1990). An important breakthrough in insect immunity research came with the first purification and biochemical characterization of several antibacterial factors from immune haemolymph (review: Boman and Hultmark, 1987). At that time, the main questions asked were: which are the responsible molecules and how do they act?

The current interest in insect immunity may be characterized by asking the question: how is the immune response induced and regulated? In order to answer these and similar questions, several laboratories have undertaken a molecular and genetic characterization of the organization and regulation of the genes responsible for insect immunity.

II. BACKGROUND

Insects are relatively well protected against invasion of foreign organisms through the structural and biochemical properties of the cuticle, which act as a barrier against invaders. Nevertheless, insects do become infected and all insects studied so far have been shown to maintain a rapidly responding immune system. The first immune reactions that occur seem to be cell mediated.

A. Cellular Reactions

The cell-mediated immune response has been studied in many insects. It comprises different types of circulating haemocytes with the capacity to phagocytose, encapsulate and form nodules of invading microorganisms. The insect haemolymph contains many different types of haemocytes that are involved in these processes, as well as in wound healing. The lack of immunological and biochemical markers has made a general classification of haemocytes difficult and there is a great diversity of haemocyte morphologies that have been described for different insect species. The most commonly described haemocyte types are called prohaemocytes, plasmatocytes, granulocytes, spherulocytes, coagulocytes and oenocytoids as reviewed in Price and Ratcliffe (1974). Gupta (1985) also includes adipohaemocytes, podocytes and vermicytes in his overview of insect haemocytes. In Lepidoptera, the plasmatocytes and the granulocytes have been found to be the main cell types involved in the immune reactions. The plasmatocytes are considered to be involved in phagocytosis, encapsulation and nodule formation, while the granulocytes are initiators of nodule formation (Götz and Boman, 1985). In the dipteran *Drosophila melanogaster*, a smaller number of haemocyte types have been described (Rizki, 1978). These are the plasmatocytes, which are believed to differentiate into podocytes and lamellocytes, and the fourth class has been called crystal cells, and might be related to the granular cells described in other insects. It has been found that certain Diptera contain a very low total number of haemocytes in their haemolymph and, thus, the immune system in these insects might be more dependent on humoral reactions, like humoral encapsulation, than on reactions that require the involvement of a large number of cells, like cellular encapsulation and nodule formation (Götz and Boman, 1985).

B. Humoral Reactions

The humoral immune response in higher insect orders like Lepidoptera, Diptera and Hymenoptera is mainly achieved by circulating factors, whose synthesis is induced upon bacterial infection (Boman and Hultmark, 1987). Some of these factors are directly antibacterial while others have an unknown function but may act as recognition factors, soluble receptors or agglutinins (Kanost et al., 1990).

Insect antibacterial factors have mainly been isolated from the haemolymph of different Lepidoptera, but also from Diptera and Hemiptera. The great advantage with using large moths for the purification of antibacterial factors from cell-free haemolymph is the size of the animal, in addition to the opportunity to turn on the genes for the immune factors in diapausing pupae without breaking the diapause. This was used in an early experiment by Faye et al. (1975) in which the specific synthesis of immune proteins was achieved by injecting bacteria into a *Hyalophora cecropia* diapausing pupa, followed by the addition of radioactive amino acids. The proteins in the haemolymph were subsequently separated in SDS-polyacrylamide gel electrophoresis and the specifically labelled proteins visualized in the gel after autoradiography, and designated immune proteins P1 to P9. Proteins P4, P5, P7 and P9 have been purified and are now renamed as haemolin, attacin, lysozyme and cecropin respectively. Homologous proteins have also been described from several other species and given alternative names, as will be indicated.

III. INSECT IMMUNE PROTEINS

A. Antibacterial Proteins

1. Lysozyme

Lysozyme was the first antibacterial protein to be isolated, first from immune haemolymph of the wax moth, *Galleria mellonella*, and the silkworm moth, *Bombyx mori* (Powning and Davidson, 1973), and subsequently identified in a number of insects (review: Kanost *et al.*, 1990). This well-known enzyme catalyses the degradation of the peptidoglycan layer of the cell wall of Gram-positive bacteria. The insect lysozymes are relatively similar to the vertebrate lysozyme of the chicken type. It is unclear if the main function of insect lysozyme is to kill bacteria, as it is only active against a few Gram-positive bacteria. Perhaps its main function is to remove bacterial cell wall products from the haemolymph, or alternatively to process the bacterial cell wall into molecules (peptidoglycans) that are recognized by the immune system and thereby to trigger the immune response.

2. Cecropins

Cecropins are the most potent antibacterial proteins isolated from immune *Hyalophora cecropia* haemolymph (Hultmark *et al.*, 1980; Steiner *et al.*, 1981). Homologous proteins have been purified from several lepidopteran species: from *Antheraea pernyi* (Qu *et al.*, 1982), from *Manduca sexta*, called bactericidin (Dickinson *et al.*, 1988), from *Bombyx mori*, called lepidopteran (Morishima *et al.*, 1990; Teshima *et al.*, 1986, 1987), from the dipteran *Sarcophaga peregrina*, called sarcotoxin I (Okada and Natori, 1985), and from a mammalian source, namely from pig intestine (Lee *et al.*, 1989). The cecropins are a family of small (4-kDa) basic proteins, synthesized as a preproprecursor molecules, which contain a signal peptide for their export into the haemolymph. The mature molecule is formed through the enzymatic cleavage by signal peptidase and peptidyl peptidase, and by amidation at the C-terminal end (Boman *et al.*, 1989). In solution, the cecropins form a random coil structure, but theoretical predictions and model building suggest that α-helical conformation is induced by the presence of organic solvents (Steiner, 1982). Nuclear magnetic resonance spectroscopy (NMR) with *Hyalophora* cecropin A confirmed that the molecule contains two α-helices interrupted by a hinge region (Holak et al., 1988). Furthermore, most of the hydrophilic residues are situated on one side of the cylindrical helix and the hydrophobic side chains are on the opposite side. Such amphipathic helices are known to interact with membranes and it has been suggested that this property of the cecropins is essential for their antibacterial action, which is probably achieved by disrupting the bacterial cell membrane (Steiner et al., 1988).

3. Attacin-like molecules

Another important group of antibacterial proteins isolated from lepidopteran and dipteran species is the attacin family, which consists of the attacins, sarcotoxin II, and the diptericins, of which the latter are much smaller but appear to share some homology with the former.

The largest protein in this family of bactericidal proteins is sarcotoxin II (28 kDa) purified from the flesh fly *Sarcophaga peregrina* (Ando et al., 1987). Three different forms of sarcotoxin II have been isolated and for one of these, sarcotoxin IIA, a cDNA clone has been isolated and molecularly characterized (Ando and Natori, 1988b).

The attacins (20 kDa), isolated from immune haemolymph of *Hyalophora cecropia*, are bactericidal proteins, structurally related to sarcotoxin II. Several different forms of attacins have been isolated (Hultmark et al., 1983) but molecular cloning has shown that there are only two different attacin genes in *Hyalophora cecropia*, encoding one basic and one neutral form of attacin (Sun et al., 1991a).

Attacins and sarcotoxin II only affect growing bacteria, by an unknown mechanism interfering with cell division. It has been suggested that the target of the attacins is the biosynthesis of the bacterial outer membrane (Engström et al., 1984). The bacteriostatic effect of attacins and sarcotoxin II was shown to be specific for a few Gram-negative bacteria (Ando and Natori, 1988a; Hultmark et al., 1983).

Diptericins are small glycine-rich proteins (9 kDa) purified from *Phormia terranovae* (Dimarcq et al., 1988). Three different forms of diptericin have been isolated from *Phormia* and one of these, diptericin A, has been studied in detail. The diptericins are active against some Gram-negative bacteria, probably by acting on the cytoplasmic membrane (Keppi et al., 1989).

4. Other antibacterial proteins

The insect defensins are a group of small cysteine-rich peptides (4 kDa) that are active against Gram-positive bacteria, presumably by attacking the bacterial cell membrane. Members of this group of peptides have been studied in two dipteran species; the *Phormia* defensin (Lambert *et al.*, 1989) and the sapecin isolated from the culture medium of a *Sarcophaga peregrina* cell line (NIH-Sape-4) (Matsuyama and Natori, 1988), as well as in a hemiptera species, namely the royalisin isolated from honey bee royal jelly (Fujiwara *et al.*, 1990).

The name defensins was given to this family of proteins due to their apparent homology to the mammalian defensins isolated from neutrophils and macrophages (Lehrer *et al.*, 1991). The first comparisons of the amino acid sequence suggested a conservation of the number and location of the cysteine residues, forming the three intramolecular S–S bridges. However, subsequent nucleotide sequence analysis and comparisons between insect and mammalian defensin genes do not support the primary hypothesis of them being related molecules.

Two families of proline-rich peptides have recently been isolated from the honey bee *Apis apis*, and named apidaecins and abecins (Casteels *et al.*, 1989, 1990).

B. Recognition Molecules

The insect immune system has the power of self/non-self discrimination, as well as recognizing and combating invading microorganisms. It is not clear how the recognition of foreign molecules is achieved, since antibodies have not been found in any invertebrates. A current hypothesis is that certain unique patterns on the surface of microorganisms are recognized by specialized molecules (Janeway, 1989). A few candidate recognition molecules have been isolated from insects and are briefly described below.

1. Haemolin

One of the most interesting candidate proteins for recognition of foreign molecules is haemolin, previously called protein P4 (Faye *et al.*, 1975). Haemolin is present in naïve haemolymph but, nevertheless, is induced to a great extent by vaccination of diapausing pupae (Andersson and Steiner, 1987; Rasmuson and Boman, 1979). During an active immune response, haemolin is the most abundant immune protein, but until recently its function was totally unknown, as haemolin does not show any antibacterial properties. The recent isolation of the haemolin gene from *Hyalophora cecropia* revealed the very exciting result that haemolin is a member of the immunoglobulin (Ig) superfamily (Sun *et al.*, 1990). Although other insect genes have been studied that code for members of the Ig superfamily, haemolin is the first insect protein belonging to the Ig superfamily *and* being part of the insect immune response. The exact function of the haemolin molecule is still not known, but it has been shown that haemolin can bind to the surface of bacteria, which suggests that haemolin might function as a soluble recognition molecule, with affinity for bacteria or bacterial cell wall components (Sun *et al.*, 1990).

2. Lectins

Lectins are multivalent proteins that bind to certain sugar residues on cell membranes and thereby lead to the aggregation of cells (Yeaton, 1981a,b). Insect haemolymph normally contains different types of lectins but a few specific lectins have been shown to be induced as part of an immune response. The M13 lectin isolated from *Manduca sexta* is one of them. It is a dimer of 36 kDa, and it was shown to interact with haemocytes and to trigger a cellular coagulation reaction (Minnick *et al.*, 1986). Another inducible lectin is the *Sarcophaga* lectin, which is made from six monomers of 30 to 32 kDa, and shows affinity for galactose and lactose (Kobayashi *et al.*, 1989; Komano *et al.*, 1980).

IV. *DROSOPHILA* AS AN INSECT IMMUNE MODEL SYSTEM

The isolation and biochemical characterization of a number of immune proteins during the last two decades were greatly supported by the use of large moths as experimental organisms. These animals provided large amounts of starting material for the purification of immune proteins and, furthermore, as described previously, the activation of an immune response in diapausing pupae enabled the specific labelling and analysis of synthesized immune proteins.

For studies on the molecular mechanisms underlying the activation and control of the insect immune system, the fruit fly *Drosophila melanogaster* is an ideally suitable model system. *Drosophila* provides a unique combination of methodology, knowledge and experience in classical and molecular genetics, and its genome is being molecularly and physiologically mapped (see Chapter 4).

A. *Drosophila* Immune Response

The first report on an immune response in *Drosophila* was published by Bakula (1970), who identified an antibacterial factor of lysosomal origin. Boman et al. (1972) described the existence of an inducible antibacterial defence system in *Drosophila* and this work was followed up by Flyg et al. (1987), who demonstrated the presence of attacin-like molecules, cecropin-like molecules, as well as a non-inducible lysozyme activity in cell-free extracts and in haemolymph from vaccinated *Drosophila* adults.

In a study by Robertson and Postlethwait (1986) it was shown that after inoculation with live bacteria, adult flies synthesized at least eight new polypeptides, ranging in size from about 5 kDa to about 75 kDa. Antibacterial activity appeared in the haemolymph 2 hours after inoculation and it was shown that the antibacterial polypeptides were synthesized *de novo*.

1. The Cecropin *locus*

In 1990, the first molecular analysis of a few of the genes encoding *Drosophila* immune proteins was published by Kylsten et al. (1990), namely the Cecropin (*Cec*) locus. Three different *Cec* genes (*CecA1*, *CecA2* and *CecB*) and two pseudogenes were cloned by homology to a cDNA clone of the *Sarcophaga* sarcotoxin IA gene (Matsumoto et al., 1986), and were identified within a small genomic region localized at 99E. Lately, a fourth cecropin-coding gene *CecC*, as well as a partly *Cec*-related gene, the *Andropin* (*Anp*) gene, have been isolated from the same chromosomal region and molecularly analysed (Samakovlis et al., 1991; Tryselius et al., 1991). All five genes are organized as two exons and a small intron; the sequence is highly conserved between the exons of the *Cec* genes, but non-homologous within the introns (Tryselius et al., 1991).

The *Drosophila Cec* genes code for a family of three small basic polypeptides, cecropins A, B and C, probably synthesized as preprepreurosor molecules, which after cleavage of the signal peptide at the N-terminal end and cleavage and amidation at the C-terminal end, form the mature exported polypeptides consisting of 39 amino acids. The amino acid sequence is highly homologous between the three *Drosophila* cecropins, as well as to the sarcotoxin I molecules isolated from *Sarcophaga peregrina* (Matsuyama and Natori, 1988; Okada and Natori, 1985).

The expression of the *Drosophila Cec* genes was studied by RNase protection analysis in extracts from embryos, larvae, pupae and adults, as well as by *in situ* hybridization on sections of larvae, pupae and adults. These experiments clearly showed that the *Cec* genes are induced as a response to bacterial infection, and that the main synthetic tissues are the fat body and haemocytes (Samakovlis *et al.*, 1990). The transcription of all four *Cec* genes can be thoroughly activated in the fat body, but the relative level of induction varies between the individual genes at different developmental stages. In short, the *CecA1* and *CecA2* genes are the main inducible *Cec* genes during larval and adult stages, while *CecB* and *CecC* transcripts are the most prominent ones after induction during pupal stages. Without pretreating the animals with bacteria or bacterial substances there was no expression of the *Cec* genes except for a weak level of expression in early pupae. This result was, however, found to be correlated with contaminated food intake, since axenically raised animals did not show any expression of the *Cec* genes without induction (Samakovlis *et al.*, 1990).

The *Andropin* (*Anp*) gene behaves very differently and was shown to be constitutively expressed in adult males, and could not be further activated by bacteria (Samakovlis *et al.*, 1991). The transcription of the *Anp* gene was, instead, induced upon mating and the transcript was exclusively found in the ejaculatory duct of adult males. The androsin molecule is not a true cecropin, but it may form amphipathic helices and it has antibacterial properties. Therefore, the authors have suggested that the function of androsin is to protect the seminal fluid and the male reproductive tract from microbial infections (Samakovlis *et al.*, 1991).

2. The diptericin gene

Molecular analysis of a cDNA clone encoding a *Drosophila* diptericin gene was recently published (Wicker *et al.*, 1990). The synthesis of the corresponding mRNA is induced after challenge with bacteria in larvae, pupae and adults. Only one *Drosophila* diptericin gene has been detected and it was mapped to the chromosomal location 56A. The diptericin amino acid sequence was deduced from the cDNA and showed 60% overall homology to the previously isolated *Phormia terranovae* diptericin molecule. The diptericin mRNA codes for a polypeptide containing a signal peptide, a proline-rich domain (P) and a glycine-rich domain (G). The presence of the P and G domains suggests that the diptericin has some evolutionary relationship to the much larger attacin and sarcotoxin II molecules (Wicker *et al.*, 1990).

3. The lysozyme locus

A family of *Drosophila* lysozyme genes has been cloned and mapped to the location 61F1-4 on the third chromosome (Kylsten *et al.*, 1992). The molecular analysis of two of these genes, the *LysD* and *LysP* genes, showed that they lack introns. In contrast to one of the previously studied insect lysozyme genes, the *Hyalophora cecropia* lysozyme gene, which was shown to be induced after injection of bacteria, the *Drosophila LysD* and *LysP* genes were expressed without any previous activation of an immune response. Instead, the transcription of the *Drosophila LysP* and *LysD* genes was found to be repressed after injection of bacteria. Transcripts from the *LysD* gene were predominantly found in an anterior section of the midgut in larvae and adults, while the *LysP* transcripts were only detected in the salivary glands of adult animals. Taking these results together, they suggest that at least the *LysD* and *Lys*P genes might not have a major function in the *Drosophila* immune defence against bacteria, but rather have a function in digesting bacteria for nutritional purposes.

4. Regulation of cecropin genes

In order to investigate the molecular mechanisms underlying the tissue-specific and inducible expression of the *Drosophila* cecropin genes we have initiated a series of experiments aiming to identify the *cis*- and *trans*-acting elements of *Cec* gene expression. We have constructed different sets of *Cec-lacZ* fusion gene constructs including promoters, upstream sequences and parts of the leader sequences linked to the reporter gene (*E. coli lacZ* gene). The expression of β-galactosidase (β-gal) is analysed both in cultured *Drosophila* cells by transient expression and in stably transformed *Drosophila* larvae and flies, after P-element-mediated transformation. Several of the *Cec-lacZ* constructs are indeed effectively activated after the induction of an immune response by lipopolysaccharide (LPS), both in the transiently transfected cells and in the transformed larvae and flies (preliminary results).

The promoter region of several of the insect immune genes contains a sequence with homology to the binding site of the mammalian transcription factor NF-κB. This binding site was first identified in the enhancer region of the immunoglobulin light kappa gene, and has since been found in promoter and enhancer elements in many mammalian genes that are involved in immune, inflammatory and acute phase responses (Leonardo and Baltimore, 1989). Recently, it was reported that the *Hyalophora cecropia* lysozyme gene contains a similar κB element that appears to bind an inducible transcription factor (Sun et al., 1991b). Also the *Hyalophora cecropia* attacin genes (Sun et al., 1991a) and the *Drosophila* cecropin genes contain elements with partial homology to κB sites (Tryselius et al., 1991). Our preliminary data indicate the existence of a nuclear factor(s) in induced *Drosophila* cell extracts that bind to these κB-like sequences (Engström, unpublished).

Acknowledgements

I would like to thank Dr Dan Hultmark and Dr Latha Kadalayil for critical reading of the manuscript. Work in the author's laboratory was supported by grants from the Swedish Natural Science Research Council, The Swedish Cancer Society, Magnus Bergvalls Stiftelse and Carl Tryggers Stiftelse.

REFERENCES

Andersson, K. and Steiner, H. (1987). *Insect Biochem.* **17** (1), 133–140.
Ando, K. and Natori, S. (1988a). *J. Biochem. (Tokyo)* **103** (4), 735–739.
Ando, K. and Natori, S. (1988b). *Biochemistry* **27** (5), 1715–1721.
Ando, K., Okada, M. and Natori, S. (1987). *Biochemistry* **26** (1), 226–230.
Bakula, M. (1970). *J. Insect Physiol.* **16**, 185–197.
Boman, H. G. and Hultmark, D. (1987). *Annu. Rev. Microbiol.* **41**, 103–126.
Boman, H. G., Nilsson, I. and Rasmuson, B. (1972). *Nature* **237** (352), 232–235.
Boman, H. G., Boman, I. A., Andreu, D., Li, Z., Merrifield, R. B., Schlenstedt, G. and Zimmerman, R. (1989). *J. Biol. Chem.* **264** (10), 5852–5860.
Boman, H. G., Faye, I., Gudmundsson, G. H., Lee, J. and Lidholm, D. A. (1991). *Eur. J. Biochem.* **201** (1), 23–31.
Casteels, P., Ampe, C., Jacobs, F., Vaeck, M. and Tempst, P. (1989). *EMBO J.* **8**, 2387–2391.
Casteels, P., Ampe, C., Riviere, L., Van Damme, J., Elicone, C., Fleming, M., Jacobs, F. and Tempst, P. (1990). *Eur. J. Biochem.* **187**, 381–386.
Dickinson, L., Russell, V. and Dunn, P. E. (1988). *J. Biol. Chem.* **263** (36), 19424–19429.
Dimarcq, J. L., Keppi, E., Dunbar, B., Lambert, J., Reichhart, J. M., Hoffmann, D., Rankine, S. M., Fothergill, J. E. and Hoffmann, J. A. (1988). *Eur. J. Biochem.* **171** (1–2), 17–29.
Engström, P., Carlsson, A., Engström, Å., Tao, Z. J. and Bennich, H. (1984). *EMBO J.* **3**, 3347–3351.
Faye, I. and Hultmark, D. (1992). In "Parasites and Pathogens of Insects" (N. Beckage, S. N. Thompson and B. A. Federici, eds). Academic Press, London. In press.
Faye, I., Pye, A., Rasmuson, T., Boman, H. G. and Boman, I. A. (1975). *Infect. Immun.* **12** (6), 1426–1438.
Flyg, G., Dalhammar, G., Rasmuson, B. and Boman, H. G. (1987). *Insect Biochem.* **17** (1), 153–160.
Fujiwara, S., Gennaro, R., Schneider, K., Przybylski, M. and Romeo, D. (1990). *J. Biol. Chem.* **265**, 11333–11337.

Gupta, A. P. (1985). *In* "Comprehensive Insect Physiology, Biochemistry and Pharmacology", 3 (G. A. Kerkut and L. I. Gilbert, eds), pp. 401–452. Pergamon Press, Oxford.

Götz, P. and Boman, H. G. (1985). *In* "Comprehensive Insect Physiology, Biochemistry and Pharmacology" (G. A. Kerkut and L. I. Gilbert, eds), pp. 453–485. Pergamon Press, Oxford.

Holak, T. A., Engström, A., Kraulis, P. J., Lindeberg, G., Bennich, H., Jones, T. A., Gronenborn, A. M. and Clore, G. M. (1988). *Biochemistry* **27** (20), 7620–7629.

Hultmark, D., Engström, A., Andersson, K., Steiner, H., Bennich, H. and Boman, H. G. (1983). *EMBO J.* **2** (4), 571–576.

Hultmark, D., Steiner, H., Rasmuson, T. and Boman, H. G. (1980). *Eur. J. Biochem.* **106** (1), 7–16.

Janeway, C. A. J. (1989). *Cold Spring Harbor Symp. Quant. Biol.* **LIV**, 1–13.

Kanost, M. R., Kawooya, J. K., Law, J. H., Ryan, R. O., Van Heusden, M. C. and Ziegler, R. (1990). *Adv. Insect Physiol.* **22**, 298–396.

Keppi, E., Pugsley, A. P., Lambert, J., Wicker, C., Dimarcq, J. L., Hoffmann, J. A. and Hoffmann, D. (1989). *Arch. Insect Biochem. Physiol.* **10** (3), 229–239.

Kobayashi, A., Hirai, H., Kubo, T., Ueno, K., Nakanishi, Y. and Natori, S. (1989). *Biochim. Biophys. Acta* **1009** (3), 244–250.

Komano, H., Mizuno, D. and Natori, S. (1980). *J. Biol. Chem.* **255** (7), 2919–2924.

Kylsten, P., Samakovlis, C. and Hultmark, D. (1990). *EMBO J.* **9** (1), 217–224.

Kylsten, P., Kimbrell, D. A., Daffre, S., Samakovlis, C. and Hultmark, D. (1992). *Mol. Gent. Genet.* **232**, 335–343.

Lackie, A. M. (1988). *Parasitology Today* **4** (4), 98–105.

Lambert, J., Keppi, E., Dimarcq, J. L., Wicker, C., Reichhart, J. M., Dunbar, B., Lepage, P., Van Dorsselaer, A., Hoffmann, J., Fothergill, J. and Hoffmann, D. (1989). *Proc. Natl. Acad. Sci. USA* **86** (1), 262–266.

Lee, J., Boman, A., Chuanxin, S., Andersson, M., Jörnvall, H., Mutt, V. and Boman, H. G. (1989). *Proc. Natl. Acad. Sci. USA* **86** (23), 9159–9162.

Lehrer, R. I., Ganz, T. and Selsted, M. E. (1991). *Cell* **64**, 229–230.

Leonardo, M. J. and Baltimore, D. (1989). *Cell* **58**, 227–229.

Matsumoto, N., Okada, M., Takahashi, H., Ming, Q. X., Nakajima, Y., Nakanishi, Y., Komano, H. and Natori, S. (1986). *Biochem. J.* **239**, 717–722.

Matsuyama, K. and Natori, S. (1988). *J. Biol. Chem.* **263** (32), 17112–17116.

Minnick, M. F., Rupp, R. A. and Spence, K. D. (1986). *Biochem. Biophys. Res. Commun.* **137**, 729–735.

Morishima, I., Suginaka, S., Ueno, T. and Hirano, H. (1990). *Comp. Biochem. Physiol. Pt B.* **95**, 551.

Okada, M. and Natori, S. (1985). *J. Biol. Chem.* **260** (12), 7174–7177.

Powning, R. F. and Davidson, W. J. (1973). *Comp. Biochem. Physiol.* **45B**, 669–681.

Price, C. D. and Ratcliffe, N. A. (1974). *Z. Zellforch. Mikrosk. Anat.* **147**, 537–549.

Qu, X., Steiner, H., Engström, Å., Bennich, H. and Boman, H. G. (1982). *Eur. J. Biochem.* **127** (1), 219–224.

Rasmuson, T. and Boman, H. G. (1979). *Insect Biochem.* **9**, 259–264.

Ratcliffe, N. A., Rowley, A. F., Fitzgerald, S. W. and Rhodes, C. P. (1985). *Int. Rev. Cytol.* **97**, 183–350.

Rizki, T. M. (1978). *In* "The Genetics and Biology of Drosophila", 2b, (M. Ashburner and T. R. F. Wright, eds) pp. 397–452. Academic Press, London.

Robertson, M. and Postlethwait, J. H. (1986). *Dev. Comp. Immunol.* **10** (2), 167–179.

Samakovlis, C., Kimbrell, D. A., Kylsten, P., Engström, Å. and Hultmark, D. (1990). *EMBO J.* **9** (9), 2969–2976.

7. Insect Immune Systems

Samakovlis, C., Kylsten, P., Kimbrell, D. A., Engström, Å. and Hultmark, D. (1991). *EMBO J.* **10** (1), 163–169.
Steiner, H. (1982). *FEBS Lett.* **137** (2), 283–287.
Steiner, H., Hultmark, D., Engström, A., Bennich, H. and Boman, H. G. (1981). *Nature* **292** (5820), 246–248.
Steiner, H., Andreu, D. and Merrifield, R. B. (1988). *Biochim. Biophys. Acta.* **939** (2), 260–266.
Sun, S., Lindström, I., Boman, H. G., Faye, I. and Schmidt, O. (1990). *Science* **250** (4988), 1729–1732.
Sun, S., Lindström, I., Lee, J. and Faye, I. (1991a). *Eur. J. Biochem.* **196**, 247–254.
Sun, S., Åsling, B. and Faye, I. (1991b). *J. Biol. Chem.* **266** (10), 6644–6649.
Teshima, T., Ueki, Y., Nakai, T. and Shiba, T. (1986). *Tetrahedron* **42**, 829–834.
Teshima, T., Nakai, T., Ueki, Y. and Shiba, T. (1987). *Tetrahedron* **43**, 4513–4518.
Tryselius, Y., Samakovlis, C., Kimbrell, D. A. and Hultmark, D. (1992). *Eur. J. Biochem.* **204**, 395–399.
Vinson, S. B. (1990). *Arch. Insect Biochem. Physiol.* **13**, 3–27.
Wicker, C., Reichhart, J., Hoffmann, D., Hultmark, D., Samakovlis, C. and Hoffmann, J. A. (1990). *J. Biol. Chem.* **265** (36), 22493–22498.
Yeaton, R. W. (1981a). *Dev. Comp. Immunol.* **5**, 535–545.
Yeaton, R. W. (1981b). *Dev. Comp. Immunol.* **5**, 391–402.

Part III.
Interactions with the Environment

8. Semiochemicals: Molecular Determinants of Activity and Biosynthesis
   J. A. PICKETT AND C. M. WOODCOCK
9. The Perception of Semiochemicals
   L. J. WADHAMS
10. Semiochemically Mediated Behaviour
    G. M. POPPY
11. Molecular Biology of Insecticide Resistance
    A. L. DEVONSHIRE, L. M. FIELD AND M. S. WILLIAMSON

8

Semiochemicals: Molecular Determinants of Activity and Biosynthesis

J. A. PICKETT AND C. M. WOODCOCK

    I. Introduction 141
    II. Molecular Determinants of Activity 141
    III. Interactions Between Different Semiochemical Types 144
    IV. Analogues of Semiochemicals 146
    V. Semiochemical Biosynthesis: New Targets for Molecular
        Genetics 147
        References 149

I. INTRODUCTION

Semiochemicals, or behaviour-controlling chemicals, include pheromones, which are involved in intraspecific communication, allomones (favouring the emitting organism) and kairomones (favouring the receiver). Semiochemicals transmitted through air, which need to be volatile, have low molecular weights and relatively high lipophilicity. However, many pheromones and other semiochemicals employed for communication through water are also low molecular weight, lipophilic compounds, e.g. the algal pheromones (Jaenicke and Boland, 1982). Although compounds as simple as ethanol can be semiochemical components, the carbon range is normally between C_5 and C_{20} with a maximum of two or three hetero atoms, generally oxygen but also sulphur or nitrogen.

II. MOLECULAR DETERMINANTS OF ACTIVITY

Semiochemicals are usually produced by polyketide, terpenoid or fatty acid pathways, but can include compounds from amino acids, polysaccharides and polyphenolics. The polyketides can have simple structures, such as the 1,3-dione aggregation pheromone of the pea and bean weevil, *Sitona lineatus* (Blight *et al.*, 1984), although this compound shows some structural sophistication through keto–enol tautomerism. Bark beetles such as the mountain pine beetle, *Dendroctonus ponderosae*, employ cyclic polyketides such as exobrevicomin (Borden, 1984). This compound contains only nine carbon atoms and two oxygens, but it has three asymmetric carbons which impart a highly specific stereochemistry. Monoterpenoids containing only ten carbon atoms commonly have asymmetric carbon atoms, for example (+)-ipsdienol and (+)-verbenone, again from the mountain pine beetle (Borden, 1984). Geometric isomerism is also a feature of the terpenoids and two components of the boll weevil (*Anthonomus grandis*) aggregation pheromone are the (*Z*) and (*E*) (*cis* and *trans*) isomers of a monocyclic monoterpene aldehyde (Tumlinson *et al.*, 1969; Dickens and Mori, 1989; Dickens and Prestwich, 1989). Some aphid sex pheromones have recently been identified as comprising monoterpenoids in the cyclopentanoid series. The nepetalactone component has three asymmetric carbon atoms giving eight isomeric possibilities, and the nepetalactol with four asymmetric carbons has 16 isomeric possibilities (Dawson *et al.*, 1990a) but for most aphids one specific isomer of each compound type is employed. In the course of identifying the stereochemistry of the nepetalactol found in many aphids, including the vetch aphid, *Megoura viciae* (Dawson *et al.*, 1987), and the greenbug, *Schizaphis graminum* (Dawson *et al.*, 1988), a 3,5-dinitrobenzoate ester was formed, allowing a crystal structure to be obtained (Fig. 1) (Dawson *et al.*, 1989b). This provides a picture of the molecular structure for the nepetalactol, showing the stereochemistry at the four chiral carbon atoms. Further diversity can be achieved, even with straight chain terpenoids, by combination with other structural types, for example when terpenoid alcohols are converted to esters, as in the case of the San José scale, *Quadraspidiotus perniciosus* (Gieselmann *et al.*, 1979).

The simplest sesquiterpene (C_{15}) pheromone component is (*E*)-β-farnesene, the alarm pheromone for many species of aphids, which is used to warn of attack by predators and parasitoids, allowing dispersal and escape (Pickett *et al.*, 1992). The simple straight chain sesquiterpenoid alcohols (*E,E*)-farnesol, (*Z,E*)-farnesol and (*E*)-nerolidol are semiochemicals employed by the two-spotted spider mite, *Tetranychus urticae*. Some very exciting chemistry has been provided by the sesquiterpenoids comprising the sex pheromone of the American cockroach, *Periplaneta americana*. Periplanone B was identified by Adams *et al.* (1979), but it was some time before the structure of periplanone A was characterized by Hauptmann *et al.* (1986). These compounds are in the germacrane

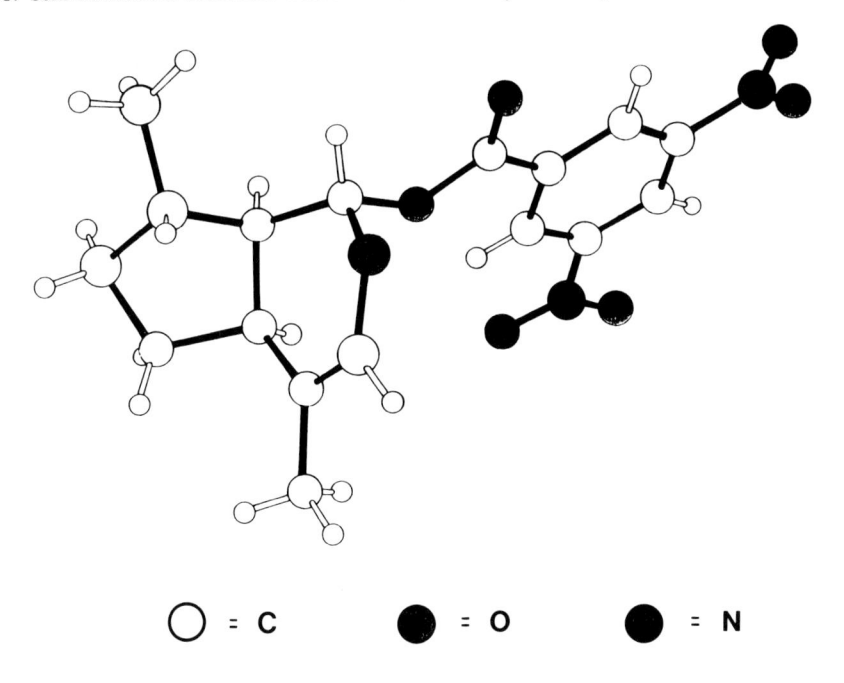

8. Semiochemicals: Molecular Determinants of Activity and Biosynthesis

○ = C   ● = O   ● = N

Fig. 1. Molecular structure of the 3,5-dinitrobenzoate of the aphid sex pheromone component (1*R*,4a*S*,7*S*,7a*R*)-nepetalactol.

series, known for having highly labile double bonds which confounded early structure elucidation. Some terpenoid pheromone compounds arise from larger molecules such as triterpenoids (C_{30}). In mammals, for example pigs, steroid metabolites, including 5α-androst-16-en-3-one, are important sex pheromone components.

The sex pheromones of Lepidoptera are almost exclusively long chain fatty acid derived structures (C_{12}–C_{18}), with some functionality such as aldehyde, alcohol or acetate. The specificity of these compounds is largely derived from the number, position and geometric isomerism of unsaturated double bonds in the chain (Arn *et al.*, 1986; Mayer and McLaughlin, 1991), e.g. for the pink bollworm, *Pectinophora gossypiella*, which employs the (*Z*,*Z*)- and (*Z*,*E*)-7,11-hexadecadienyl acetates. In Europe, the sex pheromone components of the diamondback moth, *Plutella xylostella*, are the acetate and aldehyde of the (*Z*)-11-hexadecenyl carbon chain, but in the Far East there is a sibling population for which the pheromone comprises a small percentage of the respective (*E*)-isomers (Macaulay *et al.*, 1986). Mosquitoes in the genus *Culex*, particularly *Cx. quinquefasciatus*, which is a vector of filarial diseases such as elephantiasis, also employ a long chain (C_{16}) fatty acid. However, instead of unsaturation, the functionality is dihydroxylated, with one hydroxy group forming a lactone ring with the acid group and the other hydroxy group an acetate ester. This creates two asymmetric centres and only one of the four possible isomers is active, the (5*R*,6*S*)-6-acetoxy-5-hexadecanolide (Fig. 2) (Laurence and Pickett, 1982; Laurence *et al.*, 1985; Dawson *et al.*, 1990b).

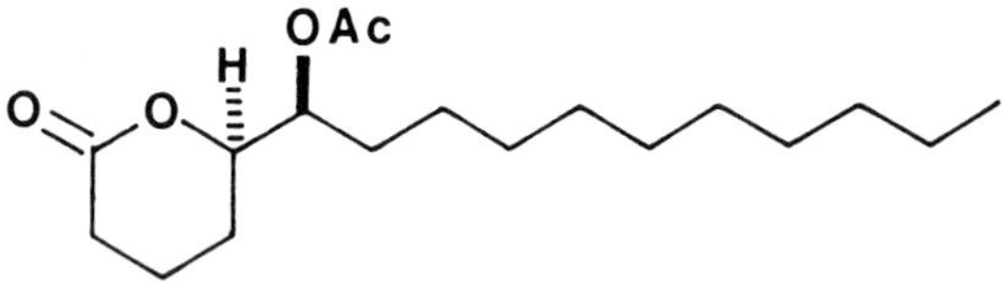

Fig. 2. (5*R*,6*S*)-6-Acetoxy-5-hexadecanolide, the oviposition pheromone of *Culex* species mosquitoes.

III. INTERACTIONS BETWEEN DIFFERENT SEMIOCHEMICAL TYPES

Of the semiochemical types, pheromones have received most attention because of their often very potent activity. Nevertheless, it is now widely appreciated that the whole subject of the chemical ecology between insects and plants, as well as between insects themselves, must be studied in order to understand chemically mediated behaviour, particularly if we are to exploit the promise of these agents in effective crop protection. Thus, the aggregation pheromone of *S. lineatus*, discussed above, is most effective in catching insects in the field when it is used together with volatiles from the legume host (e.g. *Vicia faba*), i.e. (*Z*)-3-hexen-1-ol, (*Z*)-3-hexen-1-yl acetate and linalol (Blight *et al.*, 1984; Blight and Wadhams, 1987). The mountain pine beetle also employs components from its host, i.e. *α*-pinene, myrcene and terpinolene, together with an aggregation pheromone comprising *trans*-verbenol from the female and exobrevicomin and frontalin from the male (Borden, 1984). The aphid alarm pheromone, (*E*)-*β*-farnesene, can be inhibited in its effect by (−)-*β*-caryophyllene (Fig. 3), a structurally related compound produced by plants from many different families. The aphid is thus able to distinguish farnesene, which is also present in plants, from that produced by aphids (Dawson *et al.*, 1984).

Plant defence systems, often involving the production of toxic materials, can also be employed by insects as semiochemicals. For example,

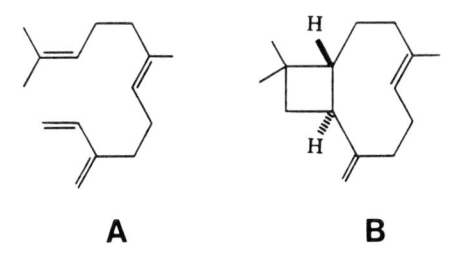

8. Semiochemicals: Molecular Determinants of Activity and Biosynthesis 145

Fig. 3. (A) (*E*)-β-farnesene, aphid alarm pheromone; (B) (−)-β-caryophyllene, alarm pheromone inhibitor.

plants in the Cruciferae (= Brassicaceae) contain glucosinolates which, on damage, are brought into contact with thioglucosidase (myrosinase) enzymes to release isothiocyanates. These broadly toxic compounds can act as repellants and antifeedants to insects not adapted to feed on crucifers, but can also be attractants for insects which specialize on these plants. Thus, in some modern cultivars of oilseed rape, disruption of the production of these defensive compounds in the vegetative parts of the plant has meant that they are more readily colonized, for example by the polyphagous peach-potato aphid, *Myzus persicae* (A. J. R. Porter, unpublished), and pollen beetles in the *Meligethes* genus (Milford *et al.*, 1989). At the same time, many crucifers are attacked by specialist feeders such as the seed weevil, *Ceutorhynchus assimilis*, and the cabbage stem flea beetle, *Psylliodes chrysocephala*, both of which have a highly developed response to glucosinolate catabolites such as the isothiocyanates (Blight *et al.*, 1989).

A number of plants produce compounds that interfere more directly with feeding, and the well-known Indian neem tree, *Azadirachta indica*, is being studied by many groups. However, the most active antifeedant component, azadirachtin (Kraus *et al.*, 1985; Broughton *et al.*, 1986), is toxic by hormonal action (Schmutterer, 1985) and its antifeedant properties may be related to this. Some plants produce compounds that are antifeedants at well below the levels required to cause physiological effects. These include the drimane antifeedants, which are cyclic sesquiterpene dialdehydes such as warburganal from the African tree *Warburgia ugandensis*, and polygodial from the temperate weed *Polygonum hydropiper*. These compounds are active against Lepidoptera and, more importantly, against sucking insects such as aphids, and can prevent plant virus transmission (Asakawa *et al.*, 1988; Griffiths *et al.*, 1989). For coleopterous pests, very high activity is found with the diterpenoid clerodane antifeedants typified by ajugarin I, which is extremely effective against the mustard beetle, *Phaedon cochleariae* (Griffiths *et al.*, 1988), and the Colorado potato beetle, *Leptinotarsa decemlineata* (Wyman *et al.*, 1989).

IV. ANALOGUES OF SEMIOCHEMICALS

Development of insecticides and pharmaceuticals from natural product leads has often involved the production of analogues with greater activity and more suitable physical and toxicological properties. Pheromones and many other semiochemical types can be extremely active and are generally of a benign toxicological nature. Nevertheless, their physical properties may prevent development as control agents. For example, the high instability of the American cockroach sex pheromone, discussed above, has meant that this potentially useful material has not yet been developed for commercial control of cockroaches.

Natural product leads for toxicants generally act within the organism, at receptors that will accept a range of chemical structures. For semiochemicals, the receptors are on the outside of the organism and are therefore subject to continual bombardment from a plethora of extraneous chemical signals. These receptors can often be highly tuned, particularly in the case of pheromones. There are one or two examples of active analogues being developed, but generally their efficacy is much less than that of the parent semiochemical.

Germacrene B shows similar activity to the American cockroach sex pheromone components, but again is extremely unstable. Bornyl acetate and some related compounds also show such activity, but this could not have been predicted by structural considerations and these analogues show differences in behavioural effects. The sex pheromone of the German cockroach, *Blattella germanica*, comprises two long chain ketones but, on reduction, the corresponding alcohols both show substantially higher activity. However, it is possible that the alcohols are natural components of the pheromone or that the ketones are precursors of the alcohols. The production of pheromone components prior to release has been demonstrated for other insect species (Pickett *et al.*, 1981; Dawson *et al.*, 1990a). Hexyl hexanoate has been identified as an attractant for *B. germanica* and conventional structure-activity studies have been used to produce a more active analogue, propyl cyclohexaneacetate (Sugawara *et al.*, 1975), but this seems to remain one of the few cases where such developments have been possible. Such studies have also been done on the lepidopterous sex pheromones. For example, the aldehyde of the diamondback moth pheromone, which is extremely unstable to aerial oxidation, can be converted into a more stable analogue, the C_{14} formate, where oxygen replaces the methylene group (Macaulay et al., 1986). However, as is the case with many such structural modifications, there is a substantial loss of activity, in this case of one order of magnitude.

One approach to producing pheromone analogues is the replacement of hydrogen atoms with fluorine (Camps et al., 1986; Briggs et al., 1986; Prestwich et al., 1988). For some of these analogues, pheromonal activity has been retained, and others show interesting antipheromonal effects. This approach has been very successful for the aphid alarm pheromone, (E)-β-farnesene, where three hydrogens in the conjugated diene part of the molecule can be replaced by fluorine with no loss of efficacy. Nishino et al. (1976) showed that removing one carbon atom to give the norfarnesene retained activity; the difluoronor compound is also highly active, but has the advantage over the three fluorine-substituted farnesene in that it is more stable and more readily synthesized (Briggs et al., 1986). Certain carbonyl analogues of (E)-β-farnesene also show activity. Since this relationship with the difluoronorfarnesene suggested that difluoromethylene could replace keto oxygen, the aldehydic oxygen of the diamondback moth sex pheromone component was replaced with the difluoromethylene group, but in this case the product was inactive. However, with the mosquito oviposition pheromone, replacement of the CH_3 group in the acetate with CF_3 gave a highly active and more volatile analogue, in spite of its high molecular weight, a typical feature of fluorine replacement. Further studies showed that replacement of as many as 17 hydrogen atoms with fluorine in the aliphatic side chain gave an even more volatile product with retention of high activity (Dawson et al., 1990b), and the structural significance of this will be discussed later (Chapter 9). Prestwich et al. (1990) have also examined high fluorine substitution in the aliphatic chain of lepidopterous sex pheromones and have found retention of activity and other interesting effects.

V. SEMIOCHEMICAL BIOSYNTHESIS: NEW TARGETS FOR MOLECULAR GENETICS

In breeding crop cultivars, it has often been necessary to reduce the production of secondary metabolites in the edible parts of the plant. However, with the advent of tissue- and season-specific promoters as tools for genetic engineering comes the possibility for improving the natural defences in the vegetative parts of crop plants, by modifying their secondary metabolism (Dawson et al., 1989a; Hallahan et al., 1992). This is the objective of a programme directed towards oilseed rape, which initially involved investigating the enzymology of biosynthetic routes to glucosinolates, particularly those conferring resistance to pests and diseases. The intention is to enhance these routes by genetic engineering. Also, attempts will be made to prevent expression of genes for pathways giving rise to compounds that attract specialist feeders. Already, catabolites from glucosinolates comprising alkenyl groups have been shown to be important in host selection by some oilseed rape-specializing insects such as *Ceutorhynchus assimilis* and *Psylliodes chrysocephala* (Blight et al., 1989). The programme will benefit by the fact that *Arabidopsis thaliana*, a crucifer with a relatively small genome and to which a considerable number of molecular biological studies are directed, seems not to produce these compounds (Hogge et al., 1988), and therefore the genes for the appropriate enzymes may be absent or not expressed in this plant.

Semiochemicals in non-crop plants are also under investigation. Initially, this research was directed at *P. hydropiper*, which produces the antifeedant (−)-polygodial from (*E,E*)-farnesyl pyrophosphate. The cyclase enzyme that converts farnesyl pyrophosphate into the drimane skeleton was the objective of these studies, since this is the key step in the biosynthesis. The work proved extremely difficult and studies on this plant have now been postponed. Instead, a collaborative programme with Professor C. J. W. Brooks at the University of Glasgow will attempt to improve production of drimenol by the Basidiomycete *Gloeophyllum odoratum*, which employs the cyclase without the complication of later oxidative stages that *P. hydropiper* uses to take drimenol through to the antifeedant dialdehydes. These oxidative stages will be investigated in the future, when the cyclase has been characterized.

Because of difficulties with the sesquiterpenoid cyclase, attention is now directed at cyclic monoterpenoids comprising aphid sex pheromones. These pheromones are produced in the hind tibiae of oviparous (sexual female) aphids by specialized cells, and emerge to the air through porous plaques situated above these cells. The aphids themselves contain very small amounts of enzymes. However, plants in the *Nepeta* genus (Labiatae) produce cyclopentanoids with the correct stereochemistry. The stereochemistry for the hop aphid (*Phorodon humuli*) sex pheromone (Fig. 4,A) is provided by *Nepeta mussinii*. Already, enzymatic and molecular biological studies at the first oxidative stage (Fig. 4,a) have led to the sequencing of a partial clone of the mixed function oxidase protein from the haem binding site towards the C-terminus (Hallahan, unpublished). Work is progressing on the later oxidative and reductive cyclization stages, with the monoterpene oxidoreductase having been purified to homogeneity (Hallahan, unpublished). This initial success has allowed the

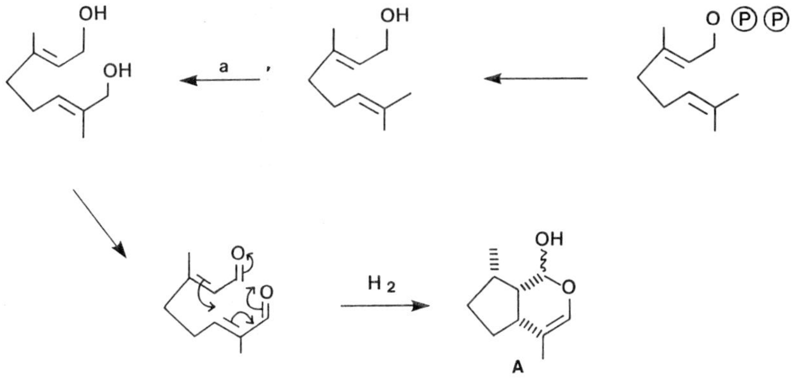

Fig. 4. Possible biosynthetic route to hop aphid, *Phorodon humuli*, sex pheromone. (The reductive cyclization is likely to involve a number of steps.)

construction of gene probes which will be used to search genomic libraries from aphids, particularly the hop aphid. The success with this mixed function oxidase system suggests that similar studies on the C_6 leafy aldehyde, produced by the enzymatic oxidation of fatty acids (Crombie *et al.*, 1991), might also elucidate pathways in plants and allow an investigation of the existence of such pathways in insects. These compounds act as semiochemicals and are used not only by herbivores to locate their host but also, at the third trophic level, by parasitoids searching for herbivores feeding on plants (Turlings *et al.*, 1990, 1991).

Biosynthesis of many lepidopterous pheromones has been investigated and a common route to these compounds has been demonstrated (Bjostad and Roelofs, 1983). This route involves desaturation, often by a Δ-11-desaturase, chain shortening and reduction through from the acyl functionality to aldehyde and alcohol, and then esterification, depending on the exact nature of the pheromone. Camps' group has elucidated the biosynthesis of the multicomponent sex pheromone of the Egyptian cotton-leaf worm, *Spodoptera littoralis* (Martinez *et al.*, 1990) and, in a collaborative programme, the (Z)-11-desaturase has now been characterized (Rodriguez *et al.*, 1992). Purification of this enzyme by use of radiolabelled inhibitor-binding studies, employing novel inhibitors developed during the course of these studies, is being attempted.

REFERENCES

Adams, M. A., Nakanishi, K., Still, W. C., Arnold, E. V., Clardy, J. and Persoons, C. J. (1979). *J. Am. Chem. Soc.* **101**, 2495–2498.

Arn, H., Tóth, M. and Priesner, E. (1986). OILB-SROP Working Group: "Use of Pheromones and other Semiochemicals in Integrated Control". Swiss Federal Research Station, Wädenswil.

Asakawa, Y., Dawson, G. W., Griffiths, D. C., Lallemand, J.-Y., Ley, S. V., Mori, K., Mudd, A., Pezechk-Leclaire, M., Pickett, J. A., Watanabe, H., Woodcock, C. M. and Zhang, Z.-N. (1988). *J. Chem. Ecol.* **14**, 1845–1855.

Bjostad, L. B. and Roelofs, W. L. (1983). *Science* **220**, 1387–1389.

Blight, M. M. and Wadhams, L. J. (1987). *J. Chem. Ecol.* **13**, 733–739.

Blight, M. M., Pickett, J. A., Smith, M. C. and Wadhams, L. J. (1984). *Naturwissenschaften* **71**, 480–481.

Blight, M. M., Pickett, J. A., Wadhams, L. J. and Woodcock, C. M. (1989). *Asp. Appl. Biol.* **23**, 329–334.

Borden, J. H. (1984). *In* "Insect Communication" (T. Lewis, ed.), pp. 123–149. Academic Press, London.

Briggs, G. G., Cayley, G. R., Dawson, G. W., Griffiths, D. C., Macaulay, E. D. M., Pickett, J. A., Pile, M. M., Wadhams, L. J. and Woodcock, C. M. (1986). *Pestic. Sci.* **17**, 441–448.

Broughton, H. B., Ley, S. V., Slawin, A. M. Z., Williams, D. J. and Morgan, E. D. (1986). *J. Chem. Soc., Chem. Commun.*, 46–47.

Camps, F., Fabriás, G. and Guerrero, A. (1986). *Tetrahedron* **42**, 3623–3629.

Crombie, L., Morgan, D. O. and Smith, E. H. (1991). *J. Chem. Soc. Perkin Trans. I*, 567–575.

Dawson, G. W., Griffiths, D. C., Pickett, J. A., Smith, M. C. and Woodcock, C. M. (1984). *Entomol. Exp. Appl.* **36**, 197–199.

Dawson, G. W., Griffiths, D. C., Janes, N. F., Mudd, A., Pickett, J. A., Wadhams, L. J. and Woodcock, C. M. (1987). *Nature* **325**, 614–616.

Dawson, G. W., Griffiths, D. C., Merritt, L. A., Mudd, A., Pickett, J. A., Wadhams, L. J. and Woodcock, C. M. (1988). *Entomol. Exp. Appl.* **48**, 91–93.

Dawson, G. W., Hallahan, D. L., Mudd, A., Patel, M. M., Pickett, J. A., Wadhams, L. J. and Wallsgrove, R. M. (1989a). *Pestic. Sci.* **27**, 191–201.

Dawson, G. W., James, N. F., Mudd, A., Pickett, J. A., Slawin, A. M. Z., Wadhams, L. J. and Williams, D. J. (1989b). *Pure Appl. Chem.* **61**, 555–558.

Dawson, G. W., Griffiths, D. C., Merritt, L. A., Mudd, A., Pickett, J. A., Wadhams, L. J. and Woodcock, C. M. (1990a). *J. Chem. Ecol.* **16**, 3019–3030.

Dawson, G. W., Mudd, A., Pickett, J. A., Pile, M. M. and Wadhams, L. J. (1990b). *J. Chem. Ecol.* **16**, 1779–1789.

Dickens, J. C. and Mori, K. (1989). *J. Chem. Ecol.* **15**, 517–518.

Dickens, J. C. and Prestwich, G. D. (1989). *J. Chem. Ecol.* **15**, 529–540.

Gieselmann, M. J., Rice, R. E., Jones, R. A. and Roelofs, W. L. (1979). *J. Chem. Ecol.* **5**, 891–900.

Griffiths, D. C., Hassanali, A., Merritt, L. A., Mudd, A., Pickett, J. A., Shah, S. J., Smart, L. E., Wadhams, L. J. and Woodcock, C. M. (1988). *In* "Proceedings Brighton Crop Protection Conference – Pests and Diseases – 1988", pp. 1041–1046. BCPC Publications, Thornton Heath.

Griffiths, D. C., Pickett, J. A., Smart, L. E. and Woodcock, C. M. (1989). *Pestic. Sci.* **27**, 269–276.

Hallahan, D. L., Pickett, J. A., Wadhams, L. J., Wallsgrove, R. M. and Woodcock, C. M. (1992). *In* "Plant Genetic Manipulation for Crop Protection" (A. Gatehouse, V. Hilder and D. Boulter, eds), pp. 215–247. CAB International, Oxford.

Hauptmann, H., Mühlbauer, G. and Sass, H. (1986). *Tetrahedron Lett.* **27**, 6189–6192.

Hogge, L. R., Reed, D. W., Underhill, E. W. and Haughn, G. W. (1988). *J. Chromatogr. Sci.* **26**, 551–556.
Jaenicke, L. and Boland, W. (1982). *Angew. Chem.* **94**, 659–669; *Angew. Chem. Int. Ed.* **94**, 643–653.
Kraus, W., Bokel, M., Klenk, A. and Pöhnl, H. (1985). *Tetrahedron Lett.* **26**, 6435–6438.
Laurence, B. R. and Pickett, J. A. (1982). *J. Chem. Soc., Chem. Commun.*, 59–60.
Laurence, B. R., Mori, K., Otsuka, T., Pickett, J. A. and Wadhams, L. J. (1985). *J. Chem. Ecol.* **11**, 643–648.
Macaulay, E. D. M., Dawson, G. W., Xun, L. and Pickett, J. A. (1986). *Asp. Appl. Biol.* **12**, 105–116.
Martinez, T., Fabriás, G. and Camps, F. (1990). *J. Biol. Chem.* **265**, 1381–1387.
Mayer, M. S. and McLaughlin, J. R. (1991). "Handbook of Insect Pheromones and Sex Attractants". CRC Press, Florida.
Milford, G. F. J., Porter, A. J. R., Fieldsend, J. K., Miller, C. A., Leach, J. E. and Williams, I. H. (1989). *Ann. Appl. Biol.* **115**, 375–380.
Nishino, C., Bowers, W. S., Montgomery, M. E. and Nault, L. R. (1976). *Appl. Ent. Zool.* **11**, 340–343.
Pickett, J. A., Williams, I. H., Smith, M. C. and Martin, A. P. (1981). *J. Chem. Ecol.* **7**, 543–554.
Pickett, J. A., Wadhams, L. J., Woodcock, C. M. and Hardie, J. (1992). *Annu. Rev. Entomol.* **37**, 67–90.
Prestwich, G. D., Sun, W.-C. and Dickens, J. C. (1988). *J. Chem. Ecol.* **14**, 1427–1439.
Prestwich, G. D., Sun, W.-C., Mayer, M. S. and Dickens, J. C. (1990). *J. Chem. Ecol.* **16**, 1761–1778.
Rodriguez, F., Hallahan, D. L., Pickett, J. A. and Camps, F. (1992). *Insect Biochem. Molec. Biol.* **22**, 143–148.
Schmutterer, H. (1985). *Z. Ang. Ent.* **100**, 468–475.
Sugawara, R., Kurihara, S. and Muto, T. (1975). *J. Insect Physiol.* **21**, 957–964.
Tumlinson, J. H., Hardee, D. D., Gueldner, R. C., Thompson, A. C., Hedin, P. A. and Minyard, J. P. (1969). *Science* **166**, 1010–1012.
Turlings, T. C. J., Tumlinson, J. H. and Lewis, W. J. (1990). *Science* **250**, 1251–1253.
Turlings, T. C. J., Tumlinson, J. H., Heath, R. R., Proveaux, A. T. and Doolittle, R. E. (1991). *J. Chem. Ecol.* **17**, 2235–2251.
Wyman, J., Feldman, J. and Griffiths, D. C. (1989). Abstract, "Symposium on Semiochemicals and Pest Control", Wageningen, October 1989, p. 33.

9

The Perception of Semiochemicals

L. J. WADHAMS

I. Introduction ... 152
II. Perception of Odours ... 152
III. Peripheral Coding of Odours ... 154
   A. Electrophysiology ... 154
   B. Pheromones ... 155
   C. Kairomones ... 158
IV. Conclusion ... 161
References ... 161

I. INTRODUCTION

Semiochemicals are a key information source utilized by most insects in interacting with mates, hosts or predators. The multicomponent nature of these chemical cues is now well established. The olfactory receptors of an insect are tuned to the detection of those compounds which comprise the chemical messages for the species.

II. PERCEPTION OF ODOURS

Insects perceive the presence of volatile semiochemicals through specialized receptor organs, the sensilla, which are located primarily on the insect's antennae. Many moths have large, often feather-like antennae which act as very effective filters, "sieving out" up to 30% of the odour molecules in the air passing over them (Kaissling, 1974, 1986a; Kanujia and Kaissling, 1985). However, antennal morphology is also determined by the lifestyle of the insect. Bark beetles, for example, which spend a large proportion of their time living inside the bark of trees, have small,

9. The Perception of Semiochemicals

unbranched club-shaped antennae and the olfactory sensilla are confined to the club (Henderson and Wadhams, 1981).

For pheromone molecules to be detected, they must first enter a sensillum and arrive at the receptor proteins in the dendritic membrane. The walls of the olfactory sensilla usually contain numerous pores (Zacharuk, 1980) and the molecules adsorbed onto the cuticular surface are assumed to enter the sensory hairs via these pores, which provide a rapid entry into the hair lumen. Sensilla contain at least three proteins for which there is biochemical evidence, a soluble binding protein present in high concentrations, low titres of sensillar enzymes such as esterases and a very scarce dendritic membrane receptor protein (Vogt, 1987).

When a pheromone enters a sensory hair it encounters a high concentration of binding protein and, at this concentration, the pheromone is solubilized into the sensillum lymph. The pheromone then moves through the lymph until it either reaches a receptor molecule or is degraded by a sensillar enzyme. The key to the effective working of this system is the multifunctional property of the binding protein, which acts as solubilizer, carrier and protector of the odorant molecule (Vogt, 1987).

Transduction mechanisms can broadly be divided into two groups, direct gating and gating of channels through a second messenger system. In the first case, the receptor and the ion channel are part of the same polypeptide complex and the binding of the agonist modulates ion channel opening. This type of system offers a very fast response but very little amplification. In insects, the high sensitivity of the olfactory cells and the delay that is found between presentation of the stimulus and the response of the cell suggests the involvement of a second messenger system. It is thought that the pheromone binds extracellularly to the receptor and initiates the transduction cascade within the cytoplasm. The activated receptor catalyses the binding of GTP to the G-protein, which in turn activates a catalytic element – in this case phospholipase C (phosphodiesterase, PDE) – which acts on a membrane lipid (phosphatidylinositol diphosphate) to produce two different second messengers.

Inositol triphosphate (IP_3) is one of the second messengers released by PDE. Work on cockroaches has shown that stimulation with the pheromone periplanone B produces a dramatic increase in IP_3 levels, which reach a maximum at about 50 ms after stimulation and thereafter decrease (Breer et al., 1990). This sort of time scale is exactly what is found in electrophysiological preparations and it appears that IP_3 accumulation is rapid enough for it to be a transmitter in olfactory excitation. The primary effects of IP_3 are mediated via Ca^+ ions and it has been shown that neurones respond to microinjections of IP_3 with transient openings of Ca^{2+}-dependent K^+ channels.

The ionic currents which generate receptor potential in pheromone-sensitive olfactory cells have been studied in moths (Kaissling, 1986b), and the data suggest that the application of pheromone causes the opening of single ion channels in the dendritic membrane. The sensillum lymph has a very high K^+ concentration (200 mM or more) and the initial depolarizing currents of signal transduction may be mediated by an inward K^+ current.

The olfactory receptor cells in the insect's antennae send their axons directly to the brain, where the first relay station is found in the glomeruli of the antennal lobe (Boeckh et al., 1984; Matsumoto and Hildebrand, 1981). In a number of insect species, it appears that information from pheromone and food odour receptors is processed through different channels. In some moths and in cockroaches, information from pheromone receptors is exclusively processed through the macroglomerulus, whereas food odour receptors contact interneurones in the smaller glomeruli.

In the antennal lobe, the information is further processed by a complicated network of interneurones. The number of output neurones leaving the antennal lobe is very small compared to the number of input antennal fibres. This convergence results not only in reduction of background noise, but also in considerable amplification of the signal. Thus, at the antennal lobe, activity of the individual interneurones can be modified at concentrations that are 100 to 1000 times less than those needed to elicit a significant change in the peripheral receptors (Boeckh et al., 1984).

The output neurones from the antennal lobe feed into the protocerebrum, where the olfactory information is integrated with inputs from the other sensory modalities, and output elements from here are then connected to the insect's motor systems (Boeckh and Ernst, 1987).

III. PERIPHERAL CODING OF ODOURS

A. Electrophysiology

Information on the nature of the peripheral receptors can be obtained using a number of electrophysiological techniques. The simplest of these is the electroantennogram (EAG), which measures an overall polarization of the antenna and reflects the response of many receptor neurons to the presentation of the stimulus. More detailed information on the nature of the olfactory neurones can be obtained by making extracellular recordings from individual sensilla. This was first achieved using sharpened tungsten microelectrodes implanted at the base of the sensilla (Boeckh, 1962). Non-polarizing electrodes can also be used, but this requires some damage to the sensillum – e.g. cutting off the tip to slip the electrode over – and can cause modifications to the sensillum lymph, affecting the hair's responsiveness. This can be overcome to a great extent by taking extreme care with the tip-cutting technique and by refinements in the composition of the saline in the recording electrode.

Using these techniques, the response profiles of individual olfactory receptors have been examined and lead to their classification as either specialist or generalist. Specialists appeared to be turned narrowly to the perception of one particular compound, whilst generalist receptors showed a response to a number of compounds. This distinction caused two hypotheses to be proposed for the mechanisms by which olfactory information is conveyed to the insect's CNS. The concept of labelled lines involves specialist receptors, each responding to one component of the pheromone. The discrimination of composition and ratios of the components in the CNS is then a simple comparison of the neural activities in the separate channels. However, the observation that generalist receptors have broad and overlapping response spectra has given rise to a second theory – that of across-fibre patterning, whereby each detectable odour has a unique pattern of activity across an array of olfactory neurones. Recognition of the odour would then require a complicated evaluation by the CNS of the activities of a whole array of olfactory receptors, so that the composition of an odour blend would be recognized when the pattern matched a template present in the CNS.

B. Pheromones

Sex pheromones are the class of semiochemicals which have been most extensively studied and their multicomponent nature in Lepidoptera is well established. Lepidopterous sex pheromones generally comprise long chain unsaturated aldehydes, alcohols or acetates. For the silkworm moth, *Bombyx mori*, recordings from the sensilla trichodea on the male antenna have shown the presence of two cell types, one of which responds to the major pheromone component, bombykol ((E,Z)-10,12-hexadecen-1-ol), whilst the other responds to the corresponding aldehyde, bombykal (Kaissling, 1971, 1986b). This latter compound, which is also produced by the female pheromone gland, suppresses the male response to bombykol. Studies in a large number of Lepidoptera have demonstrated the presence of cells which respond to individual pheromone components. Thus, in terms of the perception of pheromones, the general rule is that a pheromone blend comprising a particular number of components is perceived by an equal number of specialist receptors.

Extensive electrophysiological studies have demonstrated the key determinants of biological activity (Bestmann, 1976; Bestmann et al., 1987). These include alkene position, geometry and functionality. Somewhat surprisingly, the length of the terminal alkyl groups is critical. Modification of any of these criteria results in a very significant loss of activity, and consequently it has proved difficult to make analogues with similar activity to the parent compound. Normal isosteric replacement, such as replacing the α-methylene of an aldehyde with oxygen to give the formate, results in an order of magnitude loss of activity (Macaulay et al., 1986). However, some interesting results have been obtained by replacing hydrogen with fluorine. This is a fairly common strategy in producing biologically active compounds with modified binding properties and with different metabolic stabilities. Fluorinated analogues can function either as antagonists, blocking access of the pheromone molecule to the receptor and inhibiting responses, or as agonists, mimicking the pheromone action and producing a modified response. Replacement of the acetate hydrogens in the diamondback moth pheromone component (Z)-11-tetradecenyl acetate gives an electrophysiologically active analogue with a reduced activity (Prestwich and Streinz, 1988). Perfluorination of the alkyl chain can also produce active analogues, but again with reduced biological effects. Thus, in *Heliothis zea*, the perfluorobutyl derivative of the pheromone component (Z)-11-hexadecenyl acetate is approximately three orders of magnitude less effective than the parent compound in stimulating the receptor (Prestwich et al., 1990).

The oviposition pheromone of the mosquito *Culex quinquefasciatus* comprises the (−)-(5R,6S)-6-acetoxy-5-hexadecanolide (Laurence et al., 1985). As with the lepidopterous pheromones, shortening the alkyl chain removed activity, whilst replacement of the acetate hydrogens with fluorine gave a highly active though less stable analogue. However, greater fluorine substitution has been obtained with all the hydrogen atoms in the alkyl side chain being replaced with fluorine. Only the two methylene bridge groups were left so that the functionality would be minimally altered. Despite the fact that this molecule had a very high molecular weight, it was more volatile than the parent pheromone because molecule–molecule interactions were severely restricted. Surprisingly, the molecule retained a high level of biological activity (Dawson et al., 1990b). This, together with the studies on lepidopterous pheromones, suggests that the alkyl groups in pheromones do not have a lipophilic interaction at the receptor site, since this would be disrupted by the high fluorine substitution. Instead, it appears that the side chain has a spatial role, although even this would be modified by the larger Van der Waal's radius for fluorine.

Pheromone receptors in other insect Orders can be equally specialized and chirality can play a key role in species separation. In the bark beetle *Ips paraconfusus*, the male-produced pheromone comprising (+)-*cis*-verbenol, (+)-ipsdienol and (−)-ipsenol (Silverstein *et al.*, 1966) is perceived by separate cell types (Mustaparta, 1979, 1980; Mustaparta *et al.*, 1977). In *Ips pini*, two types of ipsenol cells exist, each specialized to one of the enantiomers. Two geographically isolated populations of this insect occur in the USA. In the western population, which uses (−)-ipsenol exclusively, the majority of the receptors are keyed to this compound. However, the eastern population, which uses a mixture of (+)- and (−)-ipsenol, possesses approximately equal numbers of receptors for the two enantiomers (Mustaparta *et al.*, 1980).

The elm bark beetle, *Scolytus scolytus*, produces two chiral pheromone components, 4-methyl-3-heptanol and α-multistriatin (Blight *et al.*, 1977, 1979, 1980). Male *S. scolytus* produce both the (−)-*threo* and (−)-*erythro* stereoisomers of the heptanol and electrophysiological studies have shown the presence of separate receptors on the antennae of both sexes for the two isomers (Wadhams *et al.*, 1982). Differences in the EAG responses to these compounds between males and females suggest that the sexes possess different proportions of the two receptor types. Separate specialized receptors have also been found that respond to the other beetle metabolite, α-multistriatin.

In *Dendroctonus* spp. bark beetles, the perception of multicomponent pheromones involves the interaction of different components with the same receptor cell. Thus, the female-produced frontalin, the host synergist pinene and the interspecific inhibitor endobrevicomin all activate the same cell on the male's antenna (Dickens and Payne, 1977; Dickens *et al.*, 1985).

The key compound in the male-produced aggregation pheromone of the pea and bean weevil, *Sitona lineatus*, is 4-methyl-3,5-heptanedione and both sexes possess specialist receptors for this compound. The dione exhibits pronounced keto–enol tautomerism and coupled gas chromatography (GC)–electrophysiological studies suggest that the olfactory cells are more responsive to the enol form (Blight, 1990).

The sex pheromones of many aphid species have been shown to comprise one or both of the biosynthetically related monoterpenes nepetalactone (I) and nepetalactol (II) (Fig. 1) (Dawson *et al.*, 1987a, 1988). The pheromone receptors of male aphids are mainly located on the secondary rhinaria of the 3rd or 4th antennal segments. Recordings from

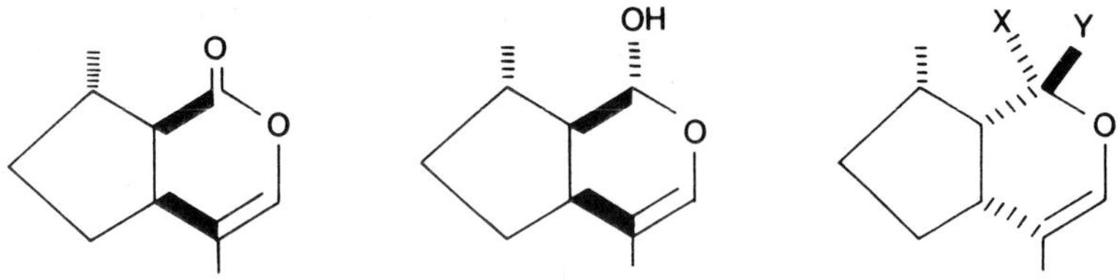

Fig. 1.

I II III

1S; X=H, Y=OH
1R; X=OH, Y=H

these show the presence of two cells tuned to the perception of either the lactone I or the lactol II (Fig. 1).

Dose–response curves show that there is a clear discrimination between the compounds and that this is maintained even at high stimulus concentrations. For most aphids, the two compounds need to be present in the correct ratio to elicit a full behavioural response (Dawson et al., 1990a; Hardie et al., 1990) and such a specialized receptor system presumably enables the aphids to distinguish clearly the conspecific ratio. The sex pheromone of the damson-hop aphid, *Phorodon humuli*, comprises a diastereomeric pair of lactols (III) (Fig. 1) (Campbell et al., 1990) and the male's secondary rhinaria again possess receptors for these compounds. *P. humuli* males also have receptors for the lactone I and lactol II, but the behavioural significance of this is not yet understood.

Compounds which act as pheromones in one species may also be involved in semiochemically mediated behaviour in other species. Initially, it was suggested that these compounds acted by competitive blocking of the pheromone receptors. However, more detailed studies, notably on bark beetles, have demonstrated that this is not the case and that specific receptors are activated by the blocking compounds.

C. Kairomones

Pheromones frequently do not act in isolation but in conjunction with other semiochemicals such as plant volatiles. This is particularly common with coleopterous aggregation pheromones, where host plant metabolites synergize the activity of the insect-produced pheromone components. In

Scolytus scolytus, the behavioural activity of the beetle metabolite 4-methyl-3-heptanol is synergized by a number of host-derived compounds, including α-cubebene and β-pinene. EAG studies showed that, relative to 4-methyl-3-heptanol, these compounds were only weakly active and although the similarity of the cubebene and pinene dose–response curves suggested that they were interacting with the same receptors, more detailed studies using single cell recording techniques demonstrated that perception of these compounds was again mediated by specialized olfactory cells. Relative EAG amplitudes were reflected in the proportions of the different cell types present on the antenna (Wadhams, 1982, 1990).

The so-called "green leaf volatiles" have frequently been implicated in semiochemically mediated behaviour in insects. The activity of the *Sitona lineatus* agggregation pheromone is strongly synergized by such host-plant-derived kairomones (Blight et al., 1984; Blight and Wadhams, 1987) and olfactory cells specialized to the perception of these compounds have been found on the antenna. Even in aphids, which possess relatively few receptors, the perception of such compounds is mediated through specialized cells. Thus, the peach-potato aphid, *Myzus persicae*, has cells on the primary rhinaria which respond selectively to (*E*)-2-hexenal.

The situation appears to be rather different in the Colorado potato beetle, *Leptinotarsa decemlineata*. Beetles respond strongly to potato plant odour by walking upwind and the green leaf volatiles are crucial in eliciting this response (Visser and Avé, 1978). The response is dependent on the ratios of the components and modification of these ratios prevents orientation of the beetles to the odour source (Thiery and Visser, 1986, 1987). Perception of the individual compounds is mediated by both specialist and generalist cells, but stimulation with binary mixtures of the green leaf volatiles showed some interesting effects. In some cells, the response to the mixture was equal to the sum of the response when the compounds were applied singly. The majority of cells, however, showed a suppression of neuronal activity and the response was smaller than the responses to the individual components. There appeared to be a correlation between the specialism of the cell and the degree of suppression, in that specialist cells responding to mixtures tended to show less excitation than their generalist counterparts. This suggests that information coding shows some separation into two channels. In one channel, the activities of the generalist receptors, which respond to mixtures by an additive effect, code for the total quantity of components in the blend, whilst the ratios of the components are coded for by the more specialized receptors which respond to mixtures by suppression (Visser and De Jong, 1988).

A somewhat analogous situation may exist in the perception of host plant volatiles by feeders on Cruciferae (= Brassicacae), such as the cabbage stem flea beetle, *Psylliodes chrysocephala*, and the seed weevil, *Ceutorhynchus assimilis*. Cruciferous plants are characterized by the presence of glucosinolates, which break down under the action of enzymes to give the corresponding volatile isothiocyanates. Although these compounds function to deter general herbivory, crucifer feeders utilize them as key components of host location behaviour. The antennae of these insects are well adapted to the perception of isothiocyanates, which appears to be mediated by both specialist and generalist cells. Electrophysiological studies on *C. assimilis* have shown that the antennae possess at least three different isothiocyanate cell types, the most abundant of which responds similarly to the 3-butenyl, 4-pentenyl and phenylethyl homologues. The other two cell types clearly discriminate between the alkenyl and phenylethyl isothiocyanates (Blight *et al.*, 1989). Although mixture effects have not been observed, it is possible that the more generalist cell type codes for quantity whilst the two specialists code for quality.

Aphids which feed on crucifers are also thought to use the isothiocyanates as part of their host location mechanism. The turnip or mustard aphid, *Lipaphis erysimi*, possesses a highly specialized receptor which responds optimally and at a very low stimulus concentration to the allyl and 3-butenyl isothiocyanates. An increase in chain length results in considerable reduction in efficacy (Dawson *et al.*, 1987b). The cabbage aphid, *Brevicoryne brassicae*, also has isothiocyanate receptors, but these are much less sensitive than those found in *L. erysimi* (Nottingham *et al.*, 1991). The high sensitivity of the *L. erysimi* cells may reflect the dual function of isothiocyanates in the behaviour of this aphid, since these compounds have also been implicated as host-derived synergists for the alarm pheromone (E)-β-farnesene (Dawson *et al.*, 1987b). In this case, the behavioural response elicited by the isothiocyanates is determined by the chemical context in which the compound is presented.

A number of non-crucifer-feeding aphids also have receptors for isothiocyanates (Nottingham *et al.*, 1991). *Phorodon humuli* and the black bean aphid, *Aphis fabae*, have cells in the proximal rhinaria which respond optimally to 4-pentenyl isothiocyanate. The response profiles of these cells are almost identical to the isothiocyanate receptors in *B. brassicae*. However, in direct contrast to *B. brassicae*, which in the laboratory is attracted to 4-pentenyl isothiocyanate, *A. fabae* is strongly repelled. This suggests that host plant location and selection may be governed not only by positive odour cues from the host plant, but also by inhibitory signals from non-host plants.

IV. CONCLUSION

Peripheral coding of odour quality of pheromones in a wide range of insect Orders is mediated through specialist receptors. However, the situation is less clear in terms of the perception of host volatiles, which is mediated by both specialist and generalist cell types. The perception of host odours by specialist receptors frequently occurs where this is a behavioural interaction between the host odour and pheromones. Where compounds are involved only in location of food or host plants, perception is often mediated through more generalist receptors, with information being transferred to the central nervous system by an across-fibre pattern. The behavioural responses elicited by compounds perceived in this manner may be more adaptive.

REFERENCES

Bestmann, H. J. (1976). *Z. Ang. Ent.* **82**, 110–111.
Bestmann, H. J., Wu, C-h, Döhla, B. and Li, K-d. (1987). *Z. Naturforsch.* **42c**, 435–441.
Blight, M. M. (1990). In "Chromatography and Isolation of Insect Hormones and Pheromones" (A. R. McCaffery and I. D. Wilson, eds), pp. 281–288. Plenum Press, New York/London.
Blight, M. M. and Wadhams, L. J. (1987). *J. Chem. Ecol.* **13**, 733–739.
Blight, M. M., Mellon, F. A., Wadhams, L. J. and Wenham, M. J. (1977). *Experientia* **33**, 845–847.
Blight, M. M., Wadhams, L. J. and Wenham, M. J. (1979). *Insect Biochem.* **9**, 525–533.
Blight, M. M., Ottridge, A. P., Wadhams, L. J., Wenham, M. J. and King, C. J. (1980). *Naturwissenschaften* **67**, 517–518.
Blight, M. M., Pickett, J. A., Smith, M. C. and Wadhams, L. J. (1984). *Naturwissenschaften* **71**, 480–481.
Blight, M. M., Pickett, J. A., Wadhams, L. J. and Woodcock, C. M. (1989). *Asp. Appl. Biol.* **23**, 329–334.
Boeckh, J. (1962). *Z. Vergleich. Physiol.* **46**, 212–248.
Boeckh, J. and Ernst, K. D. (1987). *J. Comp. Physiol.* **161**, 549–565.
Boeckh, J., Ernst, K. D., Sass, H. and Waldow, U. (1984). *J. Insect Physiol.* **30**, 15–26.
Breer, H., Boekhoff, I. and Tareilus, E. (1990). *Nature* **345**, 65–68.
Campbell, C. A. M., Dawson, G. W., Griffiths, D. C., Pettersson, J., Pickett, J. A., Wadhams, L. J. and Woodcock, C. M. (1990). *J. Chem. Ecol.* **16**, 3455–3465.
Dawson, G. W., Griffiths, D. C., Janes, N. F., Mudd, A., Pickett, J. A., Wadhams, L. J. and Woodcock, C. M. (1987a). *Nature* **325**, 614–616.
Dawson, G. W., Griffiths, D. C., Pickett, J. A., Wadhams, L. J. and Woodcock, C. M. (1987b). *J. Chem. Ecol.* **13**, 1663–1671.
Dawson, G. W., Griffiths, D. C., Merritt, L. A., Mudd, A., Pickett, J. A., Wadhams, L. J. and Woodcock, C. M. (1988). *Entomol. Exp. Appl.* **48**, 91–93.
Dawson, G. W., Griffiths, D. C., Merritt, L. A., Mudd, A., Pickett, J. A., Wadhams, L. J. and Woodcock, C. M. (1990a). *J. Chem. Ecol.* **16**, 3019–3030.

Dawson, G. W., Mudd, A., Pickett, J. A., Pile, M. M. and Wadhams, L. J. (1990b). *J. Chem. Ecol.* **16**, 1779–1789.
Dickens, J. C. and Payne, T. L. (1977). *J. Insect Physiol.* **23**, 481–489.
Dickens, J. C., Payne, T. L., Ryker, L. C. and Rudinsky, J. A. (1985). *J. Chem. Ecol.* **11**, 1359–1370.
Hardie, J., Holyoak, M., Nicholas, J., Nottingham, S. F., Pickett, J. A., Wadhams, L. J. and Woodcock, C. M. (1990). *Chemoecology* **1**, 63–68.
Henderson, N. C. and Wadhams, L. J. (1981). *Z. Ang. Ent.* **92**, 477–487.
Kaissling, K. E. (1971). *In* "Handbook of Sensory Physiology. Chemical Senses, Olfaction", **14**, Pt 1 (L. Beidler, ed.), pp. 341–431. Springer-Verlag, Berlin.
Kaissling, K. E. (1974). *In* "Biochemistry of Sensory Functions" (L. Jaenicke, ed.), pp. 243–273. Springer-Verlag, Berlin.
Kaissling, K. E. (1986a). *In* "Molecular Entomology" (UCLA Symp. Mol. Cell. Biol. New Ser. 49) (J. Law, ed.), pp. 1–11. Liss, New York.
Kaissling, K. E. (1986b). *Annu. Rev. Neurosci.* **9**, 121–145.
Kanujia, S. and Kaissling, K. E. (1985). *J. Insect Physiol.* **31**, 71–81.
Laurence, B. R., Mori, K., Otsuka, T., Pickett, J. A. and Wadhams, L. J. (1985). *J. Chem. Ecol.* **11**, 643–648.
Macaulay, E. D. M., Dawson, G. W., Xun, L. and Pickett, J. A. (1986). *Asp. Appl. Biol.* **12**, 105–116.
Matsumoto, S. G. and Hildebrand, J. G. (1981). *Proc. R. Soc. London, Ser. B* **213**, 249–277.
Mustaparta, H. (1979). *In* "Chemical Ecology: Odor Communication in Animals" (F. J. Ritter, ed.), pp. 147–158. Elsevier, Amsterdam.
Mustaparta, H. (1980). *In* "Receptors for Neurotransmitters, Hormones and Pheromones in Insects" (D. B. Satelle, L. M. Hall and J. G. Hildebrand, eds), pp. 283–298. Elsevier, Amsterdam.
Mustaparta, H., Angst, M. E. and Lanier, G. N. (1977). *J. Comp. Physiol.* **121**, 343–347.
Mustaparta, H., Angst, M. E. and Lanier, G. N. (1980). *J. Chem. Ecol.* **6**, 689–701.
Nottingham, S. F., Hardie, J., Dawson, G. W., Hick, A. J., Pickett, J. A., Wadhams, L. J. and Woodcock, C. M. (1991). *J. Chem. Ecol.* **17**, 1231–1242.
Prestwich, G. D. and Streinz, L. (1988). *J. Chem. Ecol.* **14**, 1003–1021.
Prestwich, G. D., Sun, W-c, Mayer, M. S. and Dickens, J. C. (1990). *J. Chem. Ecol.* **16**, 1761–1778.
Silverstein, R. M., Rodin, J. D. and Wood, D. L. (1966). *Science Wash. DC* **154**, 509–510.
Thiery, D. and Visser, J. H. (1986). *Entomol. Exp. Appl.* **41**, 165–172.
Thiery, D. and Visser, J. H. (1987). *J. Chem. Ecol.* **13**, 1139–1146.
Visser, J. H. and Avé, D. A. (1978). *Ent. Exp. Appl.* **24**, 738–749.
Visser, J. H. and De Jong, R. (1988). *J. Chem. Ecol.* **14**, 2005–2018.
Vogt, R. G. (1987). *In* "Pheromone Biochemistry" (G. D. Prestwich and G. J. Blomquist, eds), pp. 385–431. Academic Press, London.
Wadhams, L. J. (1982). *Z. Naturforsch.* **37C**, 947–952.
Wadhams, L. J. (1990). *In* "Chromatography and Isolation of Insect Hormones and Pheromones" (A. R. McCaffery and I. D. Wilson, eds), pp. 289–298. Plenum Press, New York.
Wadhams, L. J., Angst, M. E. and Blight, M. M. (1982). *J. Chem. Ecol.* **8**, 477–492.
Zacharuk, R. Y. (1980). *Annu. Rev. Ent.* **25**, 27–57.

10

Semiochemically Mediated Behaviour

G. M. POPPY

    I. Introduction 163
    II. The Role of Behavioural Studies in Chemical Ecology 164
    III. Aphid Sex Pheromones 165
    IV. Host Plant Volatiles 167
    V. Alarm Pheromone 169
    VI. Conclusion 170
        References 171

I. INTRODUCTION

In the obscurity of a dark chamber,
this splendid moth emits phantasmal radiations,
perhaps intermittent and reserved for the season of nuptials,
signals invisible to us, and perceptible only to those children of the night,
who may have found this means to communicate one with another,
to call one another in the darkness,
and to speak with one another.
                        (J. H. C. Fabre in Legros, 1971)

The literature of the early Aurelians abounds with anecdotes describing the extraordinary attractive powers of female Lepidoptera (Fabre, 1904). Mayer (1900) and Kellogg (1907) independently came to the conclusion that such attraction was due to a scent which was received by the antennae. The "umbrella" term semiochemical has been introduced to cover all such chemical messages between organisms (Law and Regnier, 1971). Semiochemicals by definition need to elicit a behavioural (or physiological) response in the receiving organism and thus behavioural studies are vital to the understanding of chemical ecology.

The German electrophysiologist Boeckh (1986) has suggested that insect olfaction can only be comprehensively understood by a combination of approaches and methods in a multidisciplinary effort. With modern techniques in analytical chemistry, electrophysiology and behaviour, such an approach is being used in a unique collaborative programme to study aphid chemical ecology at Rothamsted and Imperial College at Silwood Park. Recent research has shown that olfaction plays a more extensive role in the chemical ecology of aphids than previously thought (Pickett et al., 1992).

In this chapter, the role of behavioural studies in chemical ecology research will be discussed. Special emphasis will be given to the integration of behavioural studies with other techniques to provide a greater understanding of aphid chemical ecology.

II. THE ROLE OF BEHAVIOURAL STUDIES IN CHEMICAL ECOLOGY

Behavioural studies are needed at various stages in semiochemical research (Table I). It was the behavioural observations made by early workers such as Fabre, Kellogg and Mayer which paved the way for research into lepidopterous pheromones, although it was not until 1959 that the first sex pheromone was chemically identified (Butenandt et al., 1959). In spite of criticism by Kennedy (1978), the system introduced by Shorey (1973) of naming pheromones according to the behaviour they elicit is still widely used and illustrates the key role of behavioural studies. Behavioural effects of semiochemicals have largely concentrated on pheromones (intraspecies), although the roles of other semiochemical types have been substantially researched in recent years and their importance is now more generally appreciated.

The presence of scent-producing structures in male Lepidoptera was first noted at the turn of the century. However, until recently very little was known about their function due to a lack of experimental behavioural evidence (Birch et al., 1990). In *Mamestra brassicae*, behavioural observations of a "wild" strain newly established in the laboratory led to a reinvestigation of the chemicals from these male glands, to assess the possibility that they were male sex pheromones. Detailed studies showed that an "inbred" strain of these moths showed a different courtship behaviour from that of the "wild" moths, which is why no function for the chemicals had been found (Poppy, 1990). In spite of this behavioural difference, the chemical and electrophysiological investigations were relatively insensitive to such a discrepancy (Poppy, 1990).

Once an interesting observation has been made, electrophysiological

TABLE I. Stages in semiochemical research.

1. Behavioural observations
2. Chemical extraction
3. Electrophysiological investigation
4. Chemical identification/synthesis
5. Behavioural bioassays
6. Field trials

and chemical studies can be used to isolate and identify potential semiochemicals (see Table I). Behavioural bioassays can then be employed to identify and confirm the active chemicals from the candidates put forward. After these have been established, the availability of authenticated chemical samples from organic synthesis allows more sophisticated behavioural studies to be performed by eliminating unknown aspects and variations in the stimulus. The bioassays range from simple choice experiments in olfactometers to complex behavioural observations producing ethograms (e.g. courtship sequences). Such ethograms enable the researcher to "tease out" subtle roles of chemicals which may or may not be essential in the overall response. Simple choice tests, although they may not help in understanding complex behaviours, or may even miss subtle effects or interactions, can avoid the teleology and anthropomorphism found in more complex studies (Kennedy, 1972, 1978). It is for this reason that simple olfactometers are often used in applied projects.

The final stage of investigation involves field trials (see Table I). The design of such trials to eliminate other effects is crucial for the correct assessment of semiochemicals. The designing of traps may be helped by understanding the insects' behaviour in small containers and/or wind-tunnels. In moth traps, for example, knowledge of plume structure and male response to such plumes enables researchers to position traps in as effective a manner as possible (David and Birch, 1989). Eventually, a picture of the entire chemical ecology of the insect should emerge that accounts for the chemically mediated interaction throughout the various trophic levels and involving interactions with other communication modalities.

III. APHID SEX PHEROMONES

Weber (1930) cautiously suggested the existence of sex attractants among some groups of *Sternorrhyncha*. He noted the presence of pseudorhinaria on the tibiae, especially amongst host-alternating species in which finding the opposite sex presents a problem (Weber, 1931). Smith (1936) was more precise with his observations of *Hyalopterus pruni* oviparae (sexual females) being located and courted by males, which he said must be due to a signal exchange between the sexes. Pettersson (1970a, 1971) was the first to show conclusively the existence of a sex pheromone released by the hind tibiae of female *Schizaphis* species to attract males. Shortly after, similar results for *Megoura viciae* were published by Marsh (1972, 1975).

However, it was nearly 20 years before the first sex pheromone was identified as isomers of the monoterpenoids nepetalactol I and nepetalactone II (Dawson *et al.*, 1987a). Many aphid species have now been shown to have sex pheromones comprising one or both of these compounds in differing ratios (Table II). Using an olfactometer based on the design of Pettersson (1970a), the role of these chemicals in inter- and intraspecific attraction has been studied. The interspecific attraction of *Schizaphis graminum* and *M. viciae* males to different species has been behaviourally examined (Dawson *et al.*, 1990), as well as that in *Aphis fabae*, *M. viciae* and *Acyrthosiphon pisum* (Hardie *et al.*, 1990), and the level of response correlated with the known ratios of lactone and lactol. Hence, behaviour can be used to test ideas relating to the evolutionary role of sex pheromones for reproductive isolation of aphids, although other pre/postmating mechanisms may also be important for reproductive isolation (Hardie, 1991).

Short-range attraction as observed in olfactometers is not suitable for mass trapping or population surveys (Nault and Montgomery, 1977). Both Pettersson (1970a, 1971) and Marsh (1972, 1975) believed that aphids did not employ long-range attraction.

TABLE II. Composition of aphid sex pheromones.

Species		Ratio of I to II
Aphis fabae		29 : 1
Megoura viciae	(day 2–6)	4 : 1–6 : 1
	(day 7–8)	12 : 1
Acyrthosiphon pisum		1 : 1
Sitobion avenae		1 : 0
Sitobion fragariae		1 : 0
Myzus persicae		1 : 2
Schizaphis graminum		1 : 8

I, (+) − (4a*S*,7*S*,7a*R*) − nepetalactone.
II, (−) − (1*R*,4a*S*,7*S*,7a*R*) − nepetalactol.
After Pickett *et al.* (1992).

The final stage of semiochemical research involves field trials (Table I) – such field trials have recently been undertaken with aphid sex pheromones to examine their activity over longer distances. Colourless traps containing the lactone component caught significantly more male *Sitobion fragariae* than either controls or suction traps (Hardie *et al.*, 1992). In a similar trial, lactone-baited traps caught significantly more parasitoids, in the genus *Praon* (Hymenoptera), than control traps (Hardie *et al.*, 1991). Such attraction demonstrates that the aphid parasitoids can use the sex pheromone as a kairomone, thus illustrating an interaction at the third trophic level – the dual role of the lactone as a sex pheromone and a kairomone exemplifies the need for behavioural studies in order to achieve a full understanding.

The damson-hop aphid, *Phorodon humuli*, best illustrates the research approach depicted in Table I. Males were shown to be attracted to oviparae when tested in an olfactometer (Campbell *et al.*, 1990). The chemical attractant was identified, using methods described by Wadhams (1990), as the nepetalactol III comprising two of the 16 possible diastereoisomers. The lactol III was synthesized and olfactometer studies confirmed it as the sex pheromone for *P. humuli*. Traps containing the lactol III were field tested in hop gardens in the autumn, to investigate whether there was any long-range attraction which would oppose the views of Steffan (1987). Large numbers of males were attracted to the lactol traps compared to the control traps (3045 versus 210), which clearly demonstrated the role of the aphid sex pheromone in the field (Campbell *et al.*, 1990). Thus, the entire sequence of semiochemical research, from the original behavioural observations to field trials (Table I), has been demonstrated for *P. humuli*.

P. humuli also illustrates the role of host plant compounds as synergists of sex pheromones, which again can be demonstrated by behavioural studies. Myrobalan (*Prunus* spp.) bark extract on its own attracted few aphids, but in combination with the lactol sex pheromone it significantly increased the catch. This was observed in both olfactometer and field studies (Campbell *et al.*, 1990; L. Wadhams, personal communication). Such interactions are prevalent in semiochemistry, and it is only the behavioural side of research which is able to identify their occurrence and decipher their role.

IV. HOST PLANT VOLATILES

The importance of plant volatiles for host plant location and colonization is well documented in many insect families (Finch, 1980; Blight *et al.*,

1984). Early studies with aphids suggested that chemical cues, detected after visually mediated random landings, resulted in selection of the host plant (Kennedy *et al.*, 1959a,b; Müller, 1958). Kennedy *et al.* (1959a,b) believed that differential leaving rates, rather than differential arrival rates, explained aphid colonization of plants. The precise role of host plant odours in aphid host finding and selection is still uncertain, but recent integration of behavioural studies into research programmes has begun to shed some light in this area.

Attraction of the bird-cherry-oat aphid, *Rhopalosiphum padi*, to its primary host, *Prunus padus*, was noted by Pettersson (1970b) in olfactometer studies. Similar attraction to myrobalan bark extract was observed for *P. humuli* (Campbell *et al.*, 1990). The presence of antennal receptors for host plant volatiles has recently been demonstrated for *P. humuli*. The green leaf volatile (*E*)-2-hexenal and β-caryophyllene (a sesquiterpene hydrocarbon) appear to show some attraction in an olfactometer (Pettersson *et al.*, 1992), although more detailed studies are needed to elucidate their precise function.

Electrophysiological studies have shown receptors for both host and non-host plant volatiles on aphid antennae (C. M. Woodcock, personal communication). Using a linear track olfactometer, Nottingham *et al.* (1991) investigated this more fully for the more generalist feeder *Aphis fabae* and the crucifier specialists *Brevicoryne brassicae* and *Lipaphis erysimi*. The results indicate attraction of aphids to their own host plant and repulsion by some non-host plants (e.g. *A. fabae* was attracted to beans and repelled by isothiocyanates, which are components characteristic of crucifers such as the turnip, *Brassica campestris*). Masking of host plant attraction by non-host-plant volatiles was also observed and warrants further study.

As behavioural studies progress, it appears more likely that the host plant provides a "signature" which is perceived by the aphid. Therefore, there seem to be complex interactions between chemicals which may be more important than any single component, and thus only when a full understanding of the "signature" is achieved will its manipulation by molecular biology be worthwhile.

Tritrophic interactions are of interest to ecologists and evolutionists as well as insect–plant biologists. Such systems illustrate the need for sophisticated behavioural studies in order to understand the intricacies of the system. The use of host plant chemicals as cues for parasitoids was shown by Read *et al.* (1970), although wrongly interpreted. The aphid *B. brassicae* was only attractive to the parasitoid *Diaretiella rapae* when it had been on a host plant within the previous 24 hours. Read thought that this was due to host plant compounds, having "rubbed off" the plant onto the aphid's cuticle, acting as kairomones. However, it seems more likely that a myrosinase (thioglucosidase) enzyme system within *B. brassicae* (McGibbon and Alison, 1968) converts glucosinolates sequestered from the plant to the volatile isothiocyanates, which then act as parasitoid kairomones. After removal from the plant for 24 hours, the aphid produces no more isothiocyanates due to a lack of the glucosinolate precursors. However, the story becomes more complex since species which have the myrosinase enzyme (e.g. *B. brassicae* and *L. erysimi*) have fewer parasitoids than those (e.g. *M. persicae*) which do not. Hence, the isothiocyanates are playing a defensive role against generalist parasitoids, whilst providing a cue for specialist parasitoids. The isothiocyanates also act as synergists for the alarm pheromone in *L. erysimi* (Dawson *et al.*, 1987b).

V. ALARM PHEROMONE

This cornicular secretion, produced by aphids under attack, induces various types of defensive or avoidance behaviour. The sesquiterpene (E)-β-farnesene was found to be an active component of this secretion, which causes an alarm response in many economically important aphids (Table III). However, in some species (e.g. *B. brassicae* and *L. erysimi*), synthetic (E)-β-farnesene elicits only a small response in comparison with the natural cornicle secretion. This observation led to the investigation of the chemistry and electrophysiology of the *L. erysimi* system as outlined in Table I. Electrophysiology coupled with GC (Wadhams, 1990) indicated the presence of isothiocyanate receptors on the proximal primary rhinaria. Although not active alone, the isothiocyanates together with (E)-β-farnesene gave a much higher alarm response (Dawson *et al.*, 1987b).

Hops (*Humulus lupulus*), like many plants, contain the sesquiterpene (E)-β-farnesene, but it elicits no alarm response from the aphids which

TABLE III. Alarm response of some economically important aphids to (E)-β-Farnesene.

Aphid	% Response ± S.E.
Myzus persicae	99 ± 0.6
Phorodon humuli	78 ± 10.2
Aphis fabae	71 ± 5.8
Rhopalosiphum padi	47 ± 4.8
Sitobion avenae	31 ± 11.7
Lipaphis erysimi	20 ± 8.4
Brevicoryne brassicae	0

readily colonize hops. This interesting observation is explained by the fact that these plants also contain other sesquiterpenes, such as β-caryophyllene, which are known to inhibit the alarm response (Dawson *et al.*, 1984). Behavioural bioassays have shown that β-caryophyllene inhibits the alarm response in both *P. humuli* and *Myzus persicae*. In *M. persicae*, the inhibition is so great that very little response to (E)-β-farnesene is observed in a β-caryophyllene-saturated environment, and a reduced response is observed even in the presence of a predator (C. M. Woodcock, personal communication). Thus, behavioural assays are again the means by which interactions may be unravelled after chemistry and electrophysiology have indicated interesting phenomena.

VI. CONCLUSION

The prevalence of parsimony in aphid chemical ecology illustrates the need for integration of behavioural studies into any research programme. The role of isothiocyanates, for example, needs to be studied at all interaction levels before a "true" understanding of its precise function in insect–plant chemical ecology can be achieved. The role of chemical components must be fully understood before molecular biologists transform systems to alter the spectrum and the full advantages of this modern technology can be realized.

In the classic text of Shorey and McKelvey (1977), the opening chapter contained a cogent, prescient statement which has been paraphrased many times. This highlights the importance of behavioural research and it is worth repeating:

> It cannot be stressed too strongly that the key to devising efficient systems for the management of insect pests by chemically modifying their behaviour is the acquisition of an intimate knowledge of the insects' own normal use of chemicals. This important factor is too often overlooked. Once a pheromone or other behaviourally active chemical is identified, there is a tendency to feel that the research is all over, and that the chemical can be used as a bait in traps or perhaps distributed through fields, causing insect control. Rather, the identification of the chemical should open the door to more, necessary research to determine whether the normal behaviour of insects can be interfered with and manipulated to our advantage.

Acknowledgements

I wish to thank J. Hardie, J. Pickett, L. Wadhams and C. Woodcock for reading this manuscript.

REFERENCES

Birch, M. C., Poppy, G. M. and Baker, T. C. (1990). *Annu. Rev. Entomol.* **35**, 25–58.
Blight, M. M., Pickett, J. A., Smith, M. C. and Wadhams, L. J. (1984). *Naturwissenschaften* **71**, 480–481.
Boeckh, J. (1986). In "Mechanisms in Insect Olfaction" (T. L. Payne, M. C. Birch and C. E. J. Kennedy, eds), pp. 303–310. Clarendon Press, Oxford.
Butenandt, A., Beckmann, R., Stamm, D. and Hecker, E. (1959). *Z. Naturforsch.* **14b**, 283–284.
Campbell, C. A. M., Dawson, G. W., Griffiths, D. C., Pettersson, J., Pickett, J. A., Wadhams, L. J. and Woodcock, C. M. (1990). *J. Chem. Ecol.* **16**, 3455–3465.
David, C. T. and Birch, M. C. (1989). In "Insect Pheromones in Plant Protection" (A. R. Jutsom and R. F. S. Gordon, eds), pp. 17–35. Wiley, New York.
Dawson, G. W., Griffiths, D. C., Pickett, J. A., Smith, M. C. and Woodcock, C. M. (1984). *Entomol. Exp. Appl.* **36**, 197–199.
Dawson, G. W., Griffiths, D. C., Janes, N. F., Mudd, A., Pickett, J. A., Wadhams, L. J. and Woodcock, C. M. (1987a). *Nature* **325**, 614–616.
Dawson, G. W., Griffiths, D. C., Pickett, J. A., Wadhams, L. J. and Woodcock, C. M. (1987b). *J. Chem. Ecol.* **13**, 1663–1671.
Dawson, G. W., Griffiths, D. C., Merritt, L. A., Mudd, A., Pickett, J. A., Wadhams, L. J. and Woodcock, C. M. (1990). *J. Chem. Ecol.* **16**, 3019–3030.
Fabre, J. H. C. (1904). "Souvenirs Entomologiques", Vol. 8, 7th edn. Libraire Belagrave, Paris.
Finch, S. (1980). *Appl. Biol.* **5**, 67–143.
Hardie, J. (1991). *Entomol. Gener.* **16**, 249–256.
Hardie, J., Holyoak, M., Nicholas, J., Nottingham, S. F., Pickett, J. A., Wadhams, L. J. and Woodcock, C. M. (1990). *Chemoecology* **1**, 63–68.
Hardie, J., Nottingham, S. F., Powell, W. and Wadhams, L. J. (1991). *Entomol. Exp. Appl.* **61**, 97–99.
Hardie, J., Nottingham, S. F., Harrington, R., Pickett, J. A. and Wadhams, L. J. (1992) (in preparation).
Kellogg, V. L. (1907). *Biol. Bull.* **12**, 152–154.
Kennedy, J. S. (1972). *J. Aust. Entomol. Soc.* **11**, 168–176.
Kennedy, J. S. (1978). *Physiol. Entomol.* **3**, 91–98.
Kennedy, J. S., Booth, C. O. and Kershaw, W. J. S. (1959a). *Ann. Appl. Biol.* **47**, 410–423.
Kennedy, J. S., Booth, C. O. and Kershaw, W. J. S. (1959b). *Ann. Appl. Biol.* **47**, 424–444.
Law, J. H. and Regnier, F. E. (1971). *Annu. Rev. Biochem.* **40**, 533–548.
Legros, G. V. (1971). "Fabre Poet of Science." Hanson Press, New York.
Marsh, D. (1972). *Nature* **238**, 31–32.
Marsh, D. (1975). *J. Entomol. (A)* **50**, 43–64.
Mayer, A. G. (1900). *Psyche* **9**, 15–20.
McGibbon, D. B. and Alison, R. M. (1968). *N.Z. J. Sci.* **11**, 440–446.
Müller, H. J. (1958). *Entomol. Exp. Appl.* **1**, 66–72.
Nault, L. R. and Montgomery, M. E. (1977). In "Aphids as Virus Vectors" (K. F. Harris and K. Maramorosch, eds), pp. 527–545. Academic Press, New York/San Francisco/London.
Nottingham, S. F., Hardie, J., Dawson, G. W., Hick, A. J., Pickett, J. A., Wadhams, L. J. and Woodock, C. M. (1991). *J. Chem. Ecol.* **17**, 1231–1242.
Pettersson, J. (1970a). *Ent. Scand.* **1**, 63–73.
Pettersson, J. (1970b). *Lantbrhogsk Annlr.* **36**, 381–399.
Pettersson, J. (1971). *Ent. Scand.* **2**, 81–93.

Pettersson, J., Pickett, J. A., Wadhams, L. J. and Woodcock, C. M. (1992). In preparation.
Pickett, J. A., Wadhams, L. J., Woodcock, C. M. and Hardie, J. (1992). *Annu. Rev. Entomol.* **37**, 67–90.
Poppy, G. M. (1990). D. Phil. Thesis. University of Oxford, UK.
Read, D. P., Feeny, P. P. and Root, D. B. (1970). *Can. Entomol.* **102**, 1567–1578.
Shorey, H. H. (1973). *Annu. Rev. Entomol.* **18**, 349–380.
Shorey, H. H. and McKelvey, J. R. (1977). "Chemical Control of Insect Behaviour". Wiley, Interscience, Chichester.
Smith, L. M. (1936). *Hilgardia* 10, 7. Berkeley 167–211.
Steffan, A. W. (1987). *Entomol. Gener.* **12**, 235–258.
Wadhams, L. J. (1990). *In* "Chromatography and Isolation of Insect Hormones and Pheromones" (A. R. McCaffery and I. D. Wilson, eds), pp. 289–298. Plenum, New York/London.
Weber, H. (1930). "Biologie der Hemipteren". Springer, Berlin.
Weber, H. (1931). *Zeitschr. Fur Morph. oek. der Tiere* **23**, 575–577.

11

Molecular Biology of Insecticide Resistance

A. L. DEVONSHIRE, L. M. FIELD AND M. S. WILLIAMSON

I. Introduction 173
II. Target Site Resistance 174
   A. GABA Receptor/Chloride Channel Complex 174
   B. Acetylcholinesterase 175
   C. Sodium Channel 176
III. Metabolic Resistance 179
   A. Glutathione S-transferases 179
   B. Monooxygenases 179
   C. Esterases 180
   References 182

I. INTRODUCTION

Resistance to insecticides is mediated by qualitative and quantitative changes in proteins that can often be difficult to define precisely at the biochemical level. The main changes involve a limited range of detoxifying enzymes and the few insecticide target site proteins and to a lesser extent completely unknown factors that can delay the penetration of pesticide into insects.

The vast majority of commercial insecticides act at one of two targets in the nervous system. Organophosphates (OP) and carbamates interact with acetylcholinesterase (AChE) and changes in the sensitivity of this enzyme to inhibition have been well characterized biochemically. The changes are clearly due to mutant forms of the enzyme, multiple allelic forms of which have been identified in some species. The other major insecticide class, the synthetic pyrethroids, affect the function of voltage-dependent sodium channels by binding to a site distinct from those of other neurotoxins acting on the same protein. Electrophysiological and pharmacological studies have revealed a decreased sensitivity of this target to the toxic effects of pyrethroids, but the detailed biochemical basis of this is difficult to study due to the extreme lipophilicity of these ligands and the reversible nature of the interaction. Target site insensitivity has also developed widely to the cyclodienes, which act on the GABA (γ-aminobutyric acid) receptor/chloride channel, and whilst this class of insecticides has largely been superseded, there are implications for newer (e.g. avermectin) and prospective (e.g. trioxabicyclooctanes) products acting on the same target.

Three broad enzyme classes are involved in insecticide detoxification, the glutathione transferases, monooxygenases and hydrolases. Their involvement in resistance is commonly identified by increases in the characteristic metabolites they produce, often complemented by a knowledge of their cofactor (NADPH or glutathione) requirements *in vitro*. However, all three classes exist in multiple forms within each species and it is often not known whether increased activity arises from qualitative or quantitative changes in these enzyme complexes. This can often be determined by the use of model, surrogate substrates, especially for esterases and glutathione transferases, but it then becomes important to demonstrate that the activity measured plays a role in the breakdown of insecticides.

Molecular biological approaches are now elucidating the genetic changes underlying these biochemical resistance mechanisms. This chapter reviews recent progress in this area, not only for pest species, but also for *Drosophila*, which provides an important model in studies of insecticide target site genes.

II. TARGET SITE RESISTANCE

Target site insensitivity, an important mechanism by which insects develop resistance to insecticidal compounds, cannot be overcome by the use of chemical synergists and often confers cross-resistance to entire classes of insecticides, seriously limiting their effectiveness in the control of resistant populations.

A. GABA Receptor/Chloride Channel Complex

The GABA receptor/chloride channel complex has long been proposed as the primary target site for cyclodiene insecticides and although use of this class of insecticides is in decline, this receptor remains an important target for novel lead compounds. Resistance to cyclodienes has developed in about 300 species and, in all cases studied in detail, appears to be due to a single gene mechanism conferring nerve insensitivity (Georghiou, 1986). It has been proposed that this results from reduced binding affinity of insecticides for the receptor, but analysis of these interactions by conventional pharmacological techniques has been hindered by the highly lipophilic nature of the cyclodienes, which generates high levels of nonspecific binding, and by difficulties in preparing GABAergic cells from insect nervous tissue. However, the recent cloning of vertebrate GABA receptor gene sequences, and the *in vitro* expression of functional receptor/ion channel complexes from these cloned sequences in *Xenopus* oocytes (Schofield *et al.*, 1987), offer prospects for isolating and studying the homologous insect genes. Using the polymerase chain reaction (PCR) with degenerate oligonucleotide primers designed from conserved regions of the vertebrate receptor, Knipple *et al.* (1991) recently reported the amplification and characterization of GABA-receptor-like sequences from *Drosophila* and this approach should be applicable to any species with cyclodiene resistance.

An alternative approach for cloning a cyclodiene resistance gene has been employed by ffrench-Constant and Roush (1991) using a cyclodiene-resistant strain of *Drosophila melanogaster*. The elegant gene mapping techniques available in this organism were used to localize the resistance gene very precisely to a region of the polytene map (66F) on the left arm of chromosome III. This locus was cloned in a series of overlapping cosmid clones and the resistance gene identified by rescue of the sensitive phenotype following reintroduction of individual cosmid DNAs into the *Drosophila* germ line using P-element-mediated transformation (ffrench-Constant *et al.*, 1991). Sequences associated with this locus show homology to the vertebrate GABA receptor sequences, providing the most compelling evidence to date that cyclodiene resistance is indeed associated with the GABA receptor complex. One of the most desirable aspects of this approach to resistance gene cloning, which as yet is only possible in *Drosophila*, is that it requires no assumptions about the nature of the resistance gene product, since the gene is cloned purely on the basis of its phenotype and chromosomal location.

B. Acetylcholinesterase

AChE is well established as a primary target for insecticides, beginning with the first OPs in the 1940s through the carbamates and extending to at least one novel compound still under development (Murray *et al.*, 1988).

Enzyme kinetic analysis has identified insensitive forms of AChE in many species, often apparently involving just one mutant form of the enzyme (Hemingway et al., 1986), or, as in the housefly, a family of alleles each conferring a distinct pattern of sensitivity (Devonshire, 1987).

Drosophila mutants with AChE insensitivity are rare, but one showing slight malathion resistance was found to differ from the wild type gene in one base (Hall and Spierer, 1986; Bergé and Fournier, 1988) which resulted in the substitution of tyrosine for phenylalanine at position 368 (Fournier, 1991), a region removed from the established catalytic centre sequences, and any effect on the enzyme's interaction with OPs remains to be established.

Attempts to understand the genetic basis of insensitive AChE in pest species have met with limited success. A cDNA clone from *Drosophila* was used as a heterologous probe to isolate the corresponding gene from a genomic library of *Anopheles stephensi* (Hall and Malcolm, 1991), a mosquito species not yet reported to develop insensitive AChE. However, this should open the way to cloning the gene from other *Anopheles* species where this resistance mechanism is established. Our attempts to isolate housefly AChE sequences by heterologous probing with the *Drosophila* cDNA were not successful. However, using an alternative approach based on PCR we were able to amplify a small fragment (0.5 kb) directly from susceptible housefly genomic DNA. This fragment was used as a homologous probe to re-screen our libraries, and has led to the isolation of a cDNA containing around 80% of the coding sequence for the housefly AChE protein (Williamson and Devonshire, unpublished). This opens the way to identifying mutations in the gene sequences of the resistant housefly strains which alter the sequence/structure of the enzyme. The functional significance of such alterations must then be assayed by *in vitro* expression of the modified sequence (e.g. Gibney et al., 1990).

C. Sodium Channel

Mutants of the sodium channel, the target for one of the earliest and most persistent synthetic insecticides, DDT, and more recently the synthetic pyrethroids, have been under selection pressure for almost 50 years. Whilst this has had the greatest impact on pest species, leading in some cases to extreme resistance, certain *Drosophila* mutants known to be defective in sodium channel function also exhibit altered sensitivity to these compounds. Knowledge of the primary structure of the sodium channel has advanced rapidly over the last five years through the cloning of gene sequences from electric eel, rat and more recently *Drosophila*, and has led to several related coarse structural models of the channel protein based on the predicted sequences (review: Catterall, 1988). These studies have also revealed the presence of more than one type of sodium channel gene in both rat and *Drosophila* encoding different subtypes of channel mRNA. In *Drosophila* two putative sodium channel genes have been identified, termed DSC1 and *para*, both showing significant structural homologies to the vertebrate sodium channels (Salkoff *et al.*, 1987; Loughney *et al.*, 1989).

The *para* locus on the *Drosophila* X chromosome encodes a sodium channel for which a number of temperature-sensitive paralysis mutants have been identified. Recently, Hall and Kasbekar (1989) have reported the *para*^{ts} mutants show different sensitivities to both the lethal and knockdown effects of pyrethroids. In terms of kill, the resistance factors are relatively low (3–6-fold), but some mutants show very protracted knockdown compared to the wild type. Although the molecular basis of these mutants has not yet been established, a phenotypically similar conditional mutant termed *nap*^{ts} also shows marked knockdown resistance to pyrethroids and is associated with a 40% reduction in sodium channel density (Jackson *et al.*, 1984; Bloomquist *et al.*, 1989). The *nap* locus maps to chromosome 2 and is thought to down-regulate the class of sodium channels coded for by the X-linked *para* locus (Hall and Kasbekar, 1989). There is also evidence that pyrethroid resistance might be associated with the other putative sodium channel gene (*sch*, equivalent to DSC1). Pauron *et al.* (1991) have recently identified a resistance factor showing a two-fold delayed knockdown by deltamethrin which also maps to chromosome 2 in a DDT-selected strain, Tübingen-DDT. Unlike *nap*^{ts}, these flies do not exhibit a reduced density of sodium channels, but rather an apparently decreased (approximately seven-fold) affinity of deltamethrin for the sodium channel as judged by its synergistic effect on batrachotoxin binding. Partial sequencing of the *sch* gene from Tübingen-DDT and the pyrethroid-sensitive Tübingen strain revealed a single mutation that converts an aspartic acid residue to asparagine in the resistant gene. This mutation lies between segments III S5 and III S6 of the proposed channel structure, a region which some models predict is embedded in the membrane (Guy and Seetharmulu, 1986), and which appears to form the channel pore in the closely related potassium channel (Yellen *et al.*, 1991). It is therefore of interest to express these sequences in an *in vitro* assay system such as *Xenopus* oocyte to determine (1) whether the *sch* gene does encode a functional sodium channel, and (2) the effect of this mutation on the pyrethroid sensitivity of the channel.

In the housefly, knockdown resistance (kdr) is an important mechanism effective against all pyrethroids and DDT, and analogous mechanisms have been partially characterized in a number of other important pests. Several alleles of kdr have been identified in houseflies, including the *super kdr* factor which can confer 250-fold resistance to these compounds (Farnham *et al.*, 1987). These alleles map to chromosome 3 of the housefly and have been genetically isolated and characterized toxicologically in homozygous strains, providing an excellent system in which to analyse the underlying mutations at the molecular level.

Electrophysiological and biochemical studies have shown that much higher pyrethroid concentrations are required to modify the normal function of voltage-sensitive sodium channels from kdr than susceptible flies (review: Soderlund and Bloomquist, 1989). Two main hypotheses have been put forward to explain the involvement of sodium channels in this reduced sensitivity. One proposed that a reduced density of channels in the nerve membrane could produce the phenotype as is seen in nap^{ts} *Drosophila* mutants, but recent [^{3}H] saxitoxin binding experiments (Grubs *et al.*, 1988; Sattelle *et al.*, 1988) have established that the channel density is similar in the sensitive and various kdr strains, effectively excluding this model. Alternatively, kdr may be conferred by alterations in the sequence/structure of the sodium channel protein that reduces its affinity for pyrethroids. To test this hypothesis, good progress is now being made on the cloning and characterization of sodium channel gene sequences from the housefly. Using degenerate oligonucleotide primers, designed from conserved regions of the first homology domain of other cloned sodium channel sequences, Knipple *et al.* (1991) exploited PCR to amplify a segment (approximately 90 bp) of the housefly homologue of the *para* sodium channel gene. We have also isolated cloned sequences of the housefly *para* homologue by low-stringency heterologous probing of housefly cDNA libraries using a DNA probe from the *Drosophila para* gene (Williamson and Devonshire, unpublished). These cDNAs cover approximately 2 kb of 3' coding and untranslated sequence from the corresponding housefly gene and provide a starting-point from which to clone, sequence and compare the upstream gene coding regions from the susceptible and resistant (kdr) housefly strains.

In the context of understanding insecticide mode of action, the housefly system offers a major advantage over studies of vertebrate sodium channels; by focusing on sequences differing between susceptible and kdr insects, we should be able to characterize regions of the protein critical for its interaction with insecticides. This will provide much-needed insight of the topography of this key target site for a major class of insecticides, and, by comparison with data from vertebrate studies, should identify opportunities for improving insecticide selectivity.

III. METABOLIC RESISTANCE

A. Glutathione S-transferases

This group of enzymes catalyses the conjugation of glutathione with compounds having a reactive electrophilic centre, leading to the formation of a water-soluble, less reactive product. Although there are many examples of increased metabolism of insecticides or model substrates by glutathione S-transferases (GSTs) of resistant insects, few are characterized at the molecular level. However, two GST-related cDNAs from a *Drosophila* strain selected for resistance to malathion, and showing increased GST activity, have been cloned and sequenced (Cochrane *et al.*, 1991). One of these cDNAs was found to encode the small subunit of the most abundant GST and showed elevated levels of the corresponding mRNA on Northern blots of the selected line. Southern blots revealed no quantitative differences in gene copy number between the selected and control *Drosophila*, indicating that gene regulation, rather than amplification, causes overproduction of this GST.

Sequencing of GST1 cDNAs from susceptible and insecticide-resistant houseflies revealed several point mutations (Fournier, 1991) although there is no published evidence that these are involved in resistance. There was also evidence for overproduction of GST1 enzyme and its mRNA by resistant houseflies but, as in *Drosophila*, this did not involve gene amplification.

In both *Drosophila* and houseflies, there was polymorphism at this locus, and in the case of *Drosophila*, it was suggested that this might result from the presence of transposable elements.

B. Monooxygenases

The monooxygenase, or mixed function oxidase (MFO), complex involves a reductase and one or more cytochrome P450s and requires NADPH as cofactor. An increase in MFO activity is one of the most versatile mechanisms of resistance in insects, and commonly involves an increase in cytochrome P450. In housefly, cDNAs for a phenobarbital-inducible P450 protein have been cloned and sequenced (Feyereisen *et al.*, 1989). The P450 mRNA was elevated in insecticide-resistant flies as compared to a susceptible strain, and intermediate in their F_1 cross, but so far the genetic mechanism for overproduction of this P450 is not clear (Feyereisen, personal communication).

Evidence for the molecular genetic basis of differences in P450 expression as a cause of resistance has only been shown for *Drosophila* (Waters, personal communication), where resistance is associated with increased synthesis of P450 B1 which is encoded by a more abundant, but shorter, mRNA. The P450 B1 gene in susceptible insects has a long terminal repeat (LTR) at the 3' end (characteristic of a transposable element), which moves the polyadenylation signal downstream, but this LTR is absent in resistant *Drosophila*. It has been suggested that the susceptible *Drosophila* synthesize a chimeric mRNA, i.e. containing both P450 B1 and LTR sequences, and that the difference in P450 B1 expression by susceptible and resistant *Drosophila* results from down-regulation of the gene in susceptible insects as a consequence of instability of the chimeric mRNA.

C. Esterases

Insecticide detoxification by esterases has been best characterized at the molecular level in mosquitoes and aphids, where increased synthesis of enzyme results from amplification of structural genes. In aphids the massive overproduction of esterase protein (60-fold) results in detoxification of insecticidal esters by both sequestration and hydrolysis when the inhibited esterase reactivates (Devonshire and Moores, 1982) and this could also be the case for *Culex* mosquitoes where there can be up to 500-fold increase in esterase, equivalent to 15% of the insects' total proteins (Mouchès *et al.*, 1987).

In mosquitoes, there is a family of closely related esterases that can be overproduced in resistant insects. A cDNA encoding one of these enzymes, esterase B1, has been isolated from an extremely resistant strain (TemR) of *Culex pipiens quinquefasciatus* and showed a >250-fold increase in gene copy number as estimated by probing slot blots of total DNA (Mouchès *et al.*, 1986). The amplified gene in TemR was located on an abundant 2.1-kb *Eco*RI fragment as distinct from the faint 2.8-kb fragment of susceptible insects. Since these amplified esterase genes are inherited as an "allele", it appears that they are in a tandem array and that little recombination occurs (Ferrari and Georghiou, 1991).

Sequence data for the B1 gene and its deduced amino acid sequence show considerable homology with AChEs from *Drosophila* and *Torpedo*, human butyrylcholinesterase and esterases of *Heliothis*, rabbit liver and *Drosophila* (Pasteur *et al.*, 1990). Restriction analysis showed that the genes for the related esterases B2 and B3 differ from the B1 gene (Pasteur *et al.*, 1990), and that the B2 gene and its flanking DNA (approximately 20 kb) from six mosquito strains of widely diverse origin

11. Molecular Biology of Insecticide Resistance

were identical (Raymond et al., 1991). The latter result led the authors to propose that all amplified esterase B2 regions arose from a single initial event and that the world-wide presence of the amplification is due to recent migration.

The first indication that insecticide resistance might be mediated by gene amplification came when Devonshire and Sawicki (1979) proposed that the doubling in esterase content between seven progressively more resistant clones of the aphid, *Myzus persicae*, could be best explained by amplification of the esterase gene. As with mosquitoes, this was established unequivocally when a cDNA clone encoding the esterase E4 was used to show that both esterase mRNA and gene copy number are higher in resistant insects (Field et al., 1988).

Aphids with elevated E4 had amplified esterase sequences on an apparently novel 8-kb *Eco*RI fragment whereas those with a closely related elevated esterase, FE4, had a corresponding 4-kb fragment. Cloning and analyses of these fragments have shown that the amplified E4 and FE4 genes appear to differ only at their 3' ends (Field and Devonshire, 1991a). Many resistant aphid clones of diverse origin have now been examined by restriction analysis, which gave only one of two patterns correlating with the presence of amplified E4 or FE4 genes that do not co-exist (Field and Devonshire, 1991b). Thus, as in mosquitoes, there is little polymorphism at this amplified locus.

In aphids thre is also transcriptional control of the amplified esterase genes (Field et al., 1989). Some very resistant aphid clones with elevated E4 can spontaneously lose resistance and elevated enzyme. This is accompanied by loss of elevated E4 mRNA but the amplified E4 genes are retained, indicating that the "reversion" from a resistant to susceptible phenotype involves a change in gene transcription. So far the only difference observed (Field et al., 1989) between the expressed and silent E4 sequences is a change in the presence of 5-methylcytosine (5-mC) a modified base that is known to affect gene transcription in vertebrates. In aphids, expressed E4 and FE4 genes are methylated at certain sites, whereas the same sites in the unexpressed E4 genes of revertant aphids are unmethylated. This is very surprising in the light of many studies with vertebrate genes where DNA methylation has been shown to correlate with gene inactivation, and demethylation is a necessary prerequisite for transcription to occur (Doerfler, 1983).

It has recently been shown that insecticide-resistant *Myzus nicotianae*, a tobacco-feeding species, closely related to *M. persicae*, has elevated FE4 enzyme and amplified methylated FE4 genes indistinguishable from those of *M. persicae*, indicating a common molecular genetic basis of insecticide resistance (Field and Devonshire, 1991b). It is possible that the insecticide-resistant *M. nicotianae* evolved from resistant *M. persicae*, i.e. the amplified esterase genes were present before the divergence of the two species. Alternatively, as proposed for a chromosome translocation common to both species, resistance may have evolved after they separated, in which case two independent amplification events must have occurred. Since the amplified sequence in *M. persicae* and *M. nicotianae* appear to be identical, these species might be somehow "predisposed" for a particular amplification event. If so, then the same amplification event could also have occurred more than once in *M. persicae* which would explain the conserved sequences in clones of widely different origin; this would be in contrast to the migration hypothesis for insecticide-resistant mosquitoes.

REFERENCES

Bergé, J. B. and Fournier, D. (1988). "Proceedings of the XVIII International Congress Entomol." Vancouver, B.C., Canada, p. 461.
Bloomquist, J. R., Soderlund, D. M. and Knipple, D. C. (1989). *Arch. Insect Biochem. Physiol.* **10**, 293–302.
Catterall, W. A. (1988). *Science* **242**, 50–61.
Cochrane, B. J., Hargos, M., Crocquet deBelligmy, P., Holtsberg, F. and Coronella, J. (1991). *In* "Molecular Mechanisms of Resistance in Herbivorous Pests to Natural, Synthetic and Bioengineered Control Agents" (J. G. Scott and C. A. Mullin, eds). American Chemical Society Symposium Series, Washington. In press.
Devonshire, A. L. (1987). *In* "Combating Resistance to Xenobiotics – Biological and Chemical Approaches" (M. G. Ford, D. W. Hollomon, B. P. S. Khambay and R. M. Sawicki, eds), pp. 227–238. Ellis Horwood, Chichester.
Devonshire, A. L. and Moores, G. D. (1982). *Pestic. Biochem. Physiol.* **18**, 235–246.
Devonshire, A. L. and Sawicki, R. M. (1979). *Nature* **280**, 140–141.
Doerfler, W. (1983). *Annu. Rev. Biochem.* **52**, 93–124.
Farnham, A. W., Murray, A. W. A., Sawicki, R. M., Denholm, I. and White, J. C. (1987). *Pestic. Sci.* **19**, 209–220.
Ferrari, J. A. and Georghiou, G. P. (1991). *Heredity* **66**, 265–272.
Feyereisen, R., Koener, J. F., Farnsworth, D. E. and Nebert, D. W. (1989). *Proc. Natl. Acad. Sci. USA* **86**, 1465–1469.
ffrench-Constant, R. H. and Roush, R. T. (1991). *Genet. Res.* **57**, 17–21.
ffrench-Constant, R. H., Mortlock, D. P., Shaffer, C. D., MacIntyre, R. J. and Roush, R. T. (1991). *Proc. Natl. Acad. Sci. USA* **88**, 7209–7213.
Field, L. M. and Devonshire, A. L. (1991a). *In* "Resistance '91: Achievements and Developments in Combating Pesticide Resistance" (I. Denholm, A. L. Devonshire and D. W. Hollomon, eds), pp. 240–250. Elsevier, Amsterdam. In press.
Field, L. M. and Devonshire, A. L. (1991b). *In* "Molecular Mechanisms of Resistance in Herbivorous Pests to Natural, Synthetic and Bioengineered Control Agents" (J. G. Scott and C. A. Mullin, eds), American Chemical Society Symposium Series Washington. In press.
Field, L. M., Devonshire, A. L. and Forde, B. G. (1988). *Biochem. J.* **251**, 309–312.
Field, L. M., Devonshire, A. L., ffrench-Constant, R. H. and Forde, B. G. (1989). *FEBS Lett.* **243**, 323–327.

11. Molecular Biology of Insecticide Resistance

Fournier, D. (1991). In "2nd Symposium Abstracts: Insecticides – Mechanisms of Resistance" (D. Otto, H. Penzlin, H. Lyr and H. Schmutterer, eds).

Georghiou, G. P. (1986). In "Pesticide Resistance: Strategies and Tactics for Management" (National Academy of Sciences, ed.), pp. 14–43. National Academy Press, Washington D.C.

Gibney, G., Camp, S., Dionne, M., MacPhee-Quigley, K. and Taylor, P. (1990). *Proc. Natl. Acad. Sci. USA* **87**, 7546–7550.

Grubs, R. E., Adams, P. M. and Soderlund, D. M. (1988). *Pestic. Biochem. Physiol.* **32**, 217–223.

Guy, H. R. and Seetharmulu, P. (1986). *Proc. Natl. Acad. Sci. USA* **83**, 508–512.

Hall, L. M. and Kasbekar, D. P. (1989). In "Insecticide Action: from Molecule to Organism" (T. Narahashi and J. Chambers, eds), pp. 99–114. Plenum, New York.

Hall, L. M. C. and Malcolm, C. A. (1991). *Cell. Mol. Neurobiol.* **11**, 131–142.

Hall, L. M. C. and Spierer, P. (1986). *EMBO J.* **5**, 2949–2954.

Hemingway, J., Smith, C., Jayawardena, K. G. and Herath, P. R. J. (1986). *Bull. Entomol. Res.* **76**, 559–565.

Jackson, F. R., Wilson, S. D., Strichartz, G. R. and Hall, L. M. (1984). *Nature* **308**, 189–191.

Knipple, D. C., Payne, L. L. and Soderlund, D. M. (1991). *Arch. Insect Biochem. Physiol.* **16**, 45–53.

Loughney, K., Kreber, R. and Ganetzky, B. (1989). *Cell* **58**, 1143–1154.

Mouchès, C., Pasteur, N., Bergé, J. B., Hyrien, O., Raymond, M., de Saint Vincent, B. R., de Silvestri, M. and Georghiou, G. P. (1986). *Science* **233**, 778–780.

Mouchès, C., Maguin, M., Bergé, J. B., de Silvestri, M., Beyssat, V., Pasteur, N. and Georghiou, G. P. (1987). *Proc. Natl. Acad. Sci. USA* **84**, 2113–2116.

Murray, A., Siddi, G., Vietto, M., Jacobson, R. M. and Thirugnanam, M. (1988). *Br. Crop. Prot. Conf.* **1**, 73–80.

Pasteur, N., Raymond, M., Pauplin, Y., Nancé, E. and Heyse, D. (1990). In "Pesticides and Alternatives: Innovative Chemical and Biological Approaches to Pest Control" (J. E. Casida, ed.), pp. 439–447. Elsevier, Amsterdam.

Pauron, D., Amichot, M. and Bergé, J. B. (1991). In "Resistance '91: Achievements and Developments in Combating Pesticide Resistance" (I. Denholm, A. L. Devonshire and D. W. Hollomon, eds) pp. 228–239. Elsevier, Amsterdam.

Raymond, M., Callaghan, A., Fort, P. and Pasteur, N. (1991). *Nature* **350**, 151–153.

Salkoff, L., Butler, A., Wei, A., Scavarda, N., Giffen, K., Ifune, C., Goodman, R. and Mandel, G. (1987). *Science* **237**, 744–749.

Sattelle, D. B., Leech, C. A., Lummis, S. C. R., Harrison, B. J., Robinson, H. P. C., Moores, G. D. and Devonshire, A. L. (1988). In "Neurotox '88: Molecular Basis of Drug and Pesticide Action" (G. G. Lunt, ed.), pp. 563–582. Elsevier, Amsterdam.

Schofield, P. R., Darlison, M. G., Fugita, N., Burt, D. R., Stephenson, F. A., Rodriguez, H., Rhee, L. M., Ramachandran, J., Reale, V., Glencorse, T. A., Seeburg, P. H. and Barnard, E. A. (1987). *Nature* **328**, 221–227.

Soderlund, D. M. and Bloomquist, J. R. (1989). *Annu. Rev. Entomol.* **34**, 77–96.

Yellen, G., Jurman, M., Abramson, T. and MacKinnon, R. (1991). *Science* **251**, 939–942.

Part IV.
Physiological Regulation

12. Locust Adipokinetic Hormones: A Review
    M. O'SHEA AND R. C. RAYNE
13. The Structure and Functional Activity of Neuropeptides
    G. GOLDSWORTHY, G. COAST, C. WHEELER,
    O. CUSINATO, I. KAY AND B. KHAMBAY
14. Molecular Approaches to Insect Endocrinology
    L. M. RIDDIFORD

12

Locust Adipokinetic Hormones: A Review

M. O'SHEA AND R. C. RAYNE

I. Introduction 187
II. Localization and Site of Synthesis 189
III. Release and Functions 191
IV. Inactivation and Metabolism 191
V. Prohormone and Precursor Biosynthesis 194
VI. Processing *In Vivo* and *In Vitro* 198
VII. Conclusions and Future Directions 202
References 203

I. INTRODUCTION

Adipokinetic hormone (AKH) is the name given to a biological activity detectable in extracts taken from the locust corpora cardiaca (CC), major neuroendocrine structures in insects (Beenakers, 1969; Meyer and Candy, 1969). In recent years there has been a great deal of progress at the molecular level concerning these and related hormones in other insects and several primary structures are known. The CC of the locust have moreover provided us with a model preparation in which to investigate the molecular biology of neuropeptide biosynthesis.

The adipokinetic activity found in the locust CC is produced by two different but related peptide hormones called AKH-I and AKH-II. The primary structure of AKH-I is: pGlu-Leu-Asn-Phe-Thr-Pro-Asn-Trp-Gly-Thr-NH$_2$ (Stone *et al.*, 1976); the sequence of the related AKH-II peptide from *Schistocerca* is pGlu-Leu-Asn-Phe-Ser-Thr-Gly-Trp-NH$_2$ (Siegert *et al.*, 1985). In recent years structurally related hormones have been discovered in the CC of other insects. A selection of the insect hormones with the AKH signature is given in Table I.

While in the locust the adipokinetic hormones liberate diacylglycerols from the fat body into the circulation and are in general involved in the activation of lipid metabolism (Orchard, 1987), this is not the case for the similar peptides of other insect species. For example, in the cockroach, *Periplaneta americana*, the AKH-like peptides MI and MII (O'Shea *et al.*, 1984; Witten *et al.*, 1984) are involved in stimulating carbohydrate metabolism (Siegert *et al.*, 1985). In many cases two forms of an AKH-related peptide are found in the CC. The biological significance of this is not clear because only one function can be attributed to the two forms. It seems unlikely that the CC would manufacture two different peptides unless they had distinct and exclusive functions. Functional studies, however, have proven more difficult than chemistry and the gap between our knowledge of the biological activities of the AKH peptides on the one hand and their molecular structures on the other continues to widen. It has widened further recently because molecular cloning and protein sequencing techniques have been applied to the study of AKH biosynthesis (Hekimi and O'Shea, 1989a; Schulz-Aellen *et al.*, 1989; Noyes and Schaffer, 1990). This has led to the identification of the AKH precursors and novel peptides of as yet unknown function that are co-synthesized, co-stored and co-released with the AKHs. These novel peptides are parts of the AKH precursors and are named AKH precursor related peptides or APRPs (Hekimi and O'Shea, 1989b).

This chapter will add very little to our understanding of the functional significance of either the adipokinetic hormones or the co-synthesized APRPs. Our main purpose is to illustrate how useful the CC of the locust have been in allowing us to probe the molecular and cellular biology of an insect peptide hormone. We have, for example, been able conveniently to study AKH biosynthesis and AKH inactivation. Moreover, peptidergic systems may offer new potential targets for insect control (O'Shea, 1985,

TABLE I. Selected AKH family peptides.

Genus	Name	Sequence
Schistocerca	AKH-I	pGlu-Leu-Asn-Phe-Thr-Pro-Asn-Trp-Gly-Thr-NH$_2$
Schistocerca	AKH-II-S	pGlu-Leu-Asn-Phe-Ser-Thr-Gly-Trp-NH$_2$
Locusta	AKH-II-L	pGlu-Leu-Asn-Phe-Ser-Ala-Gly-Trp-NH$_2$
Periplaneta	MI	pGlu-Val-Asn-Phe-Ser-Pro-Asn-Trp-NH$_2$
Periplaneta	MII	pGlu-Leu-Thr-Phe-Thr-Pro-Asn-Trp-NH$_2$
Manduca	M-AKH	pGlu-Leu-Thr-Ser-Ser-Trp-Gly-NH$_2$
Heliothis	/H-AKH	

1986; O'Shea and Rayne, 1991; Rayne and O'Shea, 1992). The AKH system will therefore be useful for testing novel compounds which interact with the molecular machinery of peptide biosynthesis, action and inactivation.

II. LOCALIZATION AND SITE OF SYNTHESIS

The CC are the major organs of the insect neuroendocrine system that store neurohormones and release them into the circulation. As their name implies, they are intimately associated with the heart vessel and are therefore well placed to distribute hormones rapidly throughout the animal. The neurohormones secreted by the CC are synthesized by neurosecretory cells contained both in the brain and within the CC themselves. In the locust the neurosecretory cells intrinsic to the CC are clustered together in the so-called glandular lobes. The neurosecretory cells of the brain send axons to the CC through two pairs of nerves and these axons arborize in separate lobes called the storage lobes.

The adipokinetic hormones are located in and synthesized by the intrinsic neurosecretory cells of the glandular lobes (Goldsworthy et al., 1972; Hekimi and O'Shea, 1987). These are typical neurosecretory cells appearing to be short axonal-neurones with cell bodies approximately 50 μm in diameter. Several thousand of them are packed closely together in the glandular lobes and each contains large numbers of electron-dense secretory granules (Rademakers and Beenakkers, 1977; Krogh and Norman, 1977), which have been shown by differential centrifugation to contain adipokinetic activity (Stone and Mordue, 1979). Several lines of evidence show that individual neurosecretory cells of the CC make both AKH-I and AKH-II. This can be seen clearly by immunocytochemical labelling of the glandular lobe using antibodies with high specificity for AKH-I and AKH-II (Hekimi and O'Shea, 1989a). All cells of the glandular lobe in the adult locust are labelled by AKH-I-specific antibodies (Fig. 1) and all cells are also labelled with AKH-II-specific antibodies (Hekimi et al., 1991). With respect to AKH-I and AKH-II biosynthesis, therefore, the neurosecretory cells of the CC are a homogeneous population of peptide-producing cells – an extremely useful and unusual feature as most neural tissues are typically highly heterogeneous at the cellular level. The locust CC can therefore be considered to be a dedicated peptide factory, much like the ELH (egg laying hormone), producing bag cells of *Aplysia* (Scheller et al., 1982) or the insulin-producing β-cells. It is as if the glandular lobes contained a clonal cell line in which the hormone-producing genes were specifically and

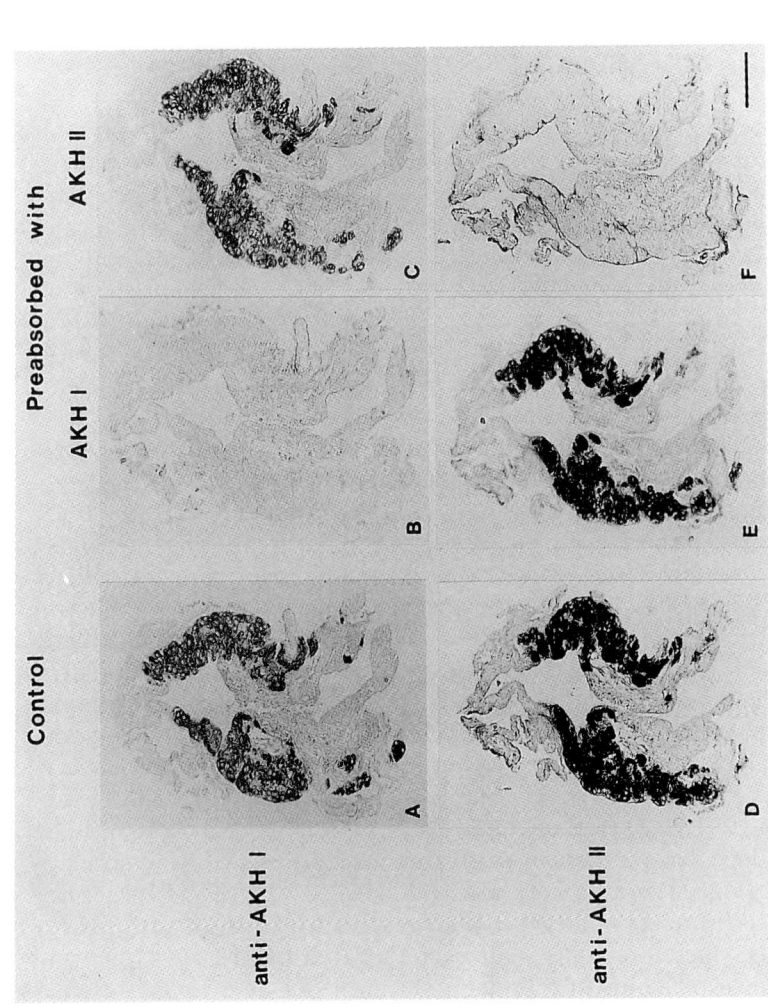

Fig. 1. Immunocytochemical evidence for the co-localization of two prohormones (pro-AKH-I, or the A-chain, and pro-AKH-II, or the B-chain) in the glandular neurosecretory cells of the CC. In (A), staining with an anti-AKH-I specific antibody reveals the neurosecretory cells of the CC. Specificity is indicated in (B) and (C) by the demonstration of a blockade of staining when the antibody is preincubated with AKH-I, but not with AKH-II. In the lower panels, a similar series of sections demonstrates the presence of AKH-II in the neurosecretory cells of the CC. In these panels, staining blockade occurs when the antibody is preabsorbed with AKH-II (F), but not with AKH-I (E).

significantly up-regulated. Unlike an artificially constructed cell line, however, the glandular lobes are a real tissue in which there is appropriate gene expression and regulation. We have exploited this system to understand the molecular and cell biology of AKH biosynthesis and its regulation.

III. RELEASE AND FUNCTIONS

The actions of peptides which share the AKH structural signature are diverse but appear in general to be concerned with increased metabolic activity. For example, some members of the AKH family are involved in stimulating carbohydrate metabolism, others act directly on muscle to favour lipid oxidation and some are cardioacceleratory. In the locust, AKH activity has primarily been studied in the adult, in which it regulates metabolism during long-distance migratory flight. Flight appears to be an important stimulus for the release of adipokinetic hormones in adult locusts and the releasing factor may be octopamine, an amine produced in the brain by some of the neurosecretory cells which project to the storage lobe of the CC (Orchard and Loughton, 1981a,b). This picture of AKH function in the adult unfortunately does not explain the need for two forms of AKH, nor does it help us understand why the AKH peptides are present in all wingless larval stages.

Concerning the physiological function of the peptides co-synthesized with the AKHs (the APRPs), we as yet have no direct evidence that they are hormones. The APRPs, however, are co-released with the AKHs (Hekimi and O'Shea, 1989b) so it seems likely that they are hormones, perhaps with an activity related to flight behaviour in the adult. An as yet untested hypothesis is that the APRPs are diuretic hormones – an idea suggested by the observation that during flight, when the locust is metabolizing lipid, metabolic water is produced. Locusts, like other insects, have impermeable cuticles so this additional water load must be removed by secretion. Indeed, it has been observed that, during long-term flight, locusts do secrete water, a physiological activity presumably provoked by a diuretic hormone. These arguments suggested to us that the APRPs are involved with water balance in some way. This hypothesis has yet to be tested.

IV. INACTIVATION AND METABOLISM

Timely and effective inactivation of neurohormones is necessary to ensure the appropriate temporal organization of the biological function. For neuropeptide-mediated functions, enzymatic degradation is an important means by which such signals are terminated. For neuropeptides in the CNS inactivation is accomplished by degradative cell surface enzymes and examples of this have been demonstrated for the inactivation of mammalian neuropeptides such as the enkephalin (Schwartz et al., 1981), substance P (Matsas et al., 1983) and neurotensin (Chabry et al., 1990), as well as for insect CNS neuropeptides (Isaac, 1987, 1988). The fate of neuropeptides such as AKH which act outside the nervous system, however, is less clear. In some cases peptide hormones are internalized by targets or other tissues where they may be degraded intracellularly (e.g. insulin) (Terris and Steiner, 1975). In other cases, cell-surface-associated peptidases have been implicated in peptide hormone degradation (Turner, 1990).

We have studied directly the metabolic fate of the two locust AKH peptides after they are released into the circulatory system. Since the AKHs possess chemically blocked N- and C-termini, enzymatic inactivation of these peptides would be expected to be initiated by an endopeptidase. In the CNS of the locust AKH-I is apparently inactivated by a membrane-bound peptidase derived from nervous tissue (Isaac, 1988). In the circulation it has been suggested that AKH is degraded by intracellular enzymes following internalization by non-neural tissues (Mordue and Stone, 1978). Our studies, however, suggest that circulating AKH is inactivated not intracellularly but by a cell surface membrane-associated endopeptidase similar to the one detected in the CNS by Isaac (1988).

By following the fate of radioactively labelled AKH-I and AKH-II we have shown that inactivation in the circulation is achieved by an endopeptidase present on the surfaces of the fat body, malpighian tubules and skeletal muscles. The enzyme cleaves the Asn to Phe bond present in both AKHs, producing peptide fragments incapable of effecting the biological actions of the AKHs. The inactive fragments of AKH-I and AKH-II C-terminal to the cleavage are then rapidly degraded *in vivo* from their free N-termini by aminopeptidases, whereas the N-terminal fragment from both hormones is relatively more stable. We were unable to detect endopeptidase activity in the haemolymph, but the haemolymph does contain aminopeptidase activity which is capable of degrading the primary C-terminal fragment following the action of the inactivating membrane-associated endopeptidase. The molecular processes involved in the inactivation of AKH-I and AKH-II in the haemolymph are summarized in Fig. 2.

By comparing the endopeptidase activity found peripherally with the neurally derived activity described by Isaac, we consider that a very

12. Locust Adipokinetic Hormones: A Review 193

AKH I metabolism

pGlu-*Leu-Asn-*Phe-Thr-Pro-Asn-*Trp-Gly-Thr-NH$_2$

↓ *Inactivation*

pGlu-*Leu-Asn (Leu I.1)   *Phe-Thr-Pro-Asn-*Trp-Gly-Thr-NH$_2$ (Trp I.1)

Metabolism of inactive fragments

Thr-Pro-Asn-*Trp-Gly-Thr-NH$_2$ (Trp I.X)

(Pro-Asn-*Trp-Gly-Thr-NH$_2$) ?

Asn-*Trp-Gly-Thr-NH$_2$ (Trp I.2)

*Trp-Gly-Thr-NH$_2$ (Trp I.2)

AKH II metabolism

pGlu-*Leu-Asn-*Phe-Ser-Thr-Gly-*Trp-NH$_2$

↓ *Inactivation*

pGlu-*Leu-Asn (Leu II.1)   *Phe-Ser-Thr-Gly-*Trp-NH$_2$ (Trp II.1)

Metabolism of inactive fragments

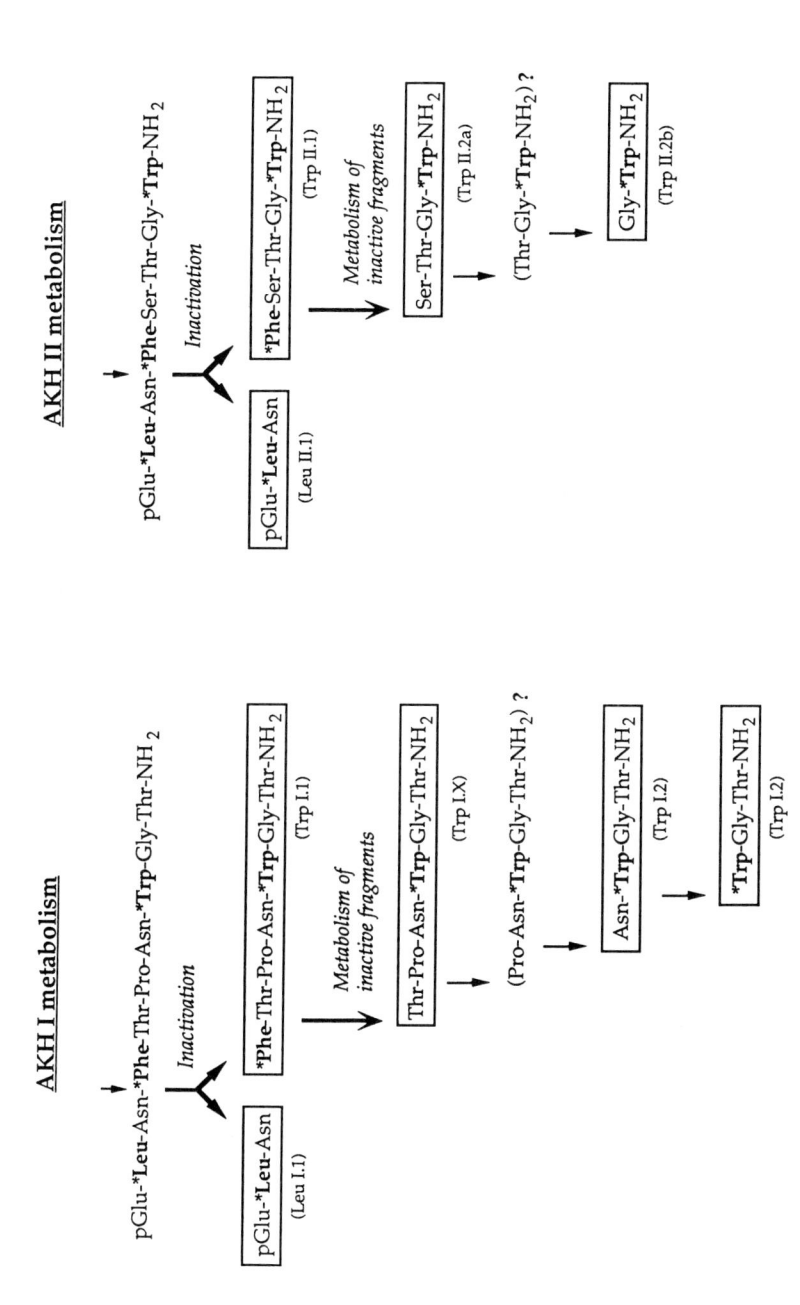

Ser-Thr-Gly-*Trp-NH$_2$ (Trp II.2a)

(Thr-Gly-*Trp-NH$_2$) ?

Gly-*Trp-NH$_2$ (Trp II.2b)

Fig. 2. Model for the inactivation and subsequent degradation of circulating AKHs. Sequences of AKH-I, AKH-II and proposed structures of their respective metabolites are depicted. Note that the initial endoproteolysis produces two fragments of each of the respective hormones; none of these fragments exhibit adipokinetic activity. The endoproteolytic step is therefore the physiologically relevant inactivating step. For further details, see Rayne and O'Shea (1992).

similar, perhaps identical, enzyme is involved in terminating neuropeptide action both in the CNS and in the circulation. Moreover, the endopeptidase activity we identify (Rayne and O'Shea, 1992) resembles mammalian endopeptidase 24.11 and may be effective against a variety of Phe-containing peptides which are N- and C-terminally protected. Our results support the view that many invertebrate and vertebrate neural hormones are inactivated by a single class of cell surface endopeptidase. It is reasonable to assume, therefore, that if a specific enzyme inhibitor were designed, a number of biological responses mediated by neuropeptides would be disrupted. Such enzyme inhibitors may therefore be insecticidal.

V. PROHORMONE AND PRECURSOR BIOSYNTHESIS

We now have a fairly complete picture of how the locust adipokinetic hormones are made (Fig. 3). The model is complicated somewhat by the fact that the immediate precursors of AKH-I and AKH-II are not linear prohormones (pro-AKH-I and pro-AKH-II) but dimeric constructs (P1, P2 and P3). This surprising (to us) and unprecedented feature of peptide hormone biosynthesis was discovered by characterizing the hormone precursors by direct protein chemistry as well as molecular cloning.

The unexpected discovery of precursors formed from two prohormone subunits led us to question the precise mode of biosynthesis of the precursor dimers. There are two ways to synthesize a dimer protein like the Ps: (1) from proteins containing more than one copy of the monomer sequence; or (2) from the oxidation of independently translated monomer prohormones. We could not distinguish between these alternatives by performing pulse-chase experiments because we could not detect potential precursor proteins large enough to contain more than one monomer chain and we have not convincingly demonstrated the existence of independent monomer chains prior to dimer formation. Thus on the basis of our cell biological and biochemical experiments, we could not decide between the two formal possibilities for the synthesis of the dimer precursors. This question was addressed using cell-free *in vitro* translation of CC mRNA and DNA sequencing of positive cDNA clones (Schulz-Aellen *et al.*, 1989).

By translating mRNA derived from the CC *in vitro* we were able to determine the size of the complete translational protein from AKH-I encoding mRNA. The size of this protein (6.8 kDa) indicated that it could *not* accommodate more than one copy of the subunit of P1. The identify of the *in vitro* translated protein was established by showing that

12. Locust Adipokinetic Hormones: A Review   195

Translation of Two mRNAs

mRNA I (~500 nt)   mRNA II (~500 nt)

Gly-Lys-Arg   Gly-Arg-Arg

Signal   AKH I   α-Chain   Signal   AKH II   β-Chain

prepro-AKH I (63 aa)   prepro-AKH II (61 aa)

Oxidation of Prohormones into Three Dimer Precursors (Ps)

pro-AKH I — Cys   pro-AKH II — Cys

A-chain (41 aa)   B-chain (39 aa)

P1 (A+A)   P2 (A+B)   P3 (B+B)

Processing to Monomer (AKHs) and Dimer (APRPs) Peptides

AKH I   APRP 1 (α + α)   AKH II   APRP 3 (β + β)

AKH I   AKH II

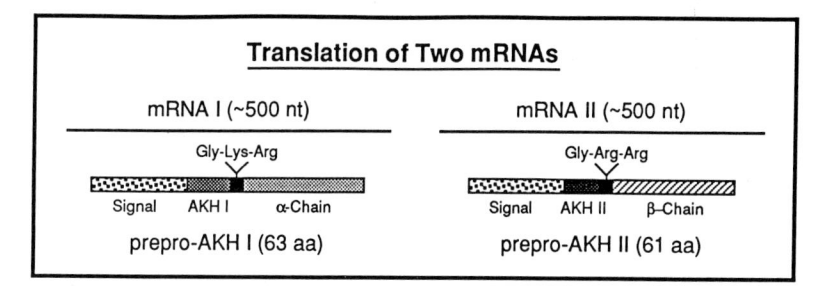

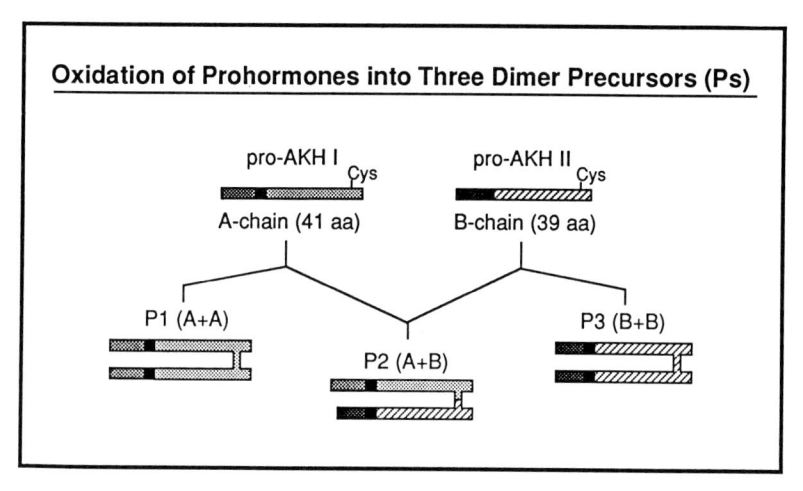

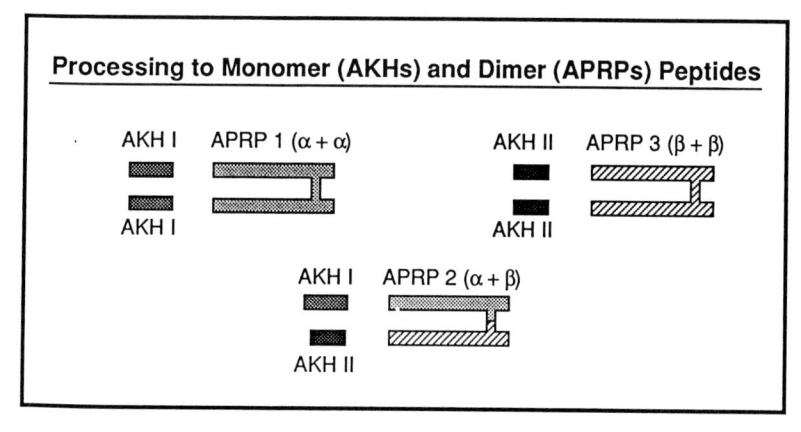

AKH I   APRP 2 (α + β)

AKH II

Fig. 3. A model of AKH-I, AKH-II and APRP biosynthesis from three dimer precursors P1, P2 and P3. nt, nucleotides; aa, amino acids. For further details, see text and O'Shea *et al.* (1990).

its translation is prevented by an antisense oligonucleotide which would specifically complement an mRNA encoding the subunit of P1. This experiment (see Fig. 3 of Schulz-Aellen *et al.* (1989)) indicated strongly that P1 is therefore produced by the oxidation of independent copies of the P1 subunit rather than by post-translational modification of a larger protein containing more than one subunit copy. Similar experiments have indicated that the heterodimer P2 is also produced by the oxidation of independently translated proteins (Hekimi *et al.*, 1991).

Confirmation that the dimeric precursors are produced from independently translated monomers is provided by sequencing of cDNA clones derived from a CC cDNA library. The sequence of the cDNA clone which translates exactly into the subunit of P1 is given in Fig. 4. The inferred protein has a molecular weight exactly corresponding to the molecular weight of the protein produced during *in vitro* translation.

During cell-free *in vitro* translation, normal post-translational or co-translational modifications do not occur. Cell-free translation therefore gives the product of genetic information alone; all other subsequent biosynthetic events must therefore depend upon the integrity of the cells and must therefore be studied within the CC. The first event which we can infer directly from a cellular biochemical experiment is that the association between the two precursor subunits (pro-AKH-I and pro-AKH-II) is an entirely random process in the endoplasmic reticulum. Random coupling of subunits would produce predictable ratios of P1, P2 and P3 dependent only on the relative amounts of the two prohormones. Actual amounts and rates of synthesis of the dimer precursors can be measured directly in pulse-labelling experiments and so the random-coupling hypothesis can be tested. Theoretical calculations and experimental measurements indicate that the predicted amounts of P1, P2 and P3 are very close to the observed amounts and the precursor ratios accurately predict product ratios (Fig. 5). Different ratios of the AKHs and APRPs can therefore be precisely regulated in the CC by differential regulation of translation of pro-AKH-I and pro-AKH-II (Hekimi *et al.*, 1991).

To summarize, the dimer precursors of AKH-I and AKH-II are formed from two prohormones called pro-AKH-I (or A-chain) and pro-AKH-II (or B-chain) which are encoded by two messenger RNAs called respectively mRNA-I and mRNA-II (Hekimi *et al.*, 1989; O'Shea *et al.*, 1990; Schulz-Aellen *et al.*, 1989). The sequence of mRNA-I translates into a prepro-AKH-I inferred protein consisting of a 22 amino acid signal peptide, followed by the 41-residue A-chain. The pro-AKH-I is AKH-I, a Gly-Lys-Arg processing site and a 28-residue peptide called the α-chain. This was followed by a stop codon in the cDNA. The protein inferred

12. Locust Adipokinetic Hormones: A Review

cDNA Clone and Deduced Pre-Pro-AKH I

G G C A A C A G G	ATG GTG CAG CGG TGC CTG GTG GTA GCC TTG
	Met Val Gln Arg Cys Leu Val Val Ala Leu

←———1

CTG GTG GTG GTG GTG GCG GCT GCC CTA TGC TCG GCG	CAG CTC AAC
Leu Val Val Val Val Ala Ala Ala Leu Cys Ser Ala	Gln Leu Asn

———— Signal Peptide ————→ 22 1 pGlu ←

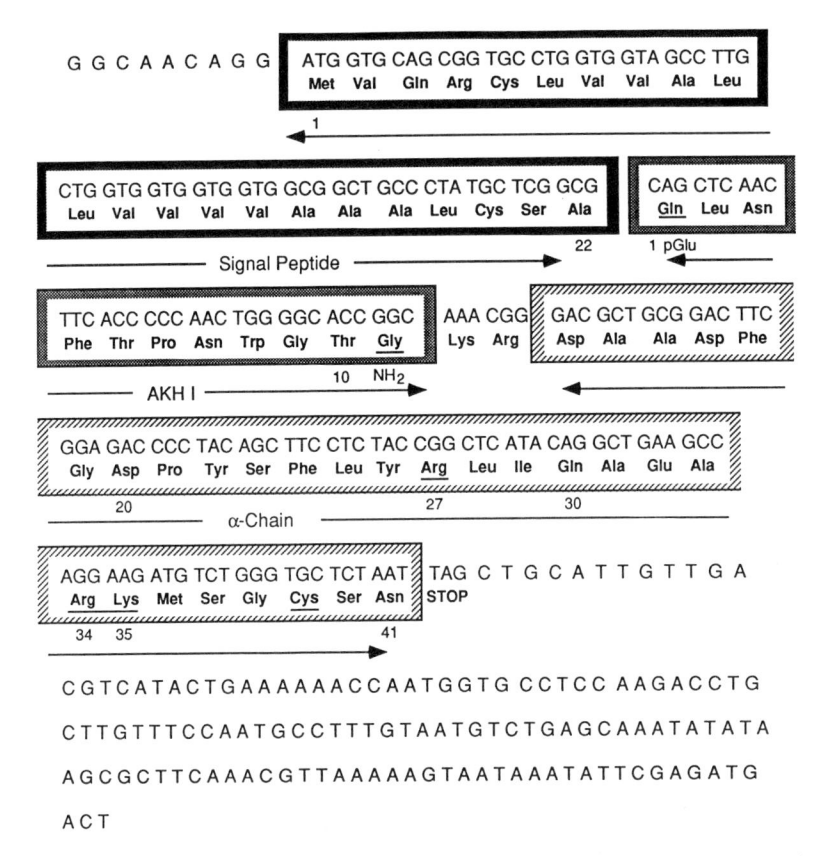

TTC ACC CCC AAC TGG GGC ACC GGC	AAA CGG	GAC GCT GCG GAC TTC
Phe Thr Pro Asn Trp Gly Thr Gly	Lys Arg	Asp Ala Ala Asp Phe

———— AKH I ———— 10 NH2 →   ←————

GGA GAC CCC TAC AGC TTC CTC TAC CGG CTC ATA CAG GCT GAA GCC
Gly Asp Pro Tyr Ser Phe Leu Tyr Arg Leu Ile Gln Ala Glu Ala

———— 20 ———— 27 30
———— α-Chain ————

AGG AAG ATG TCT GGG TGC TCT AAT	TAG C T G C A T T G T T G A
Arg Lys Met Ser Gly Cys Ser Asn	STOP

34 35   41

CGTCATACTGAAAAAACCAATGGTG CCTCC AAGACCTG

CTTGTTTCCAATGCCTTTGTAATGTCTGAGCAAATATATA

AGCGCTTCAAACGTTAAAAAGTAATAAATATTCGAGATG

ACT

Fig. 4. Nucleotide sequence of mRNA-I encoding signal peptide, AKH-I, a processing site and the α-chain of APRP-1 and APRP-2. This inferred protein sequence derived from a cloned cDNA gives the structure of prepro-AKH-I. The inferred 41 amino acid sequence starting from Gln is exactly that expected from protein sequencing of the P1 precursor of AKH-I. For further explanation see Schulz-Aellen et al. (1989).

from the mRNA-II cDNA clone indicated that a prepro-AKH-II protein also consists of a 22-residue signal peptide which is followed by pro-AKH-II. Pro-AKH-II consists of one copy of AKH-II, a processing site and a 28-residue peptide called the β-chain. The 28-residue β-peptide shares about 70% homology with the α-chain of pro-AKH-I. Amino acid sequences of purified pro-AKH-I and pro-AKH-II were determined by protein sequencing after pulse-chase experiments identified the precursors

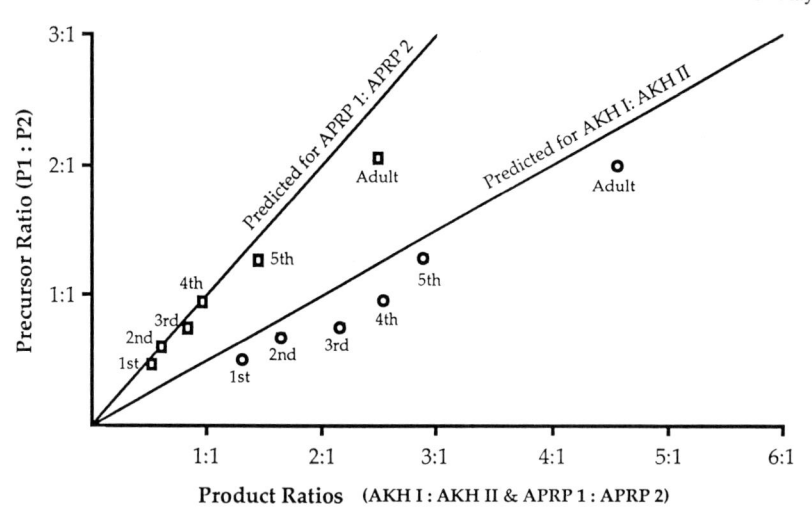

Fig. 5. Relationship between relative rates of precursor synthesis (P1 to P2) and the relative amounts of products (AKH-I to AKH-II and APRP-1 to APRP-2) throughout post-embryonic development. The slopes of these relationships, as predicted from our molecular model and our assumptions about it, are shown as solid lines. The experimentally derived data points are shown as squares (for the APRP ratios) and circles (for the AKH ratios). There is good agreement between the predicted and observed data. Explanation for this is provided in Hekimi *et al.* (1991).

of the AKHs. Inferred (from cloned cDNA) and determined (from purified precursors) sequences of the prohormones correspond exactly.

VI. PROCESSING *IN VIVO* AND *IN VITRO*

The presence of signal or leader peptide sequences upstream from the peptide hormone sequences is common in secreted proteins (Sossin *et al.*, 1989). It guides the translated protein into the lumen of the endoplasmic reticulum where post-translational processing occurs (Walter and Blobel, 1981). An early post-translational processing event is thought to be the removal of the signal peptide, probably co-translationally, transforming the preprohormone into the prohormone. Cyclization of the translated N-terminal glutamine into pyroglutamate appears to be a very early step in the AKH biosynthetic pathway and may also be co-translational. The formation of the dimeric precursors (the Ps) is by the oxidation of the single cysteine residue in the prohormones and also appears to be an

12. Locust Adipokinetic Hormones: A Review

early and rapid event because vanishingly little of the monomer prohormones can be detected in the CC. Complete processing of newly formed Ps takes about 7 hours. Recent experiments have shown that conversion of Ps to AKHs and APRPs can be blocked using the ionophore monensin (Rayne and O'Shea, unpublished). Based on the known mode of monensin action, this result indicates that proteolytic processing of Ps takes place in a post-*trans*-Golgi compartment, presumably the secretory granules. Thus the difference in the kinetics of the two stages of AKH biosynthesis (synthesis of precursor dimers and their subsequent processing) and results of the experiments with monensin suggest that the early post-translational events occur in the endoplasmic reticulum, and secondary processing events occur in secretory granules.

The formation of dimer precursors and the subsequent processing events have been studied directly using *in vitro* culture methods and the pulse-chase experimental paradigms. Moreover the structural chemistry, including protein sequencing and sizing, were performed on precursor and products (APRPs) isolated and purified from the CC following their identification by pulse-chase experiments. The molecular features of AKH prohormone and dimer precursor biosynthesis which we discovered could not have been inferred from the sequence of cDNA or genomic clones alone. Nucleotide sequences give *deduced proteins* and not the structures of the precursors explicitly. Special features of the CC allowed us to *determine* experimentally rather than deduce the precursor structures. Evidence for their dimeric forms was obtained by performing size exclusion chromatography and protein sequencing on purified precursor in reduced and native form. The dimeric structures could not have been inferred from the sequence of the cDNA alone, neither could the subsequent processing events (see below).

Pro-AKH-I (A-chain) and pro-AKH-II (B-chain) contain the AKH sequences separated by classical processing sites from the 28-residue α and β peptides. The use of this site in processing could perhaps have been correctly inferred from the nucleotide sequences of cDNA clones alone. However, the α and β peptide sequences contain additional potential processing sites (one dibasic and a single arginine, see Figs 4 and 6) and their presence might have suggested that additional peptides are produced by enzymatic processing within the α and β sequences. We know from protein sequencing, however, that these 28-residue sequences are not processed but remain intact and are present as the three dimeric constructs of APRP-1, -2 and -3.

The conversion of the P1, P2 and P3 dimers into the AKHs and APRPs requires the action of precursor convertases. Many questions remain concerning the number and types of enzymes involved in converting precursors or prohormones into biologically active peptides. For example, how many enzymes are required? How similar are the enzymes in different organisms? How is the synthesis of the convertases regulated? Do the processing enzymes involved in neuronal peptide biosynthesis resemble those which have been isolated from yeast (Fuller et al., 1988) and which appear to have substrate specificities like those which must exist in neurones?

As a first step towards isolating precursor convertases in the CC we are utilizing an *in vitro* system to reconstitute precursor processing of the AKH-I precursor, P1 (Fig. 6). In order to perform these experiments we have synthesized the complete 41 amino acid AKH-I prohormone. This small protein contains one copy of AKH-I, a Gly-Lys-Arg processing site and the 28-residue C-terminal peptide, the α-chain. This synthetic prohormone can be easily oxidized to produce the homodimer P1, the direct precursor of AKH-I and APRP-1. When incubated with homogenates of the CC which contain the precursor convertases, synthetic P1 is processed and yields fully processed APRP-1 and intermediates of AKH-I processing. The first AKH-I processing intermediate produced in this *in vitro* processing experiment is the C-terminally extended peptide AKH-I-Gly-Lys-Arg. This indicates that the first enzymatic action is an endoproteolytic cleavage of P1 between Arg and the N-terminal residue of APRP-1, Asp. The C-terminally extended AKH-I is then truncated by two stepwise carboxypeptidase activities, leading to the production of AKH-Gly-Lys and AKH-Gly. The *in vitro* conditions under which these events occur do not produce the fully processed amidated AKH-I. Experiments are currently underway to identify those *in vitro* conditions under which the amidation of AKH-Gly will occur. In other systems (Bradbury and Smyth, 1991; Kato et al., 1990; Perkins et al., 1990) it is thought that two enzymatic steps are involved in converting the C-terminal Gly into the C-terminal amide of the peptide and that amidation requires molecular oxygen, copper and ascorbic acid (Eipper et al., 1983).

By reconstituting *in vitro* the conditions for defining the properties of the prohormone or precursor convertases we hope to be able to develop assays which will allow us to purify the enzymatic activities to homogeneity. This we see as the first step in an analysis of the structural and functional characteristics of the enzymes involved in the biosynthesis of the AKH peptides. Again the special features of the CC, its cellular homogeneity in particular, make us confident that the enzymatic activities which we extract from this tissue are the *in vivo* enzymatic activities. Uncertainty on this point is a serious shortcoming of approaches to the purification of convertases in which the enzymatic activities are extracted from heterogeneous nervous tissue. Our belief that the enzymatic

12. Locust Adipokinetic Hormones: A Review

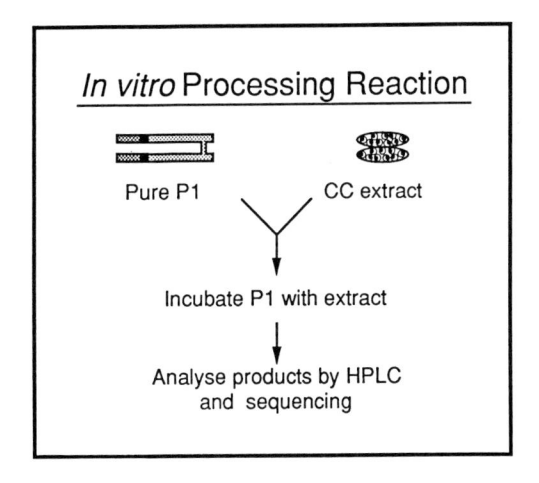

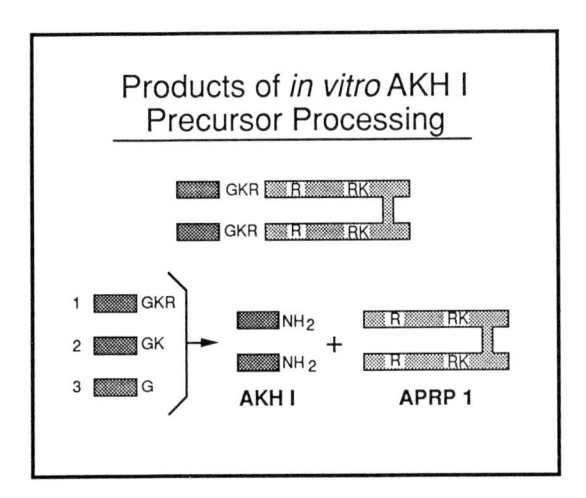

Fig. 6. Reconstitution of AKH precursor processing *in vitro* and analysis of AKH intermediates. *In vitro* processing reactions are performed as described in the upper panel. The lower panel schematically shows the peptides produced by sequential actions of prohormone convertases present in the CC extracts. Note that potential processing sites (R, RK) are not used. G, glycine; R, arginine; K, lysine.

activities which we have identified in the CC extracts are specific and likely to be the same as those used *in vivo* is strengthened by the fact that we do not see inappropriate processing *in vitro*. For example, the potential but unused processing sites in the α-chain of P1, the single Arg and the dibasic Arg-Lys, which are not digested *in vivo*, are not utilized in our reconstituted *in vitro* processing experiments (Figs 3 and 6).

VII. CONCLUSIONS AND FUTURE DIRECTIONS

The adipokinetic hormones and the tissue from which they are derived have provided an almost ideal model for analysing the cellular and molecular mechanisms of peptidergic systems. Our efforts are now focused on the regulation of AKH biosynthesis and a number of questions remain to be answered by future experiments. For example, while we know that strong differential control over AKH-I and AKH-II biosynthesis is exerted during development, we do not yet know the molecular mechanism for this regulation. At early stages of post-embryonic development pro-AKH-I and pro-AKH-II are synthesized in approximately equal amounts, but in the adult the ratio between pro-AKH-I and pro-AKH-II reaches approximately 5 : 1. This differential regulation of prohormone biosynthesis may be at the level of transcription, translation or through subsequent post-translational events. A complete answer to questions related to differential regulation will await the analysis of the structures of the AKH-I and AKH-II genes. Preliminary experiments, however, suggest strongly that some of the regulatory mechanisms are translational, suggesting that the differential expression of RNA-binding proteins may explain some of the developmental shifts in prohormone ratios. Analysis of possible transcriptional control awaits the completion of genomic cloning.

The CC, primarily because of its homogeneous cellular composition, has allowed us to perform direct experiments on the mechanisms of prohormone and precursor processing. We hope that the *in vitro* system we have developed in which most of the post-translational events can be reconstituted will eventually allow us to purify the enzymes involved in producing the AKHs and the APRPs. While this work is of fundamental importance in completing our understanding of peptide biosynthesis, it may also have some practical application. For example, it may be possible to develop specific inhibitors of these important biosynthetic enzymes and such inhibitors may be insecticidal. The development of such inhibitors will require the establishment of specific *in vitro* assays for the enzymes involved and work towards this goal is currently underway.

Much work remains to be done on the biological actions of the AKHs and the newly discovered APRPs. Although both AKH-I and AKH-II are known to be involved in lipid metabolism, it is very likely that other functions have yet to be discovered. Moreover, there are likely to be distinct functions for these two related but different hormones. No function has as yet been attributed to the APRPs. It will also be important in the future to uncover the functional significance of the dramatic shifts in peptide ratios which occur throughout postembryonic development. These shifts suggest that there is an enhanced role for AKH-II early in development but the functions of AKH-II in the pre-adult larval stages have not yet been investigated thoroughly.

The need for further experimental work on the function and pharmacology of the AKH peptides will inevitably lead to investigations of the AKH receptor. The fat body tissue in the locust is likely to be an enriched source of AKH receptor and a variety of experimental strategies could be developed towards the ultimate structural characterization of this important receptor. Such work may also have practical application since peptide mimetics, agonists and antagonists of the receptor may be potential leads towards new types of insecticides which interact with peptidergic systems.

Finally, many of the problems associated with analysing neuropeptide biosynthesis in the CC could be alleviated by the development of an immortal cell line from the neuroendocrine cells of the CC. Work is currently underway towards this goal and we expect that, if it is successful, the problems associated with analysing rare proteins of the CC which may be involved in the regulation of biosynthesis would be greatly simplified.

In conclusion, we feel that the CC has provided a very fertile ground for the study of a model insect peptidergic system. Work on this system has made important contributions to our basic knowledge of peptidergic systems and has provided fundamentally new insights into how neuropeptides are synthesized. We also look forward to exploring the possibilities that neuropeptidergic systems offer a fertile area of development of novel insecticidal compounds. Such compounds may, for example, interfere with neuropeptide biosynthesis or its regulation, neuropeptide action or neuropeptide inactivation.

REFERENCES

Beenakers, A. M. Th. (1969). *Gen. Comp. Endocrinol.* **13**, Abstract 12.
Bradbury, A. F., and Smyth, D. G. (1991). *TIBS* **16**, 112–115.
Chabry, J., Checler, F., Vincent, J.-P. and Mazella, J. (1990). *J. Neurosci.* **10** (12), 3916–3921.
Eipper, B. A., Mains, R. E. and Glembotski, C. C. (1983). *Proc. Natl. Acad. Sci. USA* **50**, 5144–5148.
Fuller, R. S., Sterne, R. E. and Thorner, J. (1988). *Annu. Rev. Physiol.* **58**, 345–362.
Goldsworthy, G. J., Johnson, R. A. and Mordue, W. (1972). *J. Comp. Physiol.* **79**, 85–96.
Hekimi, S. and O'Shea, M. (1987). *J. Neurosci.* **7**, 2773–2784.
Hekimi, S. and O'Shea, M. (1989a). *Insect Biochem.* **19**, 79–83.
Hekimi, S. and O'Shea, M. (1989b). *J. Neurosci.* **9** (3), 996–1003.

Hekimi, S., Burkhart, W., Moyer, M., Fowler, E. and O'Shea, M. (1989). *Neuron* **2**, 1363–1368.
Hekimi, A., Fischer-Lougheed, J. and O'Shea, M. (1991). *J. Neurosci.* **11**, 3246–3256.
Isaac, R. E. (1987). *Biochem. J.* **245**, 365–370.
Isaac, R. E. (1988). *Biochem. J.* **255**, 843–847.
Kato, I., Yonekura, H., Tajima, M., Yanagi, M., Yamamoto, H. and Okamoto, H. (1990). *Biochem. Biomed. Res. Commun.* **172**, 197–203.
Krogh, I. M. and Normann, T. C. (1977). *Acta Zoo.* **58**, 69–78.
Matsas, R., Fulcher, I. S., Kenny, A. J. and Turner, A. J. (1983). *Proc. Natl. Acad. Sci. USA* **80**, 3111-3115.
Mayer, R. J. and Candy, D. J. (1969). *J. Insect Physiol.* **15**, 611–620.
Mordue, W. and Stone, J. V. (1978). *In* "Comparative Endocrinology" (P. Gaillard and H. H. Boer, eds), pp. 487–490. Elsevier/North Holland Biomedical Press, Amsterdam.
Noyes, B. E. and Schaffer, M. H. (1990). *J. Biol. Chem.* **265** (1), 483–489.
Orchard, I. (1987). *J. Insect Physiol.* **33**, 451–463.
Orchard, I. and Loughton, B. G. (1981a). *Comp. Biochem. Physiol.* **68A**, 25–30.
Orchard, I. and Loughton, B. G. (1981b). *J. Neurobiol.* **12**, 143–153.
O'Shea, M. (1985). *In* "New Approaches in Insect Control" (H. von Keyserlingk, ed.) pp. 133–151. Springer-Verlag, Berlin.
O'Shea, M. (1986). *In* "Proceedings of USDA International Conference on Insect Neurochemistry and Neurophysiology", (A. B. Borkovec and D. B. Gelman, eds) pp. 3–27. Humana Press.
O'Shea, M. and Rayne, R. C. (1992). "Proc. Neurotox. 91 Conf." (in press).
O'Shea, M., Witten, J. and Schaffer, M. (1984). *J. Neurosci.* **4**, 521–529.
O'Shea, M. Hekimi, S. and Schulz-Aellen, M-F. (1990). *In* "Molecular Insect Science" (H. H. Hagedorn, J. G. Hildebrand, M. G. Kidwell and J. H. Law, eds), pp. 189–197. Plenum Press, New York.
Perkins, S. N., Husten, E. J. and Eipper, B. A. (1990). *Biochem. Biophys. Res. Commun.* **171**, 926–932.
Rademakers, L. H. P. M. and Beenakkers, A. M. Th. (1977). *Cell Tissue Res.* **180**, 155–171.
Rayne, R. C. and O'Shea, M. (1991). *Insect Biochem. Molec. Biol.*, **22**, 25–34.
Scheller, R. H., Jackson, J. F., McAllister, L. B., Schwartz, J. H., Kandel, E. R. and Axel, R. (1982). *Cell* **28**, 707–719.
Schulz-Aellen, M-F., Roulet, E., Fischer-Lougheed, J. and O'Shea, M. (1989). *Neuron* **2**, 1369–1373.
Schwartz, J-C., Malfroy, B. and De La Baume, P. (1981). *Life Sci.* **34**, 1715–1740.
Siegert, K., Morgan, P. and Mordue, W. (1985). *Biol. Chem. Hoppe-Seyler* **366**, 723–727.
Sossin, W.S., Fischer, J. M. and Scheller, R. H. (1989). *Neuron* **2**, 1407–1417.
Stone, J. V. and Mordue, W. (1979). *Gen. Comp. Endocrinol.* **39**, 543–547.
Stone, J. V., Mordue, W., Batley, K. E. and Morris, H. R. (1976). *Nature, Lond.* **263**, 207–211.
Terris, S. and Steiner, D. F. (1975). *J. Biol. Chem.* **250**, 8389–8398.
Turner, A. J. (1990). *In* "The Biology and Medicine of Signal Transduction" (Y. Nishizuka, M. Endo and C. Tanaka, eds), pp. 467–471. Raven Press, New York.
Walter, P. and Blobel, G. (1981). *J. Cell Biol.* **91**, 551–556.
Witten, J. L., Schaffer, M. H., O'Shea, M., Cook, J. C., Hemling, M. E. and Rinehart, K. L. (1984). *Biochem. Biophys. Res. Commun.* **124**, 350–358.

13

The Structure and Functional Activity of Neuropeptides

G. GOLDSWORTHY, G. COAST, C. WHEELER, O. CUSINATO,
I. KAY AND B. KHAMBAY

I. Introduction	205
II. Neuropeptides that Affect the Metabolism of the Fat Body	206
A. The Adipokinetic Hormone (AKH) Family	206
B. Achetakinins	209
III. Attempts to Predict and Determine Secondary Structure	210
A. AKH Family Peptides	212
B. Achetakinins	216
IV. N-terminal Modification of Peptides	218
A. AKH Family Peptides	218
B. Achetakinins	220
V. Insect Diuretic Peptides	221
VI. Conclusions	222
References	223

I. INTRODUCTION

In this chapter we will attempt to relate the structures of some insect neuropeptides with their biological activities, on the assumption that one of the important features of receptor recognition is the shape of the ligand. We have therefore sought to determine or predict the secondary structures of a number of insect neuropeptides from locusts and crickets for which the primary structures are known, and to recognize the features in them important for their experimentally observed activity.

One way to conceptualize hormone–receptor interactions is to regard signal molecules as comprising two major components: a site which specifies the "address", endowing high binding affinity and ligand specificity to bring about initial binding; and a second site which conveys the "message" content of the molecule to the receptor (Hruby, 1984). Although ligand–receptor interaction is undoubtedly more complex than this approach implies, the message/address concept may be useful in helping to identify key areas or features for interactions between signal molecules and their receptors.

II. NEUROPEPTIDES THAT AFFECT THE METABOLISM OF THE FAT BODY

A. The Adipokinetic Hormone (AKH) Family

The adipokinetic hormones of locusts belong to a family of arthropod neuropeptides, the AKH family (Gäde, 1990a). In the three species of locusts which have been studied, there are at least two adipokinetic peptides present in the corpora cardiaca; the decapeptide AKH-I is always present but, depending on species, there is also one of two octapeptides (Table I). In locusts, these AKH peptides mobilize diacylglycerols from the fat body during flight and stimulate oxidation of lipids by the flight muscles (Goldsworthy and Wheeler, 1989). Other effects on the metabolism of the fat body are described below. The biological actions of the AKH peptides depend presumably on interactions with specific receptors in the target tissues, but there have until recently been no direct studies of these receptors. Progress in this area of research has been severely limited by two difficulties: the lack of biologically active radiolabelled ligands of sufficiently high specific activity to enable receptor-binding analyses, and a complete lack of antagonists to any of the AKH family peptides.

Some information about receptors can be inferred from biological assay data: for most actions in locusts, AKH-I and, to a lesser extent, AKH-II, are both effective and potent agonists, but AKH-II stimulates cAMP accumulation in the fat body to a greater extent than AKH-I (Goldsworthy *et al.*, 1986a); AKH-II releases carbohydrate from the fat body, whereas AKH-I does not (Loughton and Orchard, 1981); and at saturating doses of *Locusta* AKH-II the lipid-mobilizing response in *Locusta* is truncated (Goldsworthy *et al.*, 1986a,b). According to the message/address concept, the truncated hyperlipaemic response to *Locusta* AKH-II (relative to AKH-I) could be due to this octapeptide carrying a modified or deficient message (see below) compared with the decapeptide (producing less than the maximum response), whereas reduced potency (see the activities of other octapeptides given in Table I)

13. The Structure and Functional Activity of Neuropeptides

TABLE I. Relative hyperlipaemic potencies of some members of the AKH family in locusts (mainly assayed in *Locusta migratoria* but, where indicated, also in *Schistocerca gregaria*).

Neuropeptide	Sequence and turn prediction	ED_{50} (pmol)	ED_{max} (pmol)	% Maximum response	CD evidence of β-turn
Locust AKH-I[1]	<QLNFTPNWGT-NH$_2$ (in *Schistocerca*)	1 6	3 20	100 100	Yes
[dehydro-Pro⁶]-AKH-I[2]					
Carausius HTF-II[2]	<QLTFTPNWGT-NH$_2$	1	3	100	Yes
Manduca AKH[3]	<QLTFTSSWG-NH$_2$	2	8	100	Yes
Heliothis AKH[4]		10	40	45	No
Crustacean RPCH[5]	<QLNFSPGW-NH$_2$ (in *Schistocerca*)	4.6 15	20 40	100 100	Yes
Locusta AKH-II[6]	<QLNFSAGW-NH$_2$ (in *Schistocerca*)	2 20	3 30	60 55	No
Schistocerca AKH-II[6]	<QLNFSTGW-NH$_2$ (in *Schistocerca*)	2 12	5 25	95-100 60	No
Periplaneta M-I[7]	<QVNFSPNW-NH$_2$	5	30	95-100	Yes
Periplaneta M-II[7]	<QLTFTPNW-NH$_2$	5	20	95-100	Yes
Nauphoeta-/Blaberus HTH[8]	<QVNFSPGWGT-NH$_2$ (in *Schistocerca*)	4-5 10	20 30	100 80-90	Yes
Romalea AKH-I[9]	<QVNFTPNWGT-NH$_2$	2	10	95-100	Yes
Romalea AKH-II[9]	<QVNFSTGW-NH$_2$	1	10	100	No
Gryllus AKH[10]					
Acheta AKH[11]					

[1]Stone et al. (1978); [2]Gäde and Rinehart (1987a); [3]Ziegler et al. (1986); [4]Jaffe et al. (1986); [5]Fernlund and Josefsson (1972); [6]Gäde et al. (1984, 1986), Siegert et al. (1985); [7]O'Shea et al. (1984); [8]Hayes and Keeley (1986), Gäde and Rinehart (1986); [9]Gäde et al. (1988); [10]Gäde and Rinehart (1987a,b); [11]Woodring et al. (1990), Cusinato et al. (1991).

Responses are calculated as a percentage of the maximum response to a crude extract of *Locusta* corpora cardiaca (0.02 pair equivalents/locust) tested on the same batch of locusts on the same day (data from Goldsworthy et al. (1986a,b), Wheeler et al. (1988) and unpublished observations). The final column shows if there is CD evidence for a β-turn, and underlining or overscoring indicates any predicted position.

would be interpreted as a deficiency in the address part of the molecule (Hruby, 1984). We do not favour this interpretation, because at saturating doses of *Locusta* AKH-II a further hyperlipaemic response can be elicited by injection of locust AKH-I (Goldsworthy *et al.*, 1986a). This suggests that AKH-I and AKH-II can act at different receptors in the fat body, although their high degree of cross-reactivity in all assays but that of carbohydrate release from the fat body indicates that we should not exclude the possibility that they may also stimulate common receptors.

The ready availability of synthetic preparations of the naturally occurring adipokinetic peptides from a range of insects has prompted an examination of their structure–activity relationships (Goldsworthy *et al.*, 1986a,b; Goldsworthy and Wheeler, 1986; Gäde, 1986, 1990a,b; Wheeler *et al.*, 1988; Hayes *et al.*, 1990; Hayes and Keeley, 1990). A high degree of conservatism is displayed by this family, indicating the importance of specific residues within the peptide. All have an N-terminus blocked by a pyroglutamyl residue (pGlu), an amidated C-terminus, Phe^4, and Trp^8. There are also conservative substitutions of Leu^2 or Val^2, Asn^3 or Thr^3, and Thr^5 or Ser^5. They exhibit a rather narrow range of potencies when assayed for hyperlipaemia in *Locusta* and *Schistocerca* (see Table I) or hypertrehalosaemia in *Periplaneta* (Gäde, 1986, 1990b), but a much wider range when assayed for hypertrehalosaemia in another cockroach, *Blaberus discoidalis* (Hayes and Keeley, 1990). Why the receptors in the fat body of these relatively primitive insects should appear to differ so much in their ligand preferences is unknown.

Taken at face value, the data summarized in Table I suggest that for the hyperlipaemic response in locusts, residues at both the N- and C-termini are important in relation to the "address" portion of the locust AKH-I molecule: essentially conservative substitutions at positions 2 and 3 (*Carausius* HTF-II, which is [Thr^3]-AKH-I, and *Romalea* AKH-I, which is [Val^2]-AKH-I, or the absence of the C-terminal residues Gly-Thr-NH$_2$ from [Thr^3]-AKH-I in *Periplaneta* M-II, all reduce potency (and could therefore be viewed as being concerned with the "address") while not affecting the maximum response (the "message"). Similar observations have been made for *Blaberus* HTH (Hayes and Keeley, 1990). This apparent contradiction (an address site at each end of the molecule) can be resolved in terms of our attempt to model the three-dimensional structure of locust AKH-I (see below). Goldsworthy *et al.* (1986a) argued previously that Pro^6 was important for "full" activity (and was therefore concerned with the "message"), but the full adipokinetic activities seen in *Locusta* in response to some other octapeptides which lack proline (see Table I) suggest that the proline part of the "message" can be mimicked by other amino acids. Furthermore, it could be argued that the various naturally occurring substitutions, at or on either side of residue 6, interfere with the "message" for the locust fat body, and this is why they produce truncated responses under some circumstances in either locusts or cockroaches (Table I) (Gäde, 1986, 1990b; Hayes and Keeley, 1990; Hayes et al., 1990).

In many respects the value of these potency studies on naturally occurring peptides in interpreting the "fit" requirements for receptor interaction is limited, because the effects of individual substitutions found in them are often difficult to interpret, and also because the variations in amino acid composition found in members of the family are products of evolutionary selection and are likely therefore to represent rather conservative changes. Certainly, by studying the structure–activity relationships of the AKH family it is difficult to identify with certainty specific (contiguous) areas of the molecule which are clearly concerned with the "message", or which carry the "address" (Hayes and Keeley, 1990). However, in our opinion the major problem lies elsewhere: comparisons of potency data cannot be interpreted with any confidence when we are unsure of which, and how many, populations of receptors are involved in the responses (see below).

B. Achetakinins

Holman et al. (1990) recently isolated five peptides (ranging from six to nine amino acids in length) from extracts of whole heads of *Acheta domesticus*. These myotropic peptides, which they called achetakinins, were detected by their potent stimulation of contractions of *Leucophaea* hindgut *in vitro*. In the cricket, these peptides are potent but weak diuretics *in vitro* (Coast et al., 1990), but also they promote lipid mobilization and inhibit protein synthesis in the fat body (*in vivo* and *in vitro*) in both locusts and crickets (Fig. 1). Previously, it was thought that in the AKHs a blocked N-terminus or, more specifically, the presence of a pyroglutamyl residue at this terminus was a prerequisite for lipid-mobilizing activity (Stone et al., 1978), yet none of the achetakinins are N-terminally blocked. Furthermore, short pentapeptide sequences (Phe-Xaa_1-Xaa_2-Trp-Gly-NH_2, where Xaa_1 can be Ser, His, Tyr or Asn, and Xaa_2 can be Ser or Pro) from the achetakinin C-terminal sequences show all the various activities of the parent molecules. At least eight residues had been thought necessary for lipid-mobilizing activity (Stone et al., 1978). Holman et al. have conducted a detailed analysis of the structure–activity requirements for achetakinins and related peptides in their actions on the *Leucophaea* hindgut preparation (Holman *et al.*, 1990; Nachman *et al.*, 1990). The order of potencies shown by achetakinins in the *Leucophaea* hindgut preparation appears broadly similar to that in diuretic activities against isolated malpighian tubules from crickets (Coast *et al.*, 1990), and in causing lipid mobilization or inhibiting protein synthesis in fat body from locusts and crickets (Cusinato, unpublished observations), achetakinins I, II and V usually being most active compared with achetakinins III and IV (Fig. 1). It is remarkable that receptors which interact with this family of peptides show these similarities in their structure–activity requirements, not only between different tissues, but also between different insect species. Interestingly, application of the "message/address" concept to these core pentapeptides implies either that the respective sites comprise only one or more residues, or that there is partial or total overlap between the two regions.

III. ATTEMPTS TO PREDICT AND DETERMINE SECONDARY STRUCTURE

The interpretation of structure–activity studies is limited without some concept of the shapes of the ligands. Peptides are flexible structures, and do not possess the longer-range interactions that cause a protein to adopt a stable conformation in water. However, information about the preferred conformation is generally considered relevant in interpreting structure–activity data. In short peptides the presence or absence of a β-turn is known to determine the nature of the dominant conformation (review: Rose *et al.*, 1985). They are of interest because they yield compact structures, four residues in length, and are mainly stabilized by local interactions across different parts of the peptide which may be distant in a linear (α-turns only) structure.

Fig. 1. The effects of achetakinins on lipid mobilization and inhibition of protein synthesis in locusts and crickets. In the upper set of bar charts, the effects of achetakinins (AK-I to AK-V, injected at a dose of 5 pmol) are compared with that of *Acheta* AKH (A-AKH) at a saturating dose (10 pmol) on the basis of lipid mobilization in crickets. Note that some of the achetakinins can give a maximum response (equivalent to that of A-AKH). In the middle set, the achetakinins were injected (10 pmol) into locusts. Note the truncated responses compared with that to A-AKH (10 pmol). In the lower set, the inhibition of the synthesis of haemolymph proteins (measured as a decrease in [^{3}H]leucine incorporation into haemolymph TCA-precipitable proteins *in vivo*; see Cusinato *et al.* (1991)) is shown for achetakinins injected at 5 pmol into crickets. Note again that in crickets achetakinins can be as effective as A-AKH.

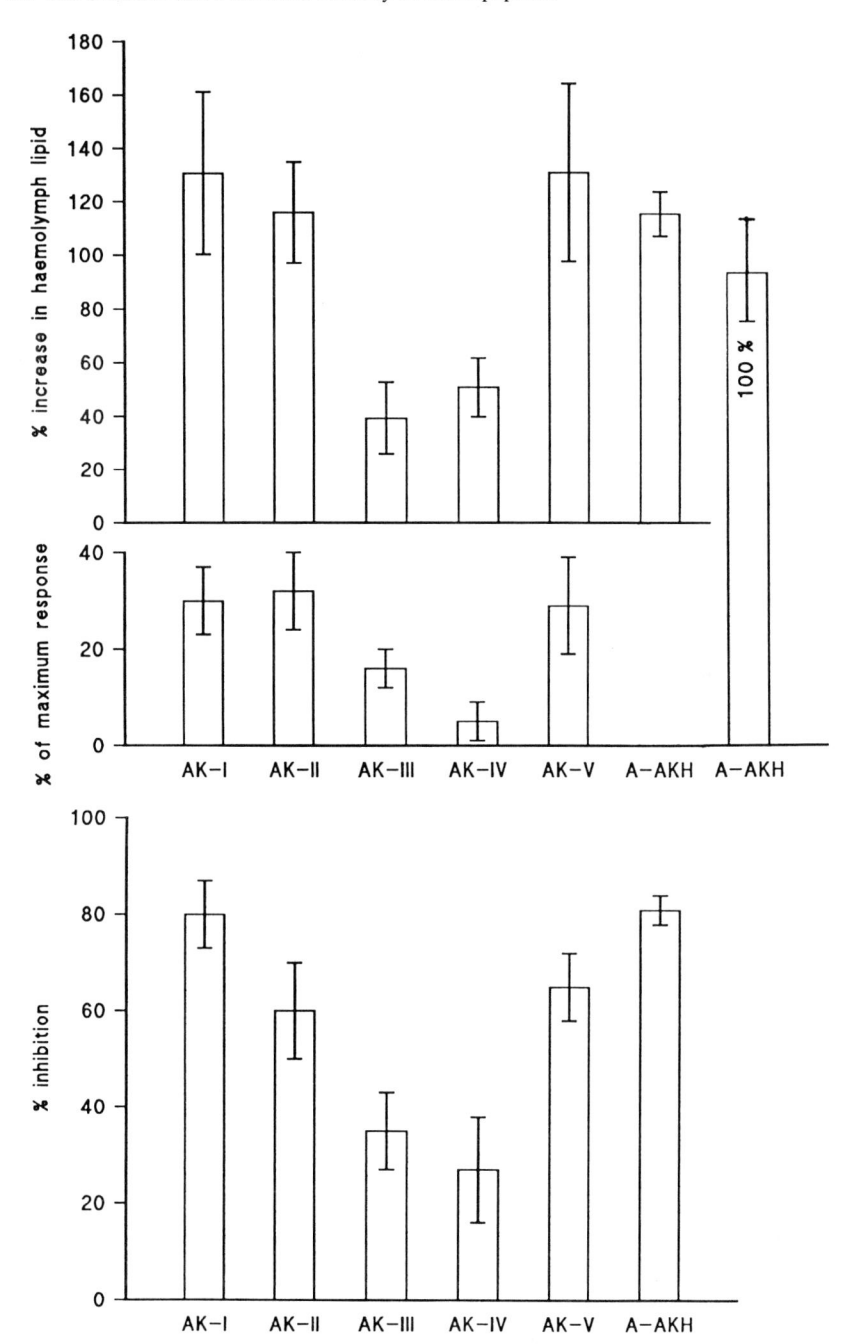

13. The Structure and Functional Activity of Neuropeptides

A. AKH Family Peptides

Stone et al. (1978) predicted that AKH-I would adopt a β-turn which would allow interaction between residues at each end of the molecule to aid stability. A β-turn predictive algorithm developed by Wilmot and Thornton (1988) has been applied to nine members of the AKH family (Table I) and all five peptides of the achetakinin family (results not shown). Type I β-turns are consistently predicted in all sequences tested, except those of *Locusta* or *Schistocerca* AKH-IIs and that of *Romalea* AKH-II (Table I).

The availability of synthetic peptides provides an opportunity to assess their secondary structures directly. A number of AKH family peptides have been subjected to CD spectroscopy (Wheeler et al., 1990). In aqueous solution at room temperature, none of them have a CD spectrum characteristic of a β-turn: the CD spectrum of *Locusta* AKH-I, for example, is relatively featureless (Fig. 2), as is true for all of the AKH family peptides we have analysed. Thus, all of the AKH peptides containing proline presented effectively identical CD spectra in water or ethanediol/water at room temperature.

On addition of SDS micelles the CD spectra of peptides containing proline change to ones characteristic of a β-turn (Fig. 2). Similar spectra indicative of β-turns are obtained after the addition of liposomes (A. F. Drake and G. J. Goldsworthy, unpublished observations). At low temperatures, all of those peptides containing proline show a CD spectral shift in the opposite direction to that after SDS treatment (Fig. 2). By contrast, peptides lacking proline present ambient spectra similar to those obtained in water alone (Fig. 3), and while these resemble to some extent the ambient-temperature CD spectrum of AKH-I, their responses to SDS are variable. *Romalea* AKH-II (Fig. 3) is relatively sensitive in the same sense as AKH-I, while the other proline-free peptides, as exemplified by *Manduca* AKH in Fig. 3, show distinctly smaller responses. The CD spectra of the *Schistocerca* and *Locusta* AKH-IIs, with and without added SDS, are almost identical to those of *Manduca* AKH.

1. A Possible Receptor-binding Model for AKH-I

Using the information from the prediction studies and CD spectroscopy, a model for one of the decapeptides, locust AKH-I, has been developed incorporating the predicted type I β-turn in position 5–8. The pyroglutamic acid and the amidated threonine were modelled by standard amino acids. This was considered to be justified, as the position of the β-turn meant that the pyroglutamic acid was not directly interacting with the

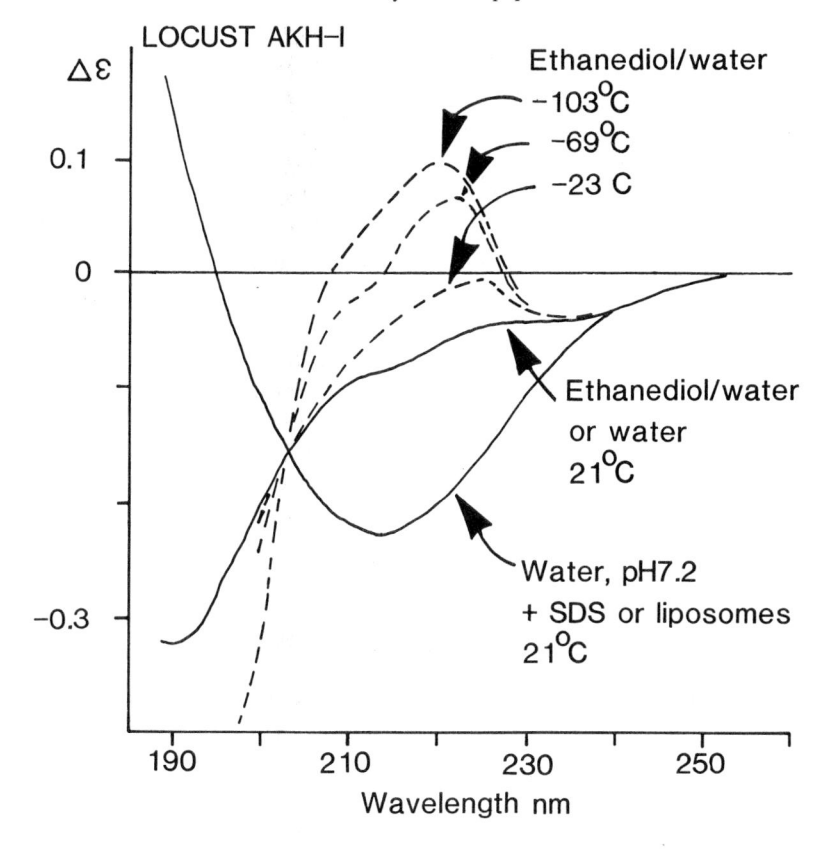

13. The Structure and Functional Activity of Neuropeptides   213

Fig. 2. The CD spectra of locust AKH-I in different solvents and at different temperatures. All other peptides listed in Table I and containing proline presented similar data.

C-terminus of the peptide. The resulting model for a possible receptor-binding conformation of locust AKH-I is shown in Fig. 4.

The main features of the energy-minimized model are a type I β-turn stabilized by two hydrogen bonds into a β-hairpin structure containing two further hydrogen bonds, and a hydrophobic cluster formed by Leu^2, Phe^4 and Trp^8 on one face of the molecule. This hydrophobic cluster may be important for binding of the locust AKH-I peptide to its receptor; the lipophilic face of the AKH-I molecule may occupy the binding site(s) on the receptor while the remainder of the molecule presents a hydrophilic face to the haemolymph.

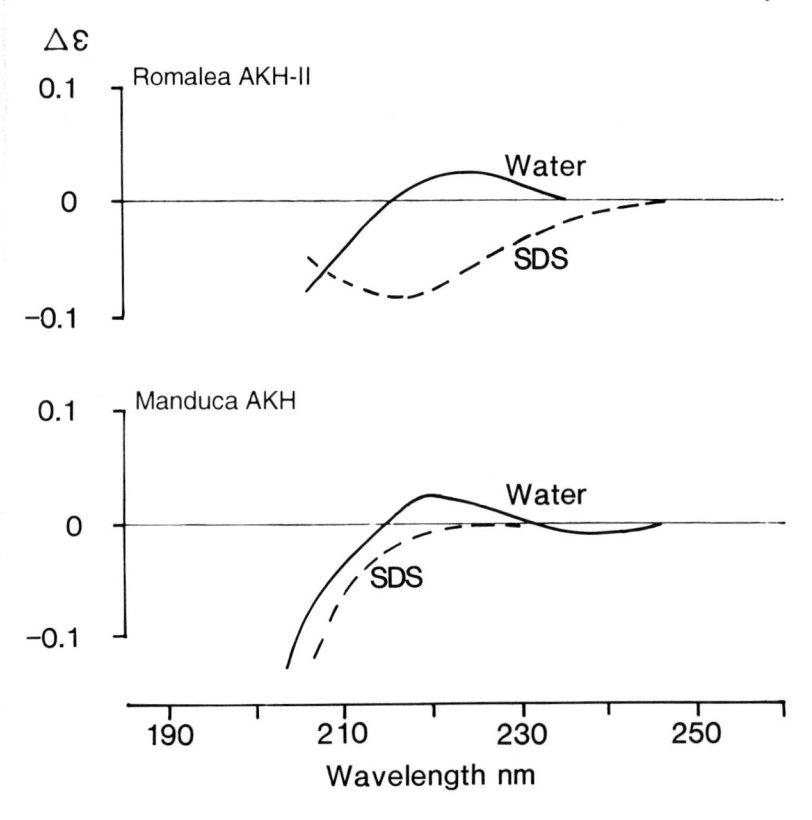

Fig. 3. The CD spectra of *Romalea* AKH-II and of *Manduca* AKH in water and in the presence of SDS micelles at pH 7.2, 21°C. The *Locusta* and *Schistocerca* AKH-IIs presented data identical to that of *Manduca* AKH.

2. Is the β-turn Important for AKH Activity?

In general, there is good agreement between the prediction studies and the CD data in the presence of SDS. Indeed, the full biological activity of the [3,4-dehydro-Pro^6]-AKH-I (Table I) is in good agreement with our CD data, but in contrast with earlier studies in which arguments (Stone *et al.*, 1978) for the possible presence and importance of the β-turn in AKH-I were partly based on the reported poor activity (Hardy and Shepherd, 1983) of this analogue, and the apparent assumption that the dehydro-proline substitution would affect the likelihood of β-turn adoption. Interestingly, while the computer prediction for *Manduca* AKH included the possibilities for turns in two positions, no evidence for this

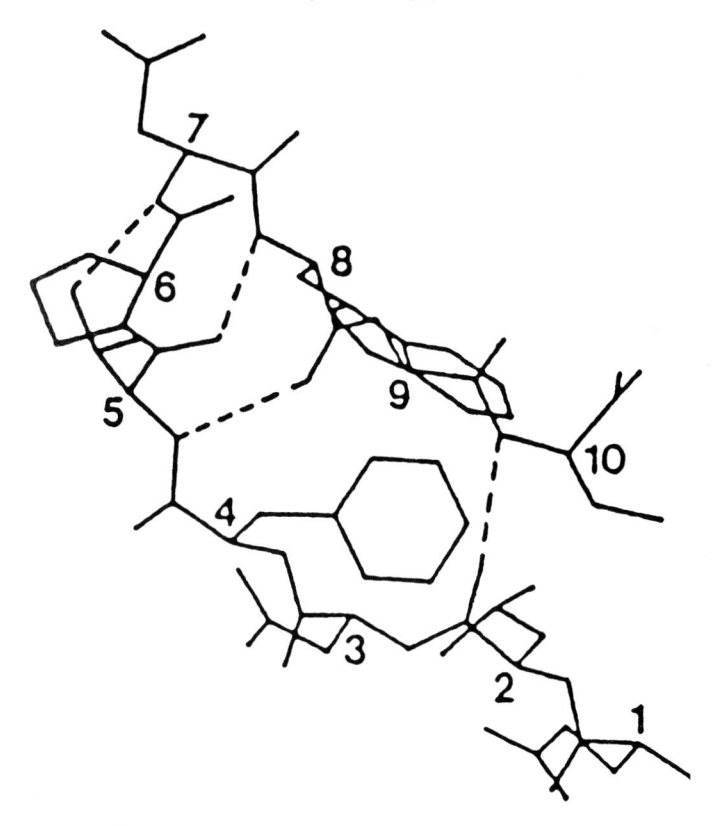

13. The Structure and Functional Activity of Neuropeptides 215

Fig. 4. A possible receptor-binding model for the locust AKH-I decapeptide. A hydrogen bond was assigned if the two electronegative atoms were less than 33 nm apart, and if the angle between the two electronegative atoms and each of their preceding main chain or side chain atoms was greater than 90°. The β-turn is shown clearly, with stabilizing hydrogen bonds (broken lines) as described in the text, and the molecule is viewed showing the hydrophobic cluster of the indole and phenyl rings and the leucyl side chain.

was obtained from the CD spectroscopy. The CD observations reported here resemble those reported earlier by Drake and colleagues (Silligardi *et al.*, 1987; Drake *et al.*, 1988; Tatham *et al.*, 1989) in studying other linear oligopeptides. The low-temperature state described for the proline-containing peptides cannot be classified as an ordinary β-turn, and its precise nature remains open to question. Presumably, those peptides which contain proline exist in free aqueous solution as an approximately 50:50 mixture of molecules in the β-turn conformation and molecules in the low-temperature state but, when bound to membranes (receptors), they show an enhanced β-turn content.

Although in general our data support the hypothesis that a β-turn is important for the biological activity of locust AKH-I (Goldsworthy et al., 1986b), whether this is an absolute requirement is uncertain (Hayes et al., 1990). Our failure to find evidence from the CD analyses for β-turn formation in the proline-free peptides does not prove that such peptides cannot or do not adopt a turn conformation when bound to their receptor(s). Further, the shifts in the CD spectra of these peptides on addition of SDS or liposomes do suggest that they could undergo a conformational change when bound to a receptor. Perhaps we have simply failed to provide for these peptides an environment which sufficiently mimics that of the receptor. On the other hand, as we have indicated earlier, we cannot assume that the biological activity of these peptides in locusts is due to interaction with a single population of receptors (see also below). The full significance of the β-turn conformation in the decapeptide AKH-I remains to be determined. Nevertheless, by causing a change in chain direction, a turn would allow the interaction of residues in different parts of the peptide, particularly the formation of hydrogen bonds as we have suggested. This latter point may be crucial in interpreting structure–activity studies as described briefly above, particularly the apparent presence of "address" sites at both the N- and C-termini; if the peptides become folded when they bind to their receptor(s) these different sites may be brought into close juxtaposition and represent an effectively contiguous area of the molecule. Above all, these results emphasize that solution conformation in this series is easily distorted, for example by micelles or temperature, and that they may be only a general guide to conformation during receptor binding.

B. Achetakinins

CD analysis of natural achetakinin parent molecules and their pentapeptide C-terminal sequences gives spectra which are not typical of type 1 β-turns, but could be interpreted as open β-turns (unpublished observations; Nachman et al., 1990). Prediction studies suggest that β-turns could comprise either residues 1 to 4, or 2 to 5. Molecular models of achetakinin core peptides, with β-turns in either position, or with two consecutive turns, suggest that the side chains of the phenylalanine and tryptophan residues are brought closely together on one face of the achetakinin molecule (Fig. 5). Such similarities in conformation, with phenyl and indole ring structures clustered on one face of the molecule, could explain why the achetakinins can mimic the action on the fat body

FSSW as a type 1 β-turn

FSSWG as two consecutive turns

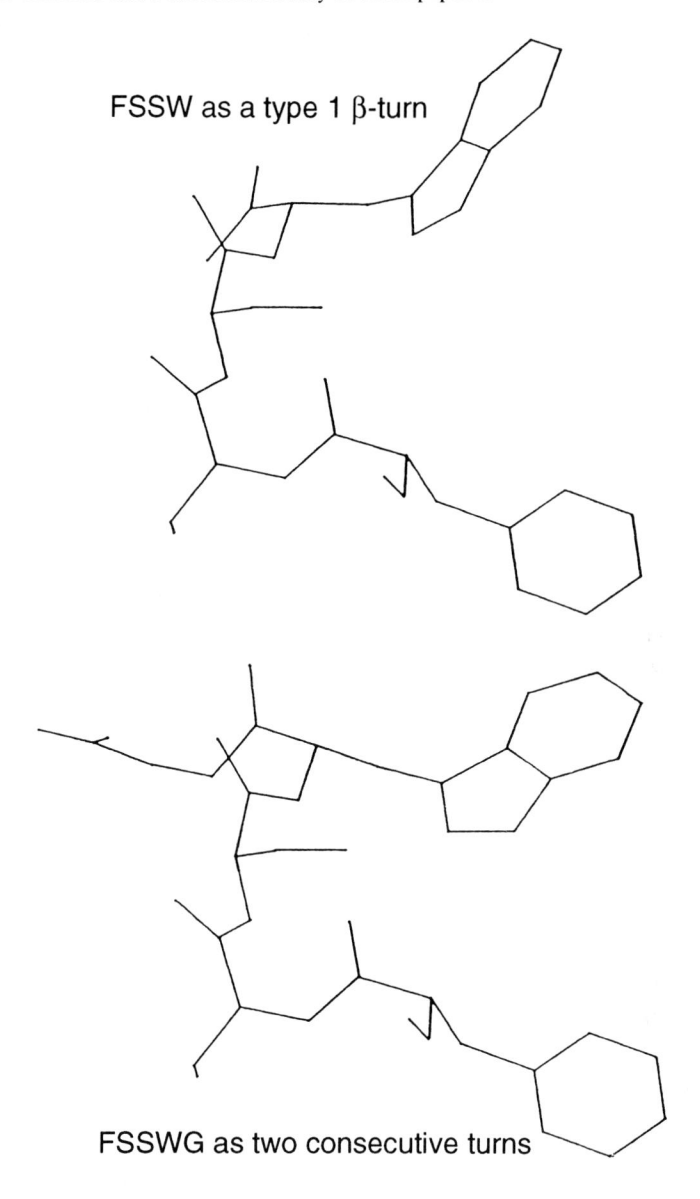

Fig. 5. Possible conformations of achetakinin III core peptide. The core peptide Phe-Ser-Ser-Trp-Gly-NH$_2$ (FSSWG) is modelled with either one or two turns (see text). Note the close juxtaposition of the indole and phenyl rings. We thank Dr Fiona Hayes for undertaking the modelling, and allowing us to use her unpublished results.

of locust decapeptide AKH-I (which shows a similar cluster, with the additional presence of a leucyl side chain – see Fig. 4).

1. Do the AKHs and the Achetakinins React at the same Receptors?

Because of the proposed similarities in secondary structure between AKHs and achetakinins, it is tempting to propose that they may work via the same or similar receptors. Caution is urged, however, in making this interpretation because we believe that in locusts the two natural adipokinetic hormones (AKH-I and AKH-II) act on the fat body via different receptors from those of achetakinins. Because achetakinins give a truncated hyperlipaemic response in locusts, we have been able to test the hypothesis that they act at different receptors from the AKH-like peptides directly. Both locust AKH-I and *Locusta* AKH-II elicit further increases in lipid mobilization in the presence of saturating doses of the achetakinin I pentapeptide C-terminal sequence peptide (Cusinato and Goldsworthy, unpublished observations). Furthermore, AKHs and achetakinins appear to work via different second messengers; recent evidence from our laboratory (Cusinato, unpublished observations) and others (W. Van Marrewijk, personal communication; D. J. Candy, personal communication) suggests that while AKH family peptides may act (primarily?) via activation of adenylyl cyclase and the production of increased levels of intracellular cAMP (Gäde and Holwerda, 1976; Spencer and Candy, 1976), they also stimulate the inositol phosphate pathway. Conversely, we are unable to show any effect of achetakinins on cAMP production in malpighian tubules (Coast *et al.*, 1990) or fat body (Kay, unpublished observations).

IV. N-TERMINAL MODIFICATION OF PEPTIDES

A. AKH Family Peptides

N-terminal modifications of AKH family neuropeptides suggest that a pyroglutamyl residue itself is not an absolute requirement but, for full biological activity (a maximal lipid-mobilization response), the N-terminus should be blocked by acetylation or the presence of a pyroglutamyl residue (Gäde and Hayes, 1990). The cricket octapeptide, *Acheta* AKH (Table I), which has been identified recently (Woodring *et al.*, 1990; Cusinato *et al.*, 1991), is identical to *Romalea* AKH-II, one of the proline-free octapeptides which can elicit a full hyperlipaemic response in locusts (Table I). When this peptide is truncated from the N-terminus and then tested for hyperlipaemic activity either with a free N-terminus or after addition of pyroglutamic acid, the resulting changes in potency (Table II) suggests that, providing the heptapeptide is blocked by a pyroglutamyl residue at its N-terminus, its C-terminal region must carry a full "address" and "message". The problem remains, however, that we are as yet still uncertain which populations of receptors are being activated in the lipid-mobilization assay. Just as was the case with the achetakinins, we find it remarkable that des-pGlu^1 *Acheta*-AKH has some activity, albeit truncated, because for those members of the AKH family tested previously, removal of the pGlu residue removes most or all activity (Stone *et al.*, 1978; Gäde, 1990b). Perhaps even more remarkable is the fact that the six-residue unblocked peptide also has hyperlipaemic activity, albeit at rather high doses. Also, it is intriguing that the addition of pyroglutamic acid to the N-terminus of the six- and seven-residue peptides appears to enhance the "message" (not just the address!) and restore "full" activity (Table II). This finding is contradictory to our previous interpretations of structure–activity relationships in the AKH family.

These studies on the structure–activity relations of *Acheta* AKH have employed both *Locusta* and *Acheta* as assay animals with very similar results so, for brevity, only the locust data are given here (Table II). Only one AKH family peptide has been identified in *Acheta*, the octapeptide, and whether or not this species also possesses another AKH peptide, perhaps a decapeptide, is unknown. It may be significant that the structure–activity relationships described here for the octapeptide *Acheta* AKH in both locusts and crickets are identical. Crickets respond well to the locust decapeptide AKH-I (Woodring *et al.*, 1989; unpublished observations), and it will be of interest to determine whether different populations of receptors are involved in the responses to octa- and decapeptides.

TABLE II. The hyperlipaemic activities of various analogues of *Acheta* AKH in locusts. For details of the assay see Table I.

Peptide sequence	ED$_{50}$ (pmol)	ED$_{max}$ (pmol)	% Maximum response
pGlu-Val-Asn-Phe-Ser-Thr-Gly-Trp-NH$_2$	0.9	10	100 ± 7
Val-Asn-Phe-Ser-Thr-Gly-Trp-NH$_2$	8.0	50	25 ± 1
pGlu-Asn-Phe-Ser-Thr-Gly-Trp-NH$_2$	6.0	20	100 ± 7
Asn-Phe-Ser-Thr-Gly-Trp-NH$_2$	53.3	130	29 ± 4

B. Achetakinins

At Birkbeck, in collaboration with our Rothamsted colleagues, we have studied the requirements for a free N-terminus in the determination of biological activity in the achetakinin I C-terminus fragment, Phe-Tyr-Pro-Trp-Gly-NH$_2$. In these preliminary studies, a different property of achetakinins has been the main focus of our attention – that of increasing fluid secretion by isolated malpighian tubules. A free amino group is not required for diuretic activity; acetylation of the free amino group of Phe[1] or addition of pyroglutamic acid yields analogues which are active at 10^{-8} M (Table III). Furthermore, the N-terminal phenylalanine has been replaced by a variety of aromatic acyl groups (Table III). Replacement by hydrocinnamic, cinnamic or phenylpropriolic acid retains activity, but our preliminary data suggest that increasing unsaturation of the acyl linker to Tyr[2] reduces activity, perhaps due to increasing its rigidity. The spacing between the phenyl ring and the Tyr[2] residue appears also to be critical for diuretic activity, because when the linker is one, two or four carbon atoms long, instead of three as in 3-phenylpropionyl-Tyr-Pro-Trp-Gly-NH$_2$, diuretic activity is lost (Table III). We have also examined briefly the effect of changing the hydrophobicity of the N-terminus (Table III).

TABLE III. Relative diuretic activity of various analogues of the achetakinin I pentapeptide terminal fragment. The data are expressed as percentages of the response of isolated malpighian tubules of *Acheta domesticus* to the parent peptide, Phe-Tyr-Pro-Trp-Gly-NH$_2$(FYPWG-NH$_2$).

Peptide or analogue	Dose	Mean	SE	n
pGlu-FYPWG-NH$_2$	10^{-8} M	64	21	8
N-acetyl-FYPWG-NH$_2$	10^{-8} M	101	30.6	8
[3-phenylpropionyl[1]]-YPWG-NH$_2$	10^{-8} M	86	14.3	10
(Hydrocinnamic acid) C$_6$H$_5$CH$_2$COOH	10^{-7} M	108	16.9	10
[3-phenyl-2-propenoic[1]]-YPWG-NH$_2$ (Cinnamic acid) C$_6$H$_5$CH=CHCOOH	10^{-8} M	56	10.8	10
[phenylpropiolic[1]]-YPWG-NH$_2$ C$_6$H$_5$C≡CCOOH	10^{-8} M	50	21.3	10
[benzoyl[1]]-YPWG-NH$_2$	10^{-8} M – 21	12.2		7
[phenylacetyl[1]]-YPWG-NH$_2$	10^{-7} M – 39	15		7
[phenylbutyric[1]]-YPWG-NH$_2$	10^{-8} M – 23	8.8		7
[methylcinnamoyl[1]]-YPWG-NH$_2$	10^{-8} M – 18	5.3		6
[methoxyphenyl-propionyl[1]]-YPWG-NH$_2$	10^{-8} M – 15	7.4		7
	10^{-8} M 43	14.6		7

Methylation of the acyl linker (methylcinnamoyl[1]-Tyr-Pro-Trp-Gly-NH$_2$) resulted in complete loss of activity at 10^{-8} M, whereas some activity was retained with a methoxy group on the phenyl ring (methoxyphenylpropionyl[1]-Tyr-Pro-Trp-Gly-NH$_2$). Methylation of the linker may therefore impose some steric constraint on the molecule, but this remains to be investigated further.

V. INSECT DIURETIC PEPTIDES

A number of attempts have been made to isolate and characterize diuretic peptides from insects over the last two decades, but findings have been contradictory and controversial (Wheeler and Coast, 1990; Spring, 1990). At Birkbeck we have recently isolated and characterized two novel diuretic peptides, one from *Acheta domesticus* (Kay *et al.*, 1991a), and one from *Locusta migratoria* (Kay *et al.*, 1991b). These peptides were identified in extracts of whole heads from the respective species using an *in vitro* bioassay based on their ability to stimulate cAMP production in conspecific malpighian tubules *in vitro*, and are potent stimulators of fluid secretion by isolated malpighian tubules. These novel peptides increase fluid secretion in a completely different manner from achetakinins (working via cAMP, and not the inositol phosphate pathway), and are structurally quite different from them, being 46-residue peptides. Importantly, however, they share about 50% sequence (Fig. 6) identity with each other, and with another insect diuretic peptide, *Manduca* diuretic hormone, isolated by Kataoka *et al.* (1989). These latter three peptides are the first members of what we predict is a family of insect diuretic peptides but, perhaps more interestingly, this insect family is itself part of a larger superfamily of CRF-like peptides which all act via cAMP and exert effects on hydromineral balance (Kataoka *et al.*, 1989; Kay *et al.*, 1991b; Coast *et al.*, 1991). Furthermore, preliminary studies suggest that the insect peptides may share some common structural elements with the vertebrate peptide members of the family: CRF, sauvagine and urotensin I stimulate fluid secretion and cAMP production in *Acheta* tubules, albeit to a limited extent (Coast *et al.*, 1991).

Detailed studies on the activities of these insect diuretic peptides are now in progress, but already we can postulate that there may be similarities between what is known of the structure–activity relationships for vertebrate CRF-like peptide members of the superfamily and the insect peptides. Structure prediction studies and CD spectroscopy suggest that the insect peptides, in common with vertebrate members of the superfamily, have prominent α-helical regions in the C-terminal half of

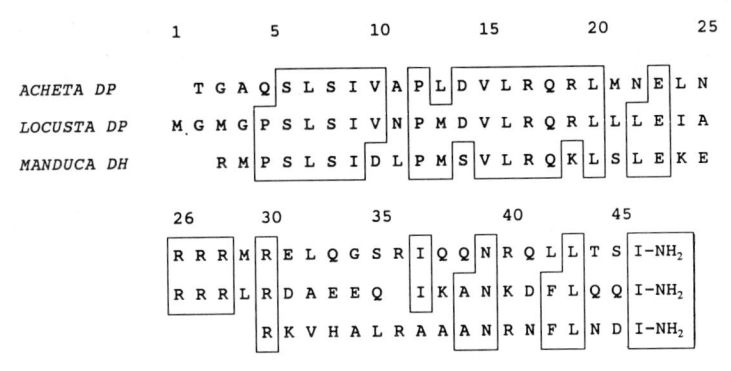

	1	5	10	15	20	25
ACHETA DP	T G A Q	S L S I V	A P	L D V L R Q R L	M N E	L N
LOCUSTA DP	M G M G	P S L S I V	N P	M D V L R Q R L	L L E	I A
MANDUCA DH	R M	P S L S I	D L P M	S V L R Q	K L S	L E	K E

	26	30	35	40	45	
R R R	M R E L Q G S R	I Q Q	N R Q L	L T S	I-NH$_2$
R R R	L R D A E E Q	I K A N K D	F L Q Q	I-NH$_2$
R K V H A L R A A	A N R N	F L N D	I-NH$_2$

Fig. 6. Comparison of amino acid sequences of *Locusta* diuretic peptide (DP), with *Acheta* DP and *Manduca* diuretic hormone. To aid alignment, a single gap has been inserted in the *Locusta* and *Manduca* sequences. Identical residues are boxed.

the molecule. Indeed, a potent antagonist of CRF, α-helical CRF-(9-41)-peptide (Rivier *et al.*, 1984), when assayed at high doses (10^{-4} M), is an antagonist of the action of *Manduca* diuretic hormone on cricket malpighian tubules (Coast *et al.*, 1991). As in vertebrate peptides, it appears likely that the C-terminal half of the insect peptides may carry the "address" and be important for receptor binding, whereas the N-terminus may carry the "message". Indeed, circumstantial evidence (Coast *et al.*, 1991) suggests that oxidation of the Met^2 and Met^{11} residues in *Manduca* diuretic hormone may reduce the maximum response of *Acheta* malpighian tubules to the peptide. Again, however, we must be cautious in drawing such conclusions until information is available about the receptors on the malpighian tubules for the various peptides.

VI. CONCLUSIONS

The likely presence of heterogeneous populations of receptors which interact with different peptides to influence the activities of insect cells makes the analysis of structure–activity relationships for insect peptides almost impossible in terms of interpreting the importance of single amino acid substitutions or the presence or absence of particular conformations. The full analysis of structure–activity relationships for insect neuropeptides must await information on their receptors, and the development of receptor-binding assays in which interactions with defined receptors can be analysed. Biologically active radiolabelled (^{125}I)-analogues of achetakinins are now being used in our laboratory for such receptor studies, and ^{125}I-labelled analogues of AKH-I and diuretic peptides are also being developed.

The realization that a range of different neuropeptides can all elicit apparently identical physiological responses in a single target tissue, and that different populations of receptors and different second messengers can be involved in these responses, is clearly important in beginning to unravel the relationships between structure and activity for insect neuropeptides. However, until detailed information on receptors for insect neuropeptides becomes available, or blockers for specific receptors are developed, application of the message/address concept in interpreting structure–activity relationships will be of limited value.

Acknowledgements

This work is supported by grants from the SERC, AFRC, and The Leverhulme Trust to GJG and GMC. Collaborative work between Birkbeck and Rothamsted is supported by the British Technology Group.

REFERENCES

Coast, G. M., Holman, G. M. and Nachman, R. J. (1990). *J. Insect Physiol.* **36**, 481–488.
Coast, G. M., Hayes, T. K., Kay, I. and Chung, J. (1991). *J. Exp. Biol.* **162**, 331–338.
Cusinato, O., Wheeler, C. H. and Goldsworthy, G. J. (1991). *J. Insect Physiol.* **37**, 461–469.
Drake, A. F., Silligardi, G. and Gibbons, W. A. (1988). *Biophys. Chem.* **31**, 143–146.
Fernlund, P. and Josefsson, L. (1972). *Science* **177**, 173–175.
Gäde, G. (1986). *Z. Naturforsch.* **41c**, 315–320.
Gäde, G. (1990a). *J. Insect Physiol.* **36**, 1–12.
Gäde, G. (1990b). *Physiol. Entomol.* **15**, 299–316.
Gäde, G. and Hayes, T. K. (1990). In "Insect Neurochemistry and Neurophysiology: 1989" (A. B. Borkovec and E. P. Masler, eds), pp. 243–246. Humana Press, New Jersey.
Gäde, G. and Holwerda, D. A. (1976). Involvement of adenosine 3′:5′-cyclic monophosphate in lipid mobilization in *Locusta migratoria*. *Insect Biochem.* **5**, 535–541.
Gäde, G. and Rinehart, K. L. (1986). *Biochem. Biophys. Res. Commun.* **141**, 774–781.
Gäde, G. and Rinehart K. L. (1987a). *Biol. Chem. Hoppe-Seyler* **368**, 67–75.
Gäde, G. and Rinehart, K. L. (1987b). *Biochem. Biophys. Res. Commun.* **149**, 908–914.
Gäde, G., Goldsworthy, G. J., Kegel, G. and Keller, R. (1984). *Physiol. Chem.* **365**, 391–398.
Gäde, G., Goldsworthy, G. J., Schaffer, M. H., Cook, J. C. and Rinehart, K. L. (1986). *Biochem. Biophys. Res. Commun.* **134**, 723–730.
Gäde, G., Hilbich, C., Beyreuther, K. and Rinehart, K. C. (1988). *Peptides* **9**, 681–688.
Goldsworthy, G. J. and Mordue, W. (1989). *Biol. Bull.* **177**, 218–224.
Goldsworthy, G. J. and Wheeler, C. H. (1986). In "Insect Neurochemistry and Neurophysiology 1986" (A. B. Borkovec and D. B. Gelman, eds), pp. 183–186. Humana Press, New Jersey.

Goldsworthy, G. J. and Wheeler, C. H. (1989). *Pest. Sci.* **25**, 85–95.
Goldsworthy, G. J., Mallison, K. and Wheeler, C. H. (1986a). *J. Insect Physiol.* **32**, 95–101.
Goldsworthy, G. J., Mallison, K., Wheeler, C. H. and Gäde, G. (1986b). *J. Insect Physiol.* **32**, 433–438.
Hardy, P. M. and Shepherd, P. W. (1983). *J. Chem. Soc. Trans.* **1**, 723–729.
Hayes, T. K. and Keeley, L. L. (1986). *In* "Insect Neurochemistry and Neurophysiology 1986" (A. B. Borkovec and D. B. Gelman, eds), pp. 195–198. Humana Press, New Jersey.
Hayes, T. K. and Keeley, L. L. (1990). *J. Comp. Physiol.* B **160**, 187–194.
Hayes, T. K., Ford, M. M. and Keeley, L. L. (1990). *In* "Insect Neurochemistry and Neurophysiology 1989" (A. B. Borkovec and E. P. Masler, eds), pp. 243–246. Humana Press, New Jersey.
Holman, G. M., Nachman, R. J. and Wright, M. S. (1990). *In* "Progress in Comparative Endocrinology" (A. W. Epple, C. G. Scanes and M. Stetson, eds), pp. 35–39. Wiley-Liss Inc., New York.
Hruby, V. J. (1984). *In* "Conformationally Directed Drug Design" (J. A. Vida and M. Gorden, eds), pp. 9–27. American Chemical Society, Washington D.C.
Jaffe, H., Raina, A. K., Riley, C. T., Fraser, B. A., Holman, G. M., Wagner, R. M. Ridgway, R. L. and Hayes, D. K. (1986). *Biochem. Biophys. Res. Commun.* **135**, 622–628.
Kataoka, H., Troetschler, R. G., Li, J. P., Kramer, S. J., Carney, R. L. and Schooley, D. A. (1989). *Proc. Natl. Acad. Sci. USA* **86**, 2976–2980.
Kay, I., Coast, G. M., Cusinato, O., Wheeler, C. H., Totty, N. F. and Goldsworthy, G. J. (1991a). *Biol. Chem. Hoppe-Seyler* **372**, 505–512.
Kay, I., Wheeler, C. H., Coast, G. M., Totty, N. F., Cusinato, O., Patel, M. and Goldsworthy, G. J. (1991b). *Biol. Chem. Hoppe-Seyler* **372**, 929–934.
Loughton, B. L. and Orchard, I. (1981). *J. Insect Physiol.* **27**, 383–385.
Nachman, R. J., Roberts, V. A., Holman, G. M. and Trainer, J. A. (1990). *In* "Progress in Comparative Endocrinology" (A. W. Epple, C. G. Scanes and M. Stetson, eds), pp. 60–66. Wiley-Liss Inc., New York.
O'Shea, M., Witten, J. and Schaffer, M. (1984). Isolation and characterization of two myoactive neuropeptides: further evidence of an invertebrate peptide family. *Neuroscience* **4**, 521–529.
Rivier, J., Rivier, C. and Vale, W. (1984). *Science* **224**, 889–891.
Rose, G. D., Gierasch, L. M. and Smith, J. A. (1985). *Adv. Protein Chem.* **37**, 1–109.
Siegert, K., Morgan, P. J. and Mordue, W. (1985). *Biol. Chem. Hoppe-Seyler* **366**, 723–727.
Silligardi, G., Drake, A. F., Mascagni, P., Neri, P., Lozzi, L., Niccolai, N. and Gibbons, W. A. (1987). *Biochem. Biophys. Res. Commun.* **143**, 1005.
Spencer, I. M. and Candy, D. J. (1976). Hormonal control of diacyglycerol mobilization from the fat body of the desert locust, *Schistocerca gregaria*. *Insect Biochem.* **6**, 289–296.
Spring, J. H. (1990). *J. Insect Physiol.* **36**, 13–22.
Stone, J. V., Mordue, W., Broomfield, C. E. and Hardy, P. M. (1978). *Eur. J. Biochem.* **89**, 195–202.
Tatham, A. S., Drake, A. F. and Shewry, P. R. (1989). *Biochem. J.* **259**, 471–476.
Van Gunsteren, W. F. and Berendsen, H. J. C. (1987). Groningen Molecular Simulation (GROMOS) Library Manual, BIOMOS, Mijenborgh 16, Groningen, The Netherlands, pp. 1–229.
Wheeler, C. H. and Coast, G. M. (1990). *J. Insect Physiol.* **36**, 23–34.
Wheeler, C. H., Drake, A. F., Wilmot, C. M., Thornton, J. M., Gäde, G. and Goldsworthy, G. J. (1990). *In* "Insect Neurochemistry and Neurophysiology: 1989" (A. B. Borkovec and E. P. Masler, eds), pp. 235–238. The Humana Press Inc.

Wheeler, C. H., Gäde, G. and Goldsworthy, G. J. (1988). *In* "Neurohormones of Invertebrates" (M. C. Thorndyke and G. J. Goldsworthy, eds), pp. 141–157. Cambridge University Press, Cambridge.
Wilmot, C. M. and Thornton, J. M. (1988). *J. Mol. Biol.* **203**, 21–232.
Woodring, J. P., Fescemeyer, H. W., Lockwood, J. A., Hammond, A. M. and Gäde, G. (1989). *Comp. Biochem. Physiol.* **92A**, 65–70.
Woodring, J., Das, S., Kellner, R. and Gäde, G. (1990). *Z. Naturforschung.* **45c**, 1176–1184.
Ziegler, R., Eckart, K., Schwarz, H. and Keller, R. (1986). *Biochem. Biophys. Res. Commun.* **133**, 337–342.

14

Molecular Approaches to Insect Endocrinology

L. M. RIDDIFORD

 I. Introduction 226
 II. Contributions of Molecular Biology to the Study of
    Developmental Insect Neuropeptides 227
    A. PTTH 227
    B. Peptides Regulating the Corpora Allata 228
    C. Eclosion Hormone 228
 III. Hormone Receptors, Transcription Factors, and the
    Regulation of Moulting and Metamorphosis 231
    A. Ecdysteroid Receptor 232
    B. Ecdysteroid-induced Transcription Factors and
       Development 233
    C. Juvenile Hormone Receptors 234
 IV. Conclusions 237
    References 237

I. INTRODUCTION

Insect moulting, metamorphosis, and reproduction are regulated by hormones, primarily ecdysteroids and juvenile hormones (JH). Neuropeptides regulate synthesis of these hormones – prothoracicotropic hormone (PTTH) stimulating synthesis and release of ecdysteroids, allatotropins and allatostatins controlling JH secretion – as well as triggering ecdysis (eclosion hormone, EH) and causing sclerotization of the new cuticle (bursicon). The age-old questions of how these hormones and neuropeptides act and how both internal and external signals can influence development are now amenable to detailed analysis. The new molecular approaches allow purification of minute amounts of peptides and of hormone receptors, isolation of their genes, and production of antibodies as well as large amounts of these peptides for biological studies. This chapter will summarize some of these new approaches and the types of information being gained.

II. CONTRIBUTIONS OF MOLECULAR BIOLOGY TO THE STUDY OF DEVELOPMENTAL INSECT NEUROPEPTIDES

Over the past few years, advances in peptide chemistry coupled with recombinant DNA technology have led to the isolation and sequencing of several of the developmental insect neuropeptides and, in some cases, the isolation of their genes. These advances have allowed the identification of the neurosecretory cells that produce these peptides and thus opened the door to a study of their regulation, both during development and by other neuronal inputs. The availability of the genes for one or a few species allows the isolation of similar genes from other insect species either by direct hybridization with genomic DNA or by use of the polymerase chain reaction (PCR). The PCR technique uses oligonucleotides based on the two ends of a conserved region of the known molecule to amplify a similar region from an unknown species (Saiki *et al.*, 1988). The neuropeptide gene can also be used to produce large amounts of the neuropeptide by recombinant DNA technology for study of its physiological and molecular mechanisms of action. Below I summarize the progress in this area relative to the developmental neuropeptides cited above. Only bursicon has not been studied with molecular techniques and thus will not be discussed further (see Reynolds (1985) for a recent review of bursicon research).

A. PTTH

Purification of the peptides in adult heads of the silkworm moth, *Bombyx mori*, that had PTTH activity in another moth species, *Samia cynthia ricini*, led to the discovery of two forms of PTTH (Nagasawa *et al.*, 1990). The smaller form, a 4-kDa peptide, is a member of the insulin-like peptide family (Nagasawa *et al.*, 1984) and has been localized by both immunocytochemistry (Mizoguchi *et al.*, 1987, 1990a) and *in situ* hybridization (Iwami, 1990) to four of the median neurosecretory cells in the *Bombyx* larval and pupal brain (Fig. 1). Curiously, it has no PTTH activity in *Bombyx* itself, and so now is called bombyxin (Mizoguchi *et al.*, 1987). Its normal physiological action in *Bombyx* has not been clarified. In *Manduca* a small form of PTTH is thought to drive the secretion of ecdysteroid by the prothoracic glands at the end of larval life to initiate metamorphosis (the so-called "wandering" or "commitment" peak of ecdysteroid) (Bollenbacher and Gilbert, 1981). Cloning of the bombyxin gene revealed that it was a member of a multigene family consisting of at least five genes (Kawakami et al., 1989; Iwami et al., 1989, 1990).

The larger 22-kDa PTTH is produced by two pairs of lateral neurosecretory cells (Agui et al., 1979; Kawakami et al., 1990; Mizoguchi et al., 1990b) (Fig. 1). The gene encodes a single polypeptide which is subsequently cleaved to produce three peptides, the largest of which is 12 kDa, which then dimerizes to form the active hormone (Kawakami et al., 1990). The physiological action of the two smaller peptides, if any, is unknown.

B. Peptides Regulating the Corpora Allata

Not as much is known about the neuropeptides that control JH biosynthesis. An allatotropin from *Manduca sexta* adults (Kataoka et al., 1989) and allatostatins from the cockroach, *Diploptera punctata* (Woodhead et al., 1989; Pratt et al., 1989, 1991), have been isolated and sequenced. The 13 amino acid *Manduca* allatotropin only stimulates the corpora allata of adult Lepidoptera; it has no activity in larval Lepidoptera or in beetles, cockroaches or grasshoppers. Recently, a 20-kDa allatotropin has been isolated from the brains of *Galleria mellonella* larvae (Bogus and Scheller, 1991), and preliminary immunocytochemical localization reveals its presence in the medial neurosecretory cells (Bogus et al., 1991). Its activity in adults has not been reported.

By contrast, there are at least five related allatostatic peptides in *Diploptera*, and these are active in both larval and adult stages as well as in other cockroach species (Pratt et al., 1990; Woodhead et al., 1989). Immunocytochemistry shows that allatostatin 1 is present in the lateral neurosecretory cells of the adult female brain which send their axons via NCCII to the corpora allata (Fig. 1) and in two pairs of medial neurosecretory cells whose axons end within the brain and probably have a neuromodulatory role (Stay et al., 1991). Preliminary studies using a photoaffinity analogue of allatostatin 1 show that the membrane fraction of the corpora allata contains two proteins (39 and 59 kDa) that specifically bind this hormone (Cusson et al., 1991).

C. Eclosion Hormone

Eclosion hormone is a 62 amino acid peptide (Marti et al., 1987; Kataoka et al., 1987; Kono et al., 1991) that acts directly on the nervous system to

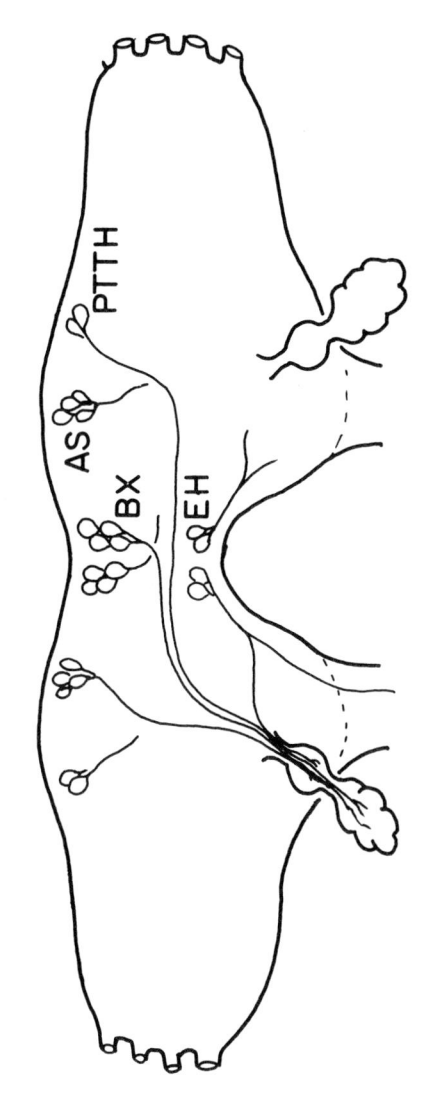

14. Molecular Approaches to Insect Endocrinology

Fig. 1. Generalized insect brain showing the identified neurosecretory cells for developmental neuropeptides with the axons projecting to the corpora cardiaca and/or corpora allata or down the ventral nerve cord. AS, allatostatin; BX, bombyxin; EH, eclosion hormone; PTTH, prothoracicotropic hormone.

trigger the ecdysis behaviour at the end of the moult in order to shed the old cuticle (Truman, 1990). Although this hormone is used by all insects, the peptide has only been isolated from two Lepidoptera, *Manduca* and *Bombyx*. The similarity of structure of the core peptide with conserved cysteines between these two moths and the isolation of the gene for the *Manduca* EH (Horodyski *et al.*, 1989) has allowed isolation of similar genes from other insect species. Using the PCR technique, we have cloned EH from *Drosophila melanogaster* (Horodyski *et al.*, 1990) and identified the EH gene in mosquitoes (*Aedes aegypti*) and beetles (*Tenebrio molitor*) (Horodyski, Riddiford and Truman, unpublished).

In *Manduca*, EH is produced by two pairs of ventromedial neurosecretory cells in the brain that send their axons down the ventral nerve cord to terminate in the neurohaemal organ and the proctodeal nerve, in the larva and the pupa (Truman and Copenhaver, 1989) (Fig. 1). For adult eclosion a second neurohaemal site, the corpora cardiaca, is also used (Truman and Riddiford, 1970; Copenhaver and Truman, 1986). EH is released directly into the neuropile where it acts on its target neurones that control ecdysis behaviour and also into the haemolymph to act on peripheral targets (Hewes and Truman, 1991). Initial immunocytochemical studies had indicated that a subset of the lateral neurosecretory cells contained EH during adult development (Copenhaver and Truman, 1986). Yet *in situ* hybridization analysis shows that EH mRNA is found only in the ventromedial cells throughout adult development; no hybridization occurs in the lateral cells at any time (Horodyski *et al.*, 1989; Riddiford *et al.*, 1991).

Early studies of EH content of the brain and ventral nervous system using a bioassay showed that EH activity increased during the intermoult, and then plateaued during the moult (Truman *et al.*, 1981). A more detailed study of the distribution of EH activity during the final larval instar up to pupation also showed a dramatic increase in total activity during the feeding phase and very little further increase during the prepupal period (Truman and Morton, 1990). To determine whether these changes were due to regulation at the level of EH mRNA, we analysed changes in EH mRNA levels by whole mount *in situ* hybridization using a digoxigenin-labelled antisense mRNA probe (Riddiford *et al.*, 1991). The digoxigenin was visualized by immunocytochemistry using an antibody conjugated with either a peroxidase substrate or with the fluorescent dye, rhodamine. The latter allowed quantification of the amount of EH mRNA in the cells by laser scanning confocal microscopy. These studies showed that the EH mRNA was present at about the same level during both the intermoult and the moult with a slight decrease a few hours (larval or pupal) or a day (adult) before EH release. After larval and pupal ecdysis the levels of EH mRNA quickly returned to normal, and neither EH nor ecdysteroid seemed to influence this level.

The EH gene can be expressed either in a baculovirus-infected cell line (Eldridge *et al.*, 1991) or in yeast (Hayashi *et al.*, 1990), thus providing large quantities of pure peptide for physiological and molecular studies. The use of such systems to produce peptide is particularly critical for the larger neuropeptides that cannot be readily made synthetically. Such peptides can then be used to isolate and study the properties of the neuropeptide receptors, an area that has received relatively little attention in insects, and also to study the intracellular signalling mechanisms utilized by these neuropeptides.

Production and secretion of an active peptide by a baculovirus-infected cell line also indicates that, at least for some insect neuropeptides, there is the potential for genetically engineering their expression at the wrong time *in vivo*, leading to possible novel methods of insect control (Keeley *et al.*, 1990). The first examples of this approach have been the use of a synthetic gene for the *Manduca* diuretic hormone (Maeda, 1989) and of the gene for *Heliothis virescens* JH esterase (Hammock *et al.*, 1990). The latter slowed growth of the cabbage looper, *Trichoplusia ni*.

III. HORMONE RECEPTORS, TRANSCRIPTION FACTORS, AND THE REGULATION OF MOULTING AND METAMORPHOSIS

Gene regulation in eukaryotes is a complex process whose mysteries are just beginning to be unravelled. Basically, in the 5' region (and sometimes also in other regions) of the gene are sequences that bind tissue-specific and various other types of DNA-binding proteins called transcription factors. The complex of binding proteins then induces a change in DNA conformation and either allows (or prevents) the interaction of the RNA polymerase II–transcription factor complex with the promoter and thus gene transcription. One class of these transcription factors contains the steroid hormone receptors which are characterized by a middle DNA-binding domain consisting of two "zinc fingers" and a C-terminal ligand-binding domain (Evans, 1988). When hormone is present, these receptors bind to a particular DNA sequence as a dimer (Beato, 1989), and both the N-terminal and C-terminal domains are involved in a complex with other nuclear factors that in turn serve to activate or inactivate the particular gene.

Regulation of gene expression in insects by 20-hydroxyecdysone (20-HE) and juvenile hormone (JH) requires the presence of intracellular receptors for these hormones which when combined with hormone may initiate or modulate gene activity in a cascading series of cellular actions. Based on a detailed study of the effects of 20-HE on the puffing of polytene chromosomes in late larval salivary glands of *Drosophila melanogaster*, Ashburner *et al.* (1974) proposed that the ecdysteroid combined with a receptor directly and immediately activated certain "early" genes whose products in turn activated a second set of "late" genes and inactivated the first set. Two of the "early" genes (E74 and E75) have now been cloned, sequenced, and shown to encode DNA-binding proteins that bind to both "early" and "late" puff sites (Burtis *et al.*, 1990; Segraves and Hogness, 1990; Thummel *et al.*, 1990; Urness and Thummel, 1990), thus fulfilling certain predictions of the model. How these interact with the various structural genes to regulate development is not yet understood.

A. Ecdysteroid Receptor

The ecdysteroid receptor gene of *Drosophila melanogaster* has recently been cloned and characterized (Koelle *et al.*, 1991). This receptor is a member of the steroid hormone receptor superfamily. An incomplete palindromic DNA sequence in the 5' promoter region of two different ecdysteroid-activated genes which binds a crude ecdysteroid–receptor complex has been identified (Cherbas *et al.*, 1991; Ozyhar *et al.*, 1991). Although there is only one gene for the ecdysteroid receptor, there are at least three different forms of the molecule generated by either alternative splicing or alternative promoters (W. Talbot and D. S. Hogness, personal communication). All three have the same DNA- and ligand-binding domains, but differ in the N-terminal domain.

Interestingly, the three forms of the receptors are differentially expressed in a cell-specific and stage-specific manner, as can be seen by the use of form-specific monoclonal antibodies in an immunocytochemical study of the developing nervous systems of both *Manduca* and *Drosophila* (J. W. Truman, W. Talbot and D. S. Hogness, personal communication). For instance, in the ventral nervous system of *Drosophila*, the B1 form of the receptor is high in most larval neurones at pupation, but then rapidly declines over the next 24 hours as the pupa is formed and adult development begins. A subgroup of these cells begins to show high levels of the A form of the ecdysteroid receptor at 16–20 hours after pupation; these high levels then persist through the remainder of adult development. These cells include the neurones that are destined to die after adult eclosion (J. W. Truman, personal communication). The remaining larval neurones show an alternative pattern, with low levels of the A form appearing in the latter half of adult development, to be joined by the B1 form just before adult eclosion. The significance of the dynamics of these changes in receptor type during development is not yet understood.

Thus, whether or not a cell can respond to the ecdysteroid present in the haemolymph is dependent on whether or not it has ecdysteroid receptors. But the type of response that it gives, i.e. the genes either activated or inactivated by ecdysteroid, then will depend on the type of receptor present. The forms of the ecdysteroid receptors can change with developmental stage and are, themselves, likely to be modulated by the hormonal milieu.

B. Ecdysteroid-induced Transcription Factors and Development

The proteins encoded by the mRNA of the two "early" puffs in *Drosophila* salivary glands are apparently transcription factors, as they have DNA-binding domains and are found in the nuclei of several tissues. The E74 gene encodes two proteins, E74A and E74B, that both contain the *ets* oncogene type of DNA-binding domain (Burtis *et al.*, 1990) but show different developmental expression patterns, with B appearing before A and in response to lower concentrations of 20-HE (Thummel *et al.*, 1990; Karim and Thummel, 1991). Also, the proteins show a tissue-specific distribution (Boyd *et al.*, 1991). The E75 gene encodes a protein E75A that is a member of the steroid hormone receptor superfamily whose ligand is unknown (Segraves and Hogness, 1990). A second encoded protein, E75B, is from an alternative promoter and lacks the first zinc finger of the DNA-binding domain. Thus, it lacks the capability to bind to DNA since the key amino acids of the oestrogen and glucocorticoid receptors that bind to the DNA are located in the first finger (Schwabe and Rhodes, 1991).

One of the benefits of recombinant DNA techniques is that one can isolate homologous genes from different organisms using sometimes rather distantly related genes. A gene with a similar response to 20-HE as that of the "early" puffs was isolated from the *Manduca* genomic library through screening with the human retinoic acid receptor (hRARα) cDNA (Palli *et al.*, 1991a, 1992). The encoded protein proved to be 65% and 20% similar to hRARα in the DNA- and ligand-binding regions respectively (Palli *et al.*, 1992). It showed higher identity (97% and 68% respectively) to a *Drosophila* member of the steroid hormone receptor superfamily, DHR3 (Koelle *et al.*, 1992), and therefore is called MHR3 (Palli *et al.*, 1992). Preliminary experiments indicate that MHR3 produced by *in vitro* transcription in a bacterial system followed by *in vitro* translation binds neither iodovinylmethoprenol (a JH analogue) nor iodoponasterone (an ecdysteroid) (Palli and Riddiford, unpublished).

Two MHR3 transcripts (3.8 and 4.5 kb) are expressed in the epidermis of the embryo, larva, and pupa at the time of the ecdysteroid rises for the moults (Palli *et al.*, 1991a, 1992). Culture experiments showed that expression was induced in day 2 4th instar larval epidermis (before the rise of ecdysteroid for the moult) by a moulting concentration of 20-HE with maximal expression by 6 hours; between 12 and 18 hours the mRNA level began to decline. The level of induction was dependent on the concentration of 20-HE and largely independent of protein synthesis. By contrast, the decline in mRNA in the continuous presence of ecdysteroid required concurrent protein synthesis. Therefore, the response of the MHR3 gene to 20-HE is similar to that of the "early" puffs in *Drosophila*.

The MHR3 gene thus encodes an ecdysteroid-inducible, DNA-binding protein that is likely to be a transcription factor involved in the cascade of gene activation and inactivation initiated by ecdysteroids at the time of the moult. Figure 2 shows our present working hypothesis for the mode of action of this factor. The gene encoding a 14-kDa cuticular protein (LCP14) (Rebers and Riddiford, 1988) is expressed during the intermoult phase of each larval instar, then suppressed by ecdysteroid during the moult. This suppression is not due to a direct action of 20-HE on the gene, but rather is dependent on protein synthesis (Hiruma *et al.*, 1991). Although LCP14 mRNA normally has a half life longer than 10 hours, in these studies 50% had disappeared after exposure to 20-HE for 6 hours (a time at which MHR3 mRNA is maximal). Thus, the kinetics of these two effects of 20-HE make MHR3 a possible candidate for the steroid-induced factor that suppresses LCP14 and other intermoult genes during the moult. Alternatively, and/or additionally, MHR3 may be important for the activation of genes involved in production of the new cuticle. Studies are underway to test these possibilities.

C. Juvenile Hormone Receptors

Juvenile hormone is a sesquiterpenoid and appears to have at least two sites of action. One is on the membrane to trigger, apparently, the phosphatidylinositol second messenger system as seen in both *Rhodnius prolixus* ovaries (Ilenchuk and Davey, 1987; Sevala and Davey, 1989) and *Drosophila* male accessory glands (Yamamoto *et al.*, 1988). The second is in the nucleus where high-affinity JH-binding sites are found in cockroaches, locusts and grasshoppers, and moths (review: Palli *et al.*, 1991a).

MOULT

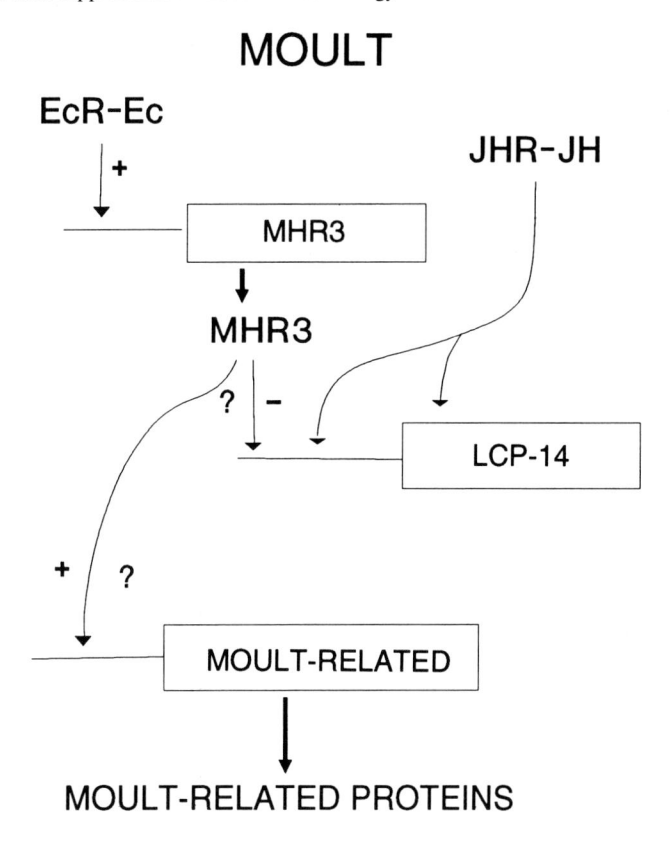

```
EcR-Ec
   |+
   ▼
 ┌─────────┐                JHR-JH
 │  MHR3   │                   |
 └─────────┘                   |
      |                        |
      ▼                        |
    MHR3                       |
    ╱ |                        |
   ?  | -                      |
      ▼         ┌─────────┐    |
                │ LCP-14  │◄───┘
                └─────────┘

  + ╱ ?
    ▼
 ┌──────────────────┐
 │  MOULT-RELATED   │
 └──────────────────┘
         |
         ▼
  MOULT-RELATED PROTEINS
```

Fig. 2. Working model for the action of ecdysteroid (Ec) and JH and the putative transcription factor MHR3 during the larval moult. See text for details.

In larval epidermis of *Manduca*, one third of the JH that enters the cells enters the nucleus, where it is bound by both a high- and a low-affinity receptor (Osir and Riddiford, 1988). Studies using biologically active, tritiated photoaffinity analogues of both the natural JHs (JH I and II) show that they are specifically bound by a 29-kDa nuclear protein (Palli *et al.*, 1990). A second 29-kDa nuclear protein specifically binds the photoaffinity analogue of methoprene. A similar protein is found in fat body.

These 29-kDa JH-binding proteins are present in the epidermis throughout the larval feeding period and in the pupa, but decrease during the ecdysteroid rise for the larval moult and disappear at the onset of metamorphosis (Palli et al., 1990, 1991b). In the latter case, in vitro experiments show that this loss is due to the action of 20-HE in the absence of JH (Palli et al., 1991b). Thus, when the cell becomes pupally committed such that it can no longer make larval cuticle, it loses its JH receptors and consequently the ability to respond to JH. The presence of these JH-binding proteins also appears to be dependent on the presence of JH during the moult. Allatectomy late during the ecdysteroid rise for the larval moult caused formation of a viable 5th instar larva which lacked the 29-kDa JH-binding protein (Palli et al., 1991b). This protein reappeared only if these larvae were treated with JH from the time of allatectomy or if they were later given JH during the pupal moult. JH treatment only during the feeding stage was ineffective.

The 29-kDa photoaffinity-labelled JH-binding nuclear protein was purified and proved to be N-terminally blocked (Prestwich, 1991). Limited digestion with the endoproteinase Lys-C provided three short, non-overlapping amino acid sequences (Touhara, Atkinson and Prestwich, unpublished). Using two-fold degenerate oligonucleotide primers based on these sequences, Palli and Riddiford (unpublished) have isolated by PCR a cDNA encoding this protein. A small region (44 amino acids) of the encoded protein shows 37% identity with the bovine interphotoreceptor retinoid binding protein (Borst et al., 1989) in the four-fold protein repeat that binds fatty acids. The complete cDNA sequence will tell us whether this is a DNA-binding protein and will allow us to synthesize the protein in vitro to determine whether it binds JH.

A 29-kDa nuclear protein with the same developmental specificity as the 29-kDa JH-binding protein binds to the LCP14 gene as determined by hybridization of this gene with a blot of the 0.5 M KCl-soluble nuclear proteins from the different stages (Palli et al., 1990). Preliminary gel retardation studies suggest that this binding may occur in both the 5' upstream regulatory region and in the first intron (Palli and Riddiford, unpublished). Our working hypothesis is that the JH–receptor complex may bind to a larval-specific intermoult gene such as LCP14 and modulate its response to ecdysteroid (Fig. 2). When ecdysteroid initiates the larval moult in the presence of JH, transcription of LCP14 is suppressed but reappears when ecdysteroid declines at the end of the moult (Hiruma et al., 1991). When JH is absent at the time of ecdysteroid action as at the onset of metamorphosis, LCP14 mRNA does not reappear. Therefore, the presence of the JH–receptor complex on the gene is thought to allow ecdysteroid-induced suppression but not the additional events that cause permanent repression.

IV. CONCLUSIONS

The molecular studies discussed here show the relatively rapid progress that molecular biology has allowed us to make in the field of insect endocrinology. We now know the structures of many of the insect neuropeptides and which cells produce them. Therefore, we can study how these cells are regulated by both hormonal and environmental signals and also their targets both within the central nervous system and peripherally. Although the second messengers for some of these neuropeptides are known, their receptors are not and neither is the transduction mechanism clear – important questions that molecular techniques can help to answer in the future.

With the intracellular hormones, ecdysteroids and juvenile hormones, the use of molecular techniques is providing us with a diversity of hormone receptors and ecdysteroid-induced transcription factors. These molecular probes not only will help us to understand how these two hormones regulate specific genes, but also should provide insight into how ecdysteroid orchestrates the moult – a cascade of cellular events first requiring the hormone, then depending on its absence. Then, a key question is how JH interacts in the first part of this cascade to prevent metamorphosis. Does it, for instance, influence the subtype of ecdysteroid receptor present or the particular transcription factors induced? Does it act directly on the stage-specific genes themselves to alter their response to ecdysteroid and, if so, how? These approaches should also in the next few years help us to understand the actions of JH on reproductive maturation where it appears to act independently of ecdysteroid and may act either intracellularly or on the membrane.

Acknowledgements

I thank Professor James W. Truman for helpful discussions and comments on this manuscript. The unpublished work in my laboratory was supported by NSF DCB88-18876 and DCB90-05202 and NIH AI12459.

REFERENCES

Agui, N., Granger, N. A., Gilbert, L. I. and Bollenbacher, W. E. (1979). *Proc. Natl. Acad. Sci. USA* **76**, 5694–5698.

Ashburner, M., Chihara, C., Meltzer, P. and Richards, G. (1974). *Cold Spring Harbor Symp. Quant. Biol.* **38**, 655–662.

Beato, M. (1989). *Cell* **56**, 335-344.
Bogus, M. I. and Scheller, K. (1991). *Zool. Jb. Physiol.* **95**, 197-208.
Bogus, M. I., Hofbauer, A., Buchner, E., Poll, S. and Scheller, K. (1991). *Sericologia* **31** (suppl.), 49.
Bollenbacher, W. E. and Gilbert, L. I. (1981) *In* "Neurosecretion. Molecules, Cells, Systems" (D. S. Farner and K. Lederis, eds), pp. 361-370. Plenum Press, New York.
Borst, D. E., Redmond, T. M., Elser, J. E., Gonda, M. A., Wiggert, B., Chader, G. J. and Nickerson, J. M. (1989). *J. Biol. Chem.* **264**, 1115-1123.
Boyd, L., O'Toole, E. and Thummel, C. S. (1991). *Development* **112**, 981-995.
Burtis, K. C., Thummel, C. S., Jones, C. W., Karim, F. D. and Hogness, D. S. (1990). *Cell* **61**, 85-99.
Cherbas, L., Lee, K. and Cherbas, P. (1991). *Genes Devel.* **5**, 120-131.
Copenhaver, P. F. and Truman, J. W. (1986). *J. Neurosci.* **6**, 1738-1747.
Cusson, M., Prestwich, G. D., Stay, B. and Tobe, S. S. (1991). Abstracts, "5th International Symposium on Juvenile Hormones", La Londe les Maures, France.
Eldridge, R., Horodyski, F. M., Morton, D. B., O'Reilly, D. R., Truman, J. W., Riddiford, L. M. and Miller, L. K. (1991). *Insect Biochem.* **21**, 341-351.
Evans, R. M. (1988). *Science* **240**, 889-895.
Hammock, B. D., Bonning, B. C., Possee, R. D., Hanzlik, T. N. and Maeda, S. (1990). *Nature* **344**, 458-460.
Hayashi, H., Nakano, M., Shibanaka, Y. and Fujita, N. (1990). *Biochem. Biophys. Res. Commun.* **173**, 1065-1071.
Hewes, R. S. and Truman, J. W. (1991). *J. Comp. Physiol. A* **168**, 697-707.
Hiruma, K., Hardie, J. and Riddiford, L. M. (1991). *Dev. Biol.* **144**, 369-378.
Horodyski, F. M., Riddiford, L. M. and Truman, J. W. (1989). *Proc. Natl. Acad. Sci. USA* **86**, 8123-8127.
Horodyski, F., Truman, J. and Riddiford, L. (1990). Abstract, "31st Annual Drosophila Conference", Asilomar, #28.123.
Ilenchuk, T. T. and Davey, K. G. (1987). *Insect Biochem.* **17**, 1085-1088.
Iwami, M. (1990). *In* "Molting and Metamorphosis" (E. Ohnishi and H. Ishizaki, eds), pp. 49-66. Japan Sci. Soc. Press, Tokyo/Springer-Verlag, Berlin.
Iwami, M., Kawakami, A., Ishizaki, H., Takahashi, S. Y., Adachi, T., Suzuki, Y., Nagasawa, H. and Suzuki, A. (1989). *Devel. Growth Diff.* **31**, 31-37.
Iwami, M., Adachi, T., Kondo, H., Kawakami, A., Suzuki, Y., Nagasawa, H., Suzuki, A. and Ishizaki, H. (1990). *Insect Biochem.* **20**, 295-303.
Karim, F. D. and Thummel, C. S. (1991). *Genes Dev.* **5**, 1067-1079.
Kataoka, H., Troetschler, R. G., Kramer, S. J., Cesarin, B. and Schooley, D. A. (1987). *Biochem. Biophys. Res. Commun.* **146**, 746-750.
Kataoka, H., Toschi, A., Li, J. P., Carney, R. L., Schooley, D. A. and Kramer, S. J. (1989). *Science* **243**, 1481-1483.
Kawakami, A., Iwami, M., Nagasawa, H., Suzuki, A. and Ishizaki, H. (1989), *Proc. Natl. Acad. Sci. USA* **86**, 6843-6847.
Kawakami, A., Kataoka, H., Oka, T., Mizoguchi, A., Kimura-Kawakami, M., Adachi, T., Iwami, M., Nagasawa, H., Suzuki, A. and Ishizaki, H. (1990). *Science* **247**, 1333-1335.
Keeley, L. L., Hayes, T. K. and Bradfield, J. Y. (1990). *In* "Insect Neurochemistry and Neurophysiology 1989" (A. B. Borkovec and E. P. Masler, eds), pp. 163-203. Humana Press, Clifton, NJ.
Koelle, M. R., Talbot, W. S., Segraves, W. A., Bender, M. T., Cherbas, P. and Hogness, D. S. (1991). *Cell* **67**, 59-77.
Koelle, M. R., Segraves, W. A. and Hogness, D. S. (1992). *Proc. Natl. Acad. Sci. USA* (in press).

Kono, T., Nagasawa, H., Isogai, A., Fugo, H. and Suzuki, A. (1991). *Insect Biochem.* **21**, 185–195.
Maeda, S. (1989). *Biochem. Biophys. Res. Commun.* **165**, 1177–1183.
Marti, T., Takio, K., Walsh, K. A., Terzi, G. and Truman, J. W. (1987). *FEBS Lett.* **219**, 415–418.
Mizoguchi, A., Ishizaki, H., Nagasawa, H., Kataoka, H., Isogai, A., Tamura, S., Suzuki, A., Fujino, M. and Kitada, C. (1987). *Mol. Cell. Endocrinol.* **51**, 227–235.
Mizoguchi, A., Hatta, M., Sato, S., Nagasawa, H., Suzuki, A. and Ishizaki, H. (1990a). *J. Insect Physiol.* **36**, 655–664.
Mizoguchi, A., Oka, T., Kataoka, H., Nagasawa, H., Suzuki, A. and Ishizaki, H. (1990b). *Devel. Growth Differ.* **32**, 591–598.
Nagasawa, H., Kataoka, H., Isogai, A., Tamura, S., Suzuki, A., Ishizaki, H., Mizoguchi, A., Fujiwara, Y. and Suzuki, A. (1984). *Science* **226**, 1344–1345.
Nagasawa, H., Kataoka, H. and Suzuki, A. (1990). In "Molting and Metamorphosis" (E. Ohnishi and H. Ishizaki, eds), pp. 33–48. Japan Sci. Soc. Press, Tokyo/Springer-Verlag, Berlin.
Osir, E. O. and Riddiford, L. M. (1988). *J. Biol. Chem.* **263**, 13812–13818.
Ozyhar, A., Strangmann-Diekmann, M., Kiltz, H-H. and Pongs, O. (1991). *Eur. J. Biochem.* **200**, 329–335.
Palli, S. R., Osir, E. O., Eng W.-S., Boehm, M. F., Edwards, M., Kulcsar, P., Ujvary, I., Hiruma, K., Prestwich, G. D. and Riddiford, L. M. (1990). *Proc. Natl. Acad. Sci. USA* **87**, 796–800.
Palli, S. R., Riddiford, L. M. and Hiruma, K. (1991a). *Insect Biochem.* **21**, 7–15.
Palli, S. R., McClelland, S., Hiruma, K., Latli, B. and Riddiford, L. M. (1991b). *J. Exp. Zool.* **260**, 337–344.
Palli, S. R., Hiruma, K. and Riddiford, L. M. (1992). *Dev. Biol.* **150**, 306–398.
Pratt, G. E., Farnsworth, D. E., Siegel, N. R., Fok, K. F. and Feyereisen, R. (1989). *Biochem. Biophys. Res. Commun.* **163**, 1243–1247.
Pratt, G. E., Farnsworth, D. E. and Feyereisen, R. (1990). *Mol. Cell. Endocrinol.* **70**, 185–195.
Pratt, G. E., Farnsworth, D. E., Fok, K. F., Siegel, N. R., McCormack, A. L., Shabanowitz, J., Hunt, D. F. and Feyereisen, R. (1991). *Proc. Natl. Acad. Sci. USA* **88**, 2412–2416.
Prestwich, G. D. (1991). *Insect Biochem.* **21**, 27–40.
Rebers, J. E. and Riddiford, L. M. (1988). *J. Mol. Biol.* **203**, 411–423.
Reynolds, S. E. (1985). In "Comprehensive Insect Physiology, Biochemistry and Pharmacology", Vol. 8 (G. A. Kerkut and L. I. Gilbert, eds), pp. 335–351. Pergamon Press, Oxford.
Riddiford, L. M., Truman, J. W., Hewes, R. and Horodyski, F. M. (1991). Abstract, "3rd International Congress on Comparative Physiology and Biochemistry", Tokyo, p. 109.
Saiki, R. K., Gelfand, D. H., Stoffel, S., Scharf, S. J., Higuchi, R., Horn, G. T., Mullis, K. B. and Erlich, H. A. (1988). *Science* **239**, 487–491.
Schwabe, J. W. R. and Rhodes, D. (1991). *TIBS* **16**, 291–296.
Segraves, W. A. and Hogness, D. S. (1990). *Genes Dev.* **4**, 204–219.
Sevala, V. L. and Davey, K. G. (1989). *Experientia* **45**, 355–356.
Stay, B., Woodhead, A. P. and Chan, K. K. (1991). Abstracts, "5th International Symposium on Juvenile Hormones", La Londe les Maures, France.
Thummel, C. S., Burtis, K. C. and Hogness, D. S. (1990). *Cell* **61**, 101–111.
Truman, J. W. (1990). In "Molting and Metamorphosis" (E. Ohnishi and H. Ishizaki, eds), pp. 67–82. Japan Sci. Soc. Press, Tokyo/Springer-Verlag, Berlin.
Truman, J. W. and Copenhaver, P. F. (1989). *J. Exp. Biol.* **147**, 457–470.

Truman, J. W. and Morton, D. B. (1990). *In* "Progress in Comparative Endocrinology" (A. Epple, C. G. Scanes and M. Stetson, eds), pp. 300–308. Wiley-Liss, New York.

Truman, J. W. and Riddiford, L. M. (1970). *Science* **167**, 1624–1626.

Truman, J. W., Taghert, P. H., Copenhaver, P. F., Tublitz, N. J. and Schwartz, L. M. (1981). *Nature* **291**, 70–71.

Urness, L. D. and Thummel, C. S. (1990). *Cell* **63**, 47–61.

Woodhead, A. P., Stay, B., Seidel, S. L., Khan, M. A. and Tobe, S. S. (1989). *Proc. Natl. Acad. Sci. USA* **86**, 5997–6001.

Yamamoto, K., Chadarevian, A. and Pellegrini, M. (1988). *Science* **239**, 916–919.

Appendix

DNA Transformation of Non-Drosophilids (Workshop Abstracts)

  I. Introduction 241
 II. Abstracts 243
     A. Transposable Genetic Elements 243
     B. Transformation Vectors: Selectable Markers 246
     C. Transformation Vectors: Promoters 247
     D. Transformation Vectors: Control of Integration 249
     E. Transformation Techniques 250
     F. Cell Transfection: *In Vitro* Studies 252
     G. Gene Expression in Insects 252
     H. Cloning and Characterization of Insect Genes 256
III. Overview 258
     A. Introduction of the Genetic Construct into the Insect . 258
     B. Stable Maintenance of the Construct and Incorporation
        into the Genome of the Insect 259
     C. Marker Genes for Distinguishing Transformed and Non-
        transformed Insects 261
     D. Control of Transgene Expression in Non-drosophilids 261
     E. Conclusions 262

I. INTRODUCTION

This Workshop, which preceded the 16th International Symposium of the Royal Entomological Society, entitled "Insect Molecular Science", was designed to provide a forum for the exploration of DNA transformation of non-drosophilid insects. It had become increasingly obvious that there was growing international interest in the exploitation of this new technology, previously used to great effect in the fruit fly *Drosophila melanogaster*, in insects of major medical, commercial or economic importance. Because of this interest, we thought that it would be worthwhile to bring together a group of like-minded people, all of whom had an interest in the DNA transformation of non-drosophilids (regardless of their organism of choice), in the hope that we could determine the progress which had been made, as well as the difficulties which still had to be faced. Most of the 30 people who took part hoped soon to be able to exploit this new technology, both for basic molecular research and for applied genetic control or improvement strategies. As might have been anticipated, however, the various presentations made it abundantly clear that we are still faced with numerous difficulties, both practical and (perhaps) ethical, which must be overcome if research centred on the transformation of non-drosophilids is to achieve its full potential. Nevertheless, it was encouraging to see evidence of the broad range of research programmes now underway and the progress which had been made in a number of these areas.

The Workshop was purposely structured to raise relevant issues under eight topic headings: Transposable genetic elements; Transformation vectors – selectable markers; Transformation vectors – promoters; Transformation vectors – control of integration; Transformation techniques; Cell transfection – *in vitro* studies; Gene expression in insects; Cloning and characterization of insect genes. These were chosen to reflect the major areas of research effort which were beginning to come together in relation to non-drosophilid transformation. It was hoped that they would also serve to guide the discussion and comment along a logical path.

Initially, there was to have been a ninth topic covering "Transgenic insects in populations". However, this was evidently a little ambitious given that most current research is not sufficiently developed for these experiments to be undertaken. Several research groups, nevertheless, are beginning to consider how the release of transgenic insects into populations should be approached. Obviously, this is an area where ethical difficulties might be encountered and these will have to be addressed, initially perhaps through a combination of theoretical modelling and carefully controlled contained release experiments. It may be that some preliminary investigations along these lines could be initiated with *Drosophila melanogaster*.

For each topic, one or more short (5–6 minute) presentations were followed by discussion and comment. The Workshop was to have concluded with a general discussion of the issues raised, together with a consensus of participants' views on research priorities. Although some of these points were raised during the Workshop, the final discussion did not materialize. This was partly through lack of time but also because, after almost 200 minutes of intensive thought and information transfer, everyone had run out of steam! It was a generally held view that not enough time was available for the range of issues raised. Sixteen presentations were scheduled and almost every presenter found it difficult to work to the 5-minute deadline. Even though two of the speakers (Ann Warren and Charles Milne) had to withdraw before the Workshop, this still left us short of time. With the benefit of hindsight, the range of material covered was such that a whole day (rather than an afternoon) could usefully have been devoted to this topic. This might also have had the advantage of leaving participants with more time for thought and discussion. Clearly, this is something for the organizers of the **next** Workshop on non-drosophilid transformation to consider.

P. EGGLESTON and J. CARLSON

II. ABSTRACTS

A. Transposable Genetic Elements

1. Sequences from Lucilia cuprina *with homology to transposable elements from* Drosophila melanogaster

H. PERKINS, K. JOHNSON AND A. HOWELLS
Division of Biochemistry and Molecular Biology, School of Life Sciences, Australian National University, Canberra, Australia

Various transposable elements from *D. melanogaster* have been used in an attempt to identify similar sequences in the Australian sheep blowfly. Subclones from the retrotransposons 297, B104, mdg3, gypsy and copia (containing conserved domains from the reverse transcriptase coding region), from the I element and from the transposons P and hobo have been used to probe Southern blots of blowfly total genomic DNA at low stringency. All probes gave positive bands but in many cases these corresponded to satellites visible on the gels and represented nonspecific binding. The 297- and B104-dervived probes gave the strongest signals after one-week exposures.

A *L. cuprina* genomic library was screened with 297-, B104-, I-element, P-element- and hobo-derived probes. Positively hybridizing plaques were obtained in all cases, the numbers of which indicated a copy number of approximately 100–200 hybridizable sequences per genome in the cases of 297, B104 and the I element, and about 30 per genome for the P element and hobo probes. A number of positively hybridizing clones from the 297, P and hobo screens have been plaque purified and are being analysed in detail. Most work so far has been done on the 297- and P-like clones.

Of the 297-like clones, over 7 kb of one of them has now been sequenced and homology has been found to the protease, RNase H and integrase domains of the 297 pol (reverse transcriptase) ORF so far.

Sequence thought to come from potential ORFs 1 and 3, however, bears little or no similarity to the 297 sequence. LTRs and potential primer-binding sites have yet to be determined. The presence of various small deletions and insertions, as well as a 22-bp and 186-bp duplication, disrupt the potential ORFs, indicating that this particular element is inactive. The different 297-like clones that have been purified have different restriction patterns and differing intensities of interclone hybridization, suggesting that they belong to several related but distinct families.

Three kilobases of sequence have been obtained from one of the P-like clones of *Lucilia* and show a fairly high degree of homology to the sequence of the canonical P element of *D. melanogaster*. Comparisons of the derived amino acid sequences of the two elements show about 50% identity throughout ORFs 1 and 2 and the beginning of ORF 3 of the P element. However, homology is less in the more distal parts of ORF 3 and barely detectable in ORF 0. The exon/intron structure also appears to be well conserved, with the exception that ORF 2 of the blowfly element is interrupted by two extra-small introns. Sequencing is incomplete at this stage so the ends of the element have not yet been determined and the nature of any terminal repeats remains unknown. Lower hybridization intensities to the P element probe, and differences in the restriction patterns of the other P-like clones, suggest that they may represent internally deleted derivatives of the full length element.

Acknowledgements

This work was supported by a Postgraduate Scholarship from the Wool Research and Development Fund of the Australian Wool Corporation.

2. *Retrotransposon reverse transcriptase-like sequences in the genome of mosquitoes*

A. WARREN AND J. M. CRAMPTON
Wolfson Unit of Molecular Genetics, Liverpool School of Tropical Medicine, Pembroke Place, Liverpool L3 5QA, UK

Surprisingly little research has been directed towards the molecular characterization of mosquito genomes despite the role these insects play as vectors of major medical importance. We are interested in the potential use of transposable genetic elements (TGEs) as tools for elucidation of the mosquito genome organization and gene isolation via transposon tagging strategies. We are further interested in the development of endogenous TGEs as DNA vectors for genome manipulation (Crampton et al., 1990). TGEs have been used successfully for molecular characterization and manipulation of the genome of a number of organisms (Daniels et al., 1985). While little research involving the use of retrotransposons (RTPs) as vectors for genome manipulation has been reported, Jacobs et al. (1988) have recently shown that modified endogenous RTPs (*Ty* elements) of yeast can be used for the reverse transcriptase-mediated transposition of foreign genes into yeast chromosomes. Although this strategy has not yet been tested in other species, it seems highly probable that manipulation of the mosquito genome may similarly be achieved using endogenous mosquito RTPs. To this end, we have initiated a search for RTPs in the genome of the mosquitoes *Aedes aegypti* and *Anopheles gambiae* which should provide us with functional elements for the manipulation of the genomes of these organisms.

Using PCR, together with oligonucleotide primers corresponding to the highly conserved amino acid regions of RTP reverse transcriptases, we have amplified DNA sequences using both *Ae. aegypti* and *An. gambiae* DNA as template. The products of the PCR reactions were blunt-end ligated into the SmaI site of M13mp9 and sequenced using the dideoxy method and Sequenase II. Considerable homology exists between the reverse transcriptase-like sequences found in the *Ae. aegypti* and *An. gambiae* genomes and other retrotransposon reverse transcriptases. The cloned PCR products have been used to identify and isolate several clones from representative genomic libraries. These genomic clones are currently being analysed to determine whether these sequences form a part of functional RTPs in the mosquito genome. We believe this should prove to be a highly effective way of rapidly identifying endogenous retrotransposons in the genomes of these and other organisms.

Acknowledgements

We wish to thank A. J. Flavell for kindly providing the oligonucleotide primers used in this study. This work was supported by the Wellcome Trust. JMC is a Wellcome Trust Senior Research Fellow in Basic Biomedical Sciences.

References

Crampton, J. M., Morris, A. C., Lycett, G. J., Warren, A. M. and Eggleston, P. (1990). *Parasitology Today* **6**, 31–36.
Daniels, S. B., Strausbaugh, L. D. and Armstrong, R. A. (1985). *Mol. Gen. Genet.* **200**, 258–265.
Jacobs, E., Dewerchin, M. and Boeke, J. D. (1988). *Gene* **67**, 259–269.

B. Transformation Vectors: Selectable Markers

1. Cloning eye colour genes from Lucilia cuprina

C. PATTERSON, A. ELIZUR, R. GARCIA, H. PERKINS AND A. HOWELLS
Division of Biochemistry and Molecular Biology, School of Life Sciences, Australian National University, Canberra, Australia

Eye colour genes are used as "conditional lethal" genes in the genetic control strategy being developed for the Australian sheep blowfly because, whereas they can be reared easily in the laboratory, they cause death in the field. Eye colour genes are also extremely useful as marker genes in attempting to develop gene transformation systems. Consequently we have attempted to clone four different eye colour genes which are essential for the synthesis of the brown pigment xanthommatin. We have had different experiences with the four different genes, so this abstract represents a synthesis of our acquired wisdom.

Our strategy has been to select regions thought likely to have been conserved in the cloned *Drosophila melanogaster* gene and to use this to probe, under conditions of reduced stringency, both Southern blotted genomic DNA from *Lucilia cuprina* and also genomic DNA libraries. Attempts have been made to clone the *L. cuprina* equivalents of *white, scarlet, vermilion* and *cinnabar*.

Of the four genes tested, two have been cloned; these are *white* and the equivalent of *scarlet* called *topaz*. With *white*, several bands are found on the Southern filters, the strongest of which correspond to satellite bands on the gels; weaker bands can be seen which in hindsight can be attributed to hybridization to the *L. cuprina white* gene. When genomic libraries were screened, positive clones were obtained; these have been plaque purified and sequence data are available for about 75% of the gene (from its 3' end). With *scarlet* the strongest bands seen on the Southern filters do not correspond to satellites and are due to hybridization to *topaz* gene sequences. Clones carrying *topaz* have been purified and sequence data are available for about 90% of the coding regions. As with *white*, the sequences appear to be less conserved at the 5' end and we have not yet been able to establish the site of transcription initiation.

Our attempts to clone the counterparts of *vermilion* and *cinnabar* have so far been unsuccessful. With *vermilion*, we can observe bands of specific hybridization on Southern filters, but all attempts to isolate positive clones from the genomic libraries have been unsuccessful. We suspect that the gene may be flanked by repeated sequences which cause fragments carrying it to clone poorly in *E. coli*. With *cinnabar* we have been unable to observe bands of specific hybridization on Southern filters, nor have we been able to detect positive clones in the one genomic library we have screened.

Acknowledgements
This work has been supported by grants from the Australian Research Council and by a research grant and a Postgraduate Research Scholarship from the Wool Research and Development Fund of the Australian Wool Corporation.

C. Transformation Vectors: Promoters

1. The expression of reporter genes in cultured cells of Aedes aegypti

G. LYCETT, J. M. CRAMPTON AND P. EGGLESTON
Wolfson Unit of Molecular Genetics, Liverpool School of Tropical Medicine, Pembroke Place, Liverpool L3 5QA, UK

To date, concerted attempts to transform non-drosophilid insects have utilized the transposable P element from *Drosophila* as a DNA vector. Such experimentation was performed with limited knowledge of the control of gene expression in non-drosophilids and was reliant on non-drosophilid gene regulation being similar to that of *Drosophila*. The inefficiency of P element transformation in alternative organisms has thus led to (1) a reappraisal of the P element constructs, (2) a search for other transposable element vectors and (3) a more detailed analysis of gene expression in insect genomes.

In this abstract we describe some of the factors that affect gene expression in the mosquito *Aedes aegypti*. In particular, those factors influencing the extent of CAT expression from the HSP70 promoter of the HSP-CAT-1 construct in an *Ae. aegypti* (MOS20) cell line. Using total reporter gene expression as a measure of transfection efficiency, preliminary analysis of several transfection techniques indicates that polybrene, calcium phosphate precipitation and Lipofectin mediate uptake of DNA into MOS20 cells. From immunohistochemical studies it was established that 5–30 per 10^5 MOS20 cells were synthesizing CAT protein at detectable levels following polybrene transfection. To maximize CAT activity and assay sensitivity, various parameters affecting heat-shock-induced gene expression were characterized. These parameters included temperature and duration of heat shock, post-transfection time and duration of recovery post-heat-shock. In agreement with published data concerning *Ae. albopictus* cell transfection, optimal induction of the *Ae. aegypti* analogue to HSP70 occurs at 41°C.

In addition, we have compared the inducibility of different length *Drosophila* HSP70 promoter sequences. Our findings suggest that the 1.1 kb promoter sequence of HSP-CAT-1 yields greater transfected cell reporter gene activity than the truncated (440-bp) sequence present on the P element vector pUChsneo. Furthermore, we have demonstrated that *Drosophila* promoters from the actin 5C gene and the E74 gene are actively and constitutively expressed in *Ae. aegypti* cells. Further work is now directed towards characterizing the activity of other eukaryotic promoters in this mosquito cell line.

Acknowledgements

This work was supported by the MRC, Wellcome Trust and Lister Institute of Preventive Medicine. JMC is a Wellcome Trust Senior Research Fellow in Basic Biomedical Sciences and PE is a Lister Institute Research Fellow.

2. *Regulated expression from the* Drosophila *metallothionein promoter in mosquito cells*

M. KOVACH, J. CARLSON AND B. BEATY

Department of Microbiology, Colorado State University, Fort Collins, CO 80523, USA

Expression from the *Drosophila* metallothionein promoter (MT) was examined in mosquito cells. A plasmid, pMT-1, that contains the MT promoter (Johansen *et al.*, 1989) transcribing the β-galactosidase gene, was constructed. pMT-1 was introduced into C6/36 *Aedes albopictus* cells via lipofection (BRL). Stable transformants were achieved by co-transfection of pMT-1 with a second plasmid carrying the selectable hygromycin marker.

Slot blot hybridization with the labelled β-galactosidase gene was used to estimate the copy number of the plasmid in the pMT-1 transformant. Approximately 600 copies of the β-galactosidase gene were present per cell. The plasmid containing the selectable marker was present at about 65 copies per cell.

A wide spectrum of heavy metals were tested for their inducibility of the MT promoter. Copper sulphate was the most efficient inducer of the promoter in mosquito cells, resulting in an 8–30-fold increase in enzyme activity. Induction could be achieved with concentrations as low as 25 μM and up to 300 μM. Concentrations of 400 μM and above were toxic to the cells. Non-induced cells did not express β-galactosidase. Minimal induction was achieved with cadmium chloride, with optimal conditions resulting in 1–2-fold induction.

The work described provides an inducible and controllable expression system in mosquito cells, and could be beneficial in developing *in vivo* molecular strategies for control of arthropod-borne diseases.

References
Johansen, H., Vander-Straten, A., Sweet, R., Otto, E., Mavoni, G. and Rosenberg, M. (1989). *Genes Dev.* **3**, 882–889.

D. Transformation Vectors: Control of Integration

1. FLP-mediated recombination in the vector mosquito, Aedes aegypu

A. MORRIS AND A. JAMES
Department of Molecular Biology and Biochemistry, University of California, Irvine, California 92717, USA

Low-frequency illegitimate recombination events such as those isolated from previous attempts to transform mosquitoes could be utilized effectively if the integrated sequence served as a target for a heterologous high-frequency recombination system. We report the activity in mosquito embryos of a yeast recombinase, FLP, acting on a specific target DNA sequence, FRT, isolated from the yeast 2-μm plasmid. In a series of experiments, plasmids containing the FLP recombinase under the control of the *Drosophila melanogaster* heat shock 70 promoter were co-injected with target plasmids containing FRT sites. FLP-mediated recombination was detected between (1) target sites located on separate plasmids resulting in the formation of dimers or higher order multimers and (2) target sites located on the dimers reformed in (1), leading to resolution of the dimers to their original monomeric forms. Synthetic FRT sites were also used and gave rise to similar results to those obtained using the FRT sites originally isolated from the yeast 2-μm plasmid. This successful demonstration of yeast FLP recombinase activity within the mosquito embryo suggests a possible future application of this system in establishing transformed lines of mosquitoes to further the molecular analysis of this important vector and for use in control strategies.

E. Transformation Techniques

1. Stable silkworm moth embryo transformation by DNA injection

S. MAHALINGAM, Y. BANNO and K. IATROU
Department of Medical Biochemistry, Faculty of Medicine, University of Calgary, 3330 Hospital Drive NW, Calgary, Alberta T2N 4N1, Canada

We have developed a plasmid vector that promotes transgene chromosomal integration into the germ line of injected embryos of the domesticated silkworm moth *Bombyx mori* at high frequency. Preblastoderm stage embryos injected with this vector were allowed to develop to maturity, crossed to each other and sacrificed after mating. PCR analyses of residual gonad DNA demonstrated that over 70% of these adults carried the injected vector DNA in their gonads and that all positive individuals contained the injected vector sequences as integral parts of their chromosomes.

Sibling crosses amongst the F_1 progeny of injected individuals were carried out and the gonads of the F_1 generation (parents of F_2) were similarly analysed by PCR to determine whether the injected vector persisted in the F_1. Of 32 F_1 individuals analysed from a cross between two injected parents, 28 (14 males and 14 females) were found to be positive while 4 females were found to be negative. This transmission rate suggests that chromosomal integration of the vector sequences occurs prior to, or at the time of, establishment of the germ cell progenitors. Quantitative PCR analyses revealed that, under the specific injection conditions employed in our experiments, F_1 progeny carry between one and five integrated copies of the vector sequences per haploid genome.

Analysis of genomic DNA of F_2 progeny by PCR and Southern hybridization are currently in progress to determine the mechanism of integration and the stability of integrated sequences. The applicability of our method for embryo transformation to other lepidopteran and non-lepidopteran insect species will be discussed.

Acknowledgements

This work has been supported by the Canadian National Network of Centres of Excellence for "Biotechnology for Insect Pest Management" (Insect Biotech Canada) and the Medical Research Council of Canada.

2. Sperm-mediated honey bee transformation

C. MILNE
Department of Entomology, Program in Genetics and Cell Biology, Washington State University, Pullman, WA 99164-6432, USA

Considering reports of sperm-mediated gene transfer in rabbits, sea urchins, mice and chickens, it is worthwhile to determine if insect sperm can transfer genes, and honey bees are the ideal insect to examine this transformation technique. This was investigated by inseminating honey bee queens with sperm that had been incubated with a 1-kb linear DNA fragment, and assaying the DNA of G_0 progeny by the polymerase chain reaction. Preliminary evidence indicates that DNA incubated with bee sperm (1) was present in about 30% of the progeny, (2) appeared to be in the genome, (3) only occurred in a fraction of the cells, and (4) was probably not transferred by incorporation into the sperm genome.

If honey bee sperm can transfer genes, then the sperm should physically bind DNA. Honey bee sperm was incubated with ^{32}P-labelled DNA, and the sperm bound DNA (1) relatively rapidly, (2) in large amounts (up to 1000 molecules/sperm), (3) on the outside of the sperm, (4) relatively tightly, and (5) to nucleic-acid-specific sites. Sperm-bound DNA retained its size and structure and did not appear to integrate into the sperm genome.

These preliminary results indicate that honey bee sperm can transfer genes into the genome. Considering the elaborate techniques often employed to transfer genes, this method represents a substantial breakthrough in gene transfer technology in terms of its ease, success rate and broad applicability to other insects. This technique may work in other insects for which instrumental insemination with sperm has not been perfected by placing a concentrated DNA-containing solution in the reproductive tracts of females before natural mating to treat the sperm *in vivo*.

F. Cell Transfection: *In Vitro* Studies

1. Stable transformation of a mosquito cell line results in extraordinarily high copy numbers of plasmid

T. MONROE, M. KOVACH, M. MUHLMANN, J. CARLSON, J. BEDFORD AND B. BEATY

Department of Microbiology, Colorado State University, Fort Collins, CO 80523, USA

Stable incorporation of high copy numbers (10^4–10^5) of a plasmid vector was achieved in a cell line derived from the *Aedes albopictus* mosquito. Plasmid sequences were readily observed by ethidium bromide staining of cellular DNA after restriction endonuclease digestion and agarose gel electrophoresis. The plasmid was demonstrated by *in situ* hybridization to be present both in large arrays integrated in metaphase chromosomes and in minute and double minute replicating elements. Karyotype analysis of subclones of high copy number lines reveals stable maintenance of large extrachromosomal arrays of the plasmid. The original as well as modified versions of the plasmid were rescued by transformation of *E. coli* using total cellular DNA. Southern blot analyses of some of the recovered plasmids indicate the presence of mosquito-derived sequences.

G. Gene Expression in Insects

1. In vivo *heterologous gene expression in the mosquito* Aedes aegypti

I. COMLEY, J. M. CRAMPTON AND P. EGGLESTON
Wolfson Unit of Molecular Genetics, Liverpool School of Tropical Medicine, Pembroke Place, Liverpool L3 5QA, UK

We present investigations into the expression of heterologous genes in the developing embryos, larvae and adults of the mosquito *Aedes aegypti*, Bangkok strain. The DNA is introduced into the embryo 90–120 minutes post-oviposition by means of microinjection. The injected material is in the form of supercoiled plasmid DNA which has been purified on a caesium chloride gradient. Plasmids used consist of a construct of the chloramphenicol acetyl transferase (CAT) gene under the control of different transcriptional promoters. Any transient expression facilitated by these constructs can be detected as CAT activity. This provides a very serviceable test for gene expression as, firstly, the mosquito possesses no endogenous CAT activity and, secondly, any CAT activity present in the microinjected organism can be detected by a very simple assay. This consists of lysing a number of the organisms and using the resulting supernatant in an enzyme reaction whereby ^{14}C-labelled chloramphenicol is acetylated. The resultant products are extracted into an organic solvent and separated by thin layer chromatography. CAT activity is easily distinguished as the presence of the higher mobility labelled product on the autoradiograph of the TLC plate. The promoter used currently is the hsp70 heat shock promoter from *Drosophila*. This is of interest not only because of the conserved nature of heat shock promoters but also, more importantly, because of the fact that correct transcription and translation of CAT has been demonstrated in mosquito cell culture using this hsp70 promoter (see Abstract, G. Lycett, this Workshop). The expression of the hsp70/CAT construct has been tested at different times post-injection and for different durations of heat shock. In this way we have demonstrated that the embryo is able to recognize the *Drosophila* promoter and to produce functional CAT. This then provides us with a suitable *in vivo* system to assess the effectiveness of characterized promoters and to determine the functionality of putative promoter sequences in the mosquito system.

Acknowledgements

This work was supported by the Research Development Fund of Liverpool University, the Wellcome Trust and the Lister Institute of Preventive Medicine. JMC is a Wellcome Trust Senior Research Fellow in Basic Biomedical Sciences and PE is a Lister Institute Research Fellow.

2. *Expression from two* Drosophila *promoters in embryos of* Locusta migratoria

S. MATHI, V. WALKER AND G. WYATT
Department of Biology, Queens University, Kingston, K7L 3N6, Canada

We have demonstrated transient expression of genes injected into embryos of the African migratory locust. Freshly deposited eggs (early cleavage embryos) are easily injected without dechorionation or desiccation. Eggs were injected with circular or linear plasmids containing the *Drosophila* hsp70 promoter and the chloramphenicol acetyl transferase (CAT) reporter gene (hsp-cat), or with circular plasmid containing the *Drosophila* copia promoter fused to CAT (copia-cat). Southern blot analysis showed that the hsp-cat plasmid persisted extrachromosomally for at least 8 days after injection. There was no evidence for plasmid replication. After injection of hsp-cat, CAT enzyme activity was detected at varying levels in 6–8% of the surviving injected embryos on days 3 and 9. Embryos injected with copia-cat, assayed on day 3, had a greater frequency but no higher level of expression. If optimization of conditions can achieve higher frequencies of expression, the system is promising for analysis of other promoters.

Acknowledgement
Supported by NSERC of Canada.

3. CAT protein production in mosquitoes and mosquito cells using expression vectors derived from Sindbis virus

K. OLSON, J. CARLSON AND B. BEATY
Department of Microbiology, Colorado State University, Fort Collins, CO 80523, USA

We have used Sindbis (SIN) virus expression vectors, developed by C. M. Rice, to express CAT protein in mosquitoes and mosquito cells. The pTRCAT expression vector transcribes a self-replicating RNA containing the non-structural coding sequences of SIN clone TOTO 1002, but has the viral structural genes removed and replaced with the CAT gene. Expression vectors pTE/3'2J/CAT, pTOTO/3'2J/CAT, and pTE/5'2J/CAT transcribe RNA containing SIN virus structural and non-structural genes as well as the CAT gene. The genomes of each of these vectors contain two subgenomic RNA promoters which transcribe mRNA for the structural genes (26S RNA) and the CAT gene. The genomic RNAs generate infectious recombinant virus within the appropriate cell lines.

TRCAT RNA has been transfected into C6/36 (*Aedes albopictus*) and ATC-10 (*Aedes aegypti*) cells using Lipofectin. CAT protein has been detected, although at low levels, in both cell lines using an indirect immunofluorescent assay. Conomic RNAs transcribed from pTE/5'2J/CAT and pTE/3'2J/CAT have been transfected into BHK-21 cells to generate virus. The virus in turn has been used to infect C6/36 cells and both *Aedes aegypti* and *Aedes triseriatus* adult mosquitoes by intrathoracic inoculations. CAT protein has been detected in both C6/36 cells and mosquitoes. These expression vectors should prove useful for introducing novel genes into mosquitoes.

4. Zinc finger motifs and silkworm moth chorion gene transcription factors

J. DREVET AND K. IATROU
Department of Medical Biochemistry, Faculty of Medicine, University of Calgary, 3330 Hospital Drive NW, Calgary, Alberta T2N 4N1, Canada

Band shift and DNA footprinting assays have been used to identify proteins which bind specifically to the shared promoters of high-cysteine (Hc) chorion gene pairs, which are expressed during late choriogenesis. These assays resulted in the identification of two DNA-binding proteins, BCFI and BCFII, whose promoter recognition/binding sequences are conserved in all Hc genes and whose appearance in follicular nuclei coincides with the activation of the Hc genes *in vivo*.

The sequence of the binding site of the major chorion factor, BCFI, contains a perfect match to the consensus sequence of the erythroid-specific transcription factors known collectively as the "GATA-1" family factors. We have used degenerate oligonucleotides to PCR amplify from *Bombyx mori* genomic DNA a sequence (ZF1) which, upon conceptual translation, shows extensive homology with the DNA-binding domain of the GATA-1 family of zinc finger proteins. Using this amplified product as a probe, we have isolated and characterized a 4.3 kb genomic clone (ZF2) which contains GATA-1-like zinc finger motifs separated by an intron. Sequence comparisons as well as Southern analysis suggest that a multigene family encoding this class of zinc finger proteins exists in *B. mori*.

RT-PCR amplifications of follicular RNAs using primers that encompass the zinc finger domain of ZF2 resulted in the isolation of two cDNA sequences, ZF2S and ZF2L (for short and long ZF2 cDNAs), which are present in follicular but not in silk gland cell nuclear extracts, and which differ from each other by the presence (or absence) of a stretch of 14 amino acids. Considering that only one ZF2 gene was found in *B. mori* genomic DNA under stringent hybridization conditions, these results suggest that ZF2 expression may be regulated via differential splicing. Moreover, the RT-PCR experiments suggest that ZF2S cDNA is developmentally regulated since, although it can be detected during early choriogenesis, it accumulates maximally in late choriogenic follicles, the stage at which Hc gene expression takes place. ZF2S and ZF2L were subcloned into expression vectors and over-expressed in *E. coli* cells. Band shift assays using bacterial protein extracts have shown that both over-expressed fusion proteins are able to bind specifically to the BCFI recognition sequences of Hc genes. Antibody production is now in progress to find out whether ZF2S and ZF2L represent two variant forms of the chorion transcription factor BCFI.

Acknowledgement

This work has been supported by the Medical Research Council of Canada.

H. Cloning and Characterization of Insect Genes

1. Basic studies for genetic manipulation in Ceratitis capitata *Wied (Diptera, Tephritidae)*

A. MALACRIDA, G. GASPERI, L. BARUFFI, C. TORTI AND C. GUGLIELMINO
Departimento di Biologia Animale, Universita di Pavia, 1–27100 Pavia, Italy

The medfly *Ceratitis capitata* is the target of a determined effort to produce a genetic sexing system for use in SIT control of this species. Both for the classical and molecular approach to genetic sexing, considerable progress has been made in the following topics:

1. The analysis of the complexity and organization of the genome of this species.
2. The genetic, biochemical and molecular analyses of the *Adh* loci which are of potential use in selection.
3. The failure so far to achieve transformation in the medfly using P elements has necessitated a more fundamental approach: the systematic search for hybrid dysgenic symptoms between different populations.
4. The analysis of the genetic structure of natural populations.

2. The Drosophila melanogaster *lysozyme locus: gene organization and pattern of expression*

S. D. PEREIRA
Department of Microbiology, University of Stockholm, S-106 91, Stockholm, Sweden

We have identified seven expressed genes in the lysozyme locus of *Drosophila melanogaster*. By comparison of their DNA sequences they were classified into four different types: LysD (including LysB, LysC, LysD and LysE), LysP, LysS and LysX. These four types of lysozyme exhibit between 67% and 82% homology. They also have different expression patterns: the LysD gene is expressed during all feeding stages of development, in the anterior midgut of larvae and adults; the LysP gene is only expressed in the salivary glands of adults; the LysS gene is only active in the third larval instar; and the LysX gene is expressed from late third larval instar to adult. Interestingly the LysD, P and S genes are repressed when bacteria are injected into the haemocoel, whereas expression of the LysX gene does not change. The LysD and P genes might have a function in the digestion of bacteria present in the food rather than an active role in immune defence. We have identified two lysozymes in the haemolymph of the fruit flies that might be involved in their immune defence. The results achieved so far offer a basis for further studies on gene regulation mechanisms as well as evolution.

3. Physical mapping of dipteran genomes using a high-resolution in situ protocol

W. PROCUNIER
Department of Entomology, Michigan State University, E. Lansing, Michigan, 48824-1115, USA

A major hurdle in isolating and characterizing genes of interest from many insect taxa is the lack of detailed genetic linkage maps containing polymorphic segments (sequences) which can serve as genetic reference points. One reason for this is that many insect species are either univoltine or have special biological attributes that make them difficult to rear in the laboratory. The construction of a physical map might be facilitated by first determining the position of particular genes directly, before undertaking any genetic crosses, by the direct *in situ* hybridization of labelled DNA probes to specific sites on the polytene chromosomes.

Most recently, we have been successful in demonstrating that heterologous single and multiple copy genes from *Drosophila* can be *in situ* hybridized to the polytene chromosomes of taxa in both the higher and lower diptera. Using an avidin/biotinylated horse radish peroxidase/ diaminobenzidine system, the polytene chromosome preparations were of sufficiently high quality for direct banding comparisons without recourse to photographing nuclei before adding probe or having to use image analysis.

This technique has the potential for broad scientific and practical applications. Physical maps would be useful for comparing linkage relationships between taxa, assessing phylogenetic relationships, mapping quantitative trait genes and facilitating the localization and sequencing of other genes of biological interest, including those involved in development and insecticide resistance.

This work was carried out in collaboration with Jim Smith, Jeff Feder and Guy Bush and we are grateful for the support of Tom Friedman.

Acknowledgements
This research was supported in part by a Michigan State Biotechnology Research Centre grant to Guy Bush, William Procunier and Jim Smith.

III. OVERVIEW

P. EGGLESTON[1] and J. CARLSON[2]
[1]*Wolfson Unit of Molecular Genetics, Liverpool School of Tropical Medicine, Liverpool L3 5QA, UK*
[2]*Department of Microbiology, Colorado State University, Fort Collins, CO 80523, USA*

The generation of transformed insects will require, to a considerable extent, the use of innovative approaches and the integration of both new and existing technology. The majority of attempts to date have tried to adapt the widely used P element transformation technique of *Drosophila* to insects of major medical, commercial or economic importance. Thus far, P element transposition has been demonstrated only in *Drosophila* although excision activity mediated by P transposase has been demonstrated in mammalian tissue culture cells and in yeast. Reports of stable transformation in mosquitoes appear to have occurred by rare illegitimate recombination events not mediated by P element transposase. It would appear, therefore, that the P element does not function as a truly autonomous transposon and may require other host-encoded factors present only in the Drosophilidae. Whether this is, in fact, the case must await further scientific investigation. In relation to the present state of research on the transformation of non-drosophilids, it is clear that at least four critical barriers need to be surmounted for the generation of stably transformed insects: (1) introduction of the genetic construct into the insect; (2) stable maintenance of the construct and incorporation into the genome of the insect; (3) identification of suitable marker genes that allow transformed and non-transformed insects to be distinguished; and (4) identification of appropriate promoters that control expression of the introduced gene in the transformed insects. Papers presented at this Workshop were of relevance to all of these aspects of the generation of transgenic insects.

A. Introduction of the Genetic Construct into the Insect

Introduction of DNA constructs into the genomes of non-drosophilid insects has been pursued both in cell culture and in the living insect. Both approaches have their own advantages. Cultured cells offer a more controlled environment in which to develop transformation vector constructs and to investigate gene expression. The genetic constitution of cultured cells, however, is undoubtedly different from that of the differentiated organism from which they were derived. Clearly, the ultimate goal of most research in this area must be the stable transformation of living insects even though presentations at this Workshop clearly demonstrate the importance of cell transfection in the pursuit of this goal.

Most attempts to introduce DNA constructs into living insects have focused on direct embryo microinjection as performed routinely in *Drosophila*. The DNA is injected posteriorly (although this appears not to be critical) and prior to pole cell formation so that the introduced DNA is taken up into the germ line primordia. Various modifications to the standard *Drosophila* protocol have been developed to account for differences in the physiology of the insects involved. Papers in this Workshop report on the introduction of transformation vector constructs through microinjection of the embryos of mosquitoes (Comley *et al.*; Morris and James), locusts (Mathi *et al.*) and silkworm moths (Mahalingam *et al.*). In some cases, expression of the introduced DNA has been demonstrated through the use of reporter gene fusions. An alternative and novel approach is the sperm-mediated transformation of honey bees (Milne). The sperm appears not to incorporate the foreign DNA but to act as a vector for its introduction into the female gamete. An approach that has not been extensively explored is the use of viruses to introduce genes into organisms via the process of infection as can be done with retroviruses in mammals. Baculoviruses have been successfully exploited for high levels of expression of genes in Lepidoptera; however, they have not been adapted for continuous long-term expression in insects. One of the drawbacks of the recombinant virus approach is the lack of well-characterized insect viruses with either a DNA genome or a DNA replication intermediate as in the case of retroviruses. The abstract by Olson *et al.* in this Workshop demonstrates the expression of a foreign gene by an infectious recombinant sindbis virus in cells and intrathoracically inoculated adults of the mosquitoes *Aedes albopictus* and *Aedes aegypti*. Although the virus is maintained as a persistent infection in mosquitoes, the RNA genome and the possibility of transmission to alternative hosts limit its use to transient-expression-type experiments.

B. Stable Maintenance of the Construct and Incorporation into the Genome of the Insect

Stable maintenance of the introduced genetic construct requires both that it be incorporated into the recipient germ line (such that it becomes subject to the normal hereditary machinery) and that it is incorporated at a genomic site which will allow normal gene expression. These issues have been addressed by several reports in this Workshop. One important approach is the identification of transposable elements from insect genomes which might form the basis of high-efficiency transformation vectors, as with the P element in *Drosophila*, Perkins *et al.*, working with *Lucilia cuprina*, used low-stringency hybridization with cloned *Drosophila* transposons as probes to identify both retrotransposon-like and P-element-like sequences. An interesting point to note is that these sequences were positively identified through library screens and were not detected by initial genomic Southern blots. Other approaches include the use of PCR to identify reverse transcriptase-like sequences in the genomes of *Aedes aegypti* and *Anopheles gambiae* (Warren and Crampton). The potential for development of these putative transposable elements into transformation vectors is unknown but will be of considerable interest in the near future. Incorporation of recombinant DNA constructs into insect genomes by mechanisms other than transposition was addressed by other reports. The abstract by Morris and James reports encouraging preliminary studies in embryos of the mosquito *Aedes aegypti* demonstrating both intra- and intermolecular recombination between plasmid molecules mediated by the FLP-FRT site-specific recombination system from yeast. This system has been in use in both *Drosophila* and mammalian cells and thus may allow the reproducible integration of constructs at a specific site in the genomes of many organisms.

Integration by homologous recombination was reported in silkworm moth embryos (Mahalingam *et al.*). Plasmid constructs carrying a highly repetitive sequence from the genome of *Bombyx mori* were shown to integrate into the genome by a PCR-based assay in a high percentage of injected embryos. The efficiency of integration appeared to be much higher than that observed in other insects to date and may reflect the dynamics of recombination in this particular insect.

The majority of the insect cell culture experiments reported at this Workshop involved the study of transient gene expression. However, stable transformation of *Aedes albopictus* mosquito cell cultures was reported by Monroe *et al.* Extremely high copy numbers (10^4/cell) of the transforming plasmid were found to be maintained as large integrated arrays and/or self-replicating chromosome-like elements in these cells.

C. Marker Genes for Distinguishing Transformed and Non-transformed Insects

An important aspect of the transformation procedure for any organism is the ability to distinguish transformed from non-transformed individuals. In *Drosophila*, this is routinely achieved through the use of visible phenotypic mutations (primarily cloned eye colour genes) such that transformed individuals are readily distinguished by an altered eye colour phenotype. This facility is not yet available in other insects, where eye colour genes have still to be cloned and characterized. Most progress to date has been achieved with *Lucilia cuprina* (Patterson *et al.*) where both *white* and the homologue of *scarlet*, known as *topaz*, have been cloned. These may soon be available for *Lucilia* transformation studies.

Where such phenotypic markers are not available, as in most non-drosophilid insects, antibiotic resistance genes are routinely used to identify transformed individuals. Most studies have focused on the antibiotics G418 and hygromycin, which appear to be universally toxic but which can be inactivated by appropriate bacterial resistance genes. A number of problems have been identified in the use of this approach, most of them relating to the efficiency with which the resistance genes are expressed (Lycett *et al.*; Monroe *et al.*; and see next section). As an alternative approach, the alcohol dehydrogenase gene is being investigated as a selectable marker in the medfly, *Ceratitis capitata* (Malacrida *et al.*). In addition (and as demonstrated in *Drosophila*) putative transformed progeny can be screened without a selectable marker either through rapid squash-blots or by exploiting PCR (Mahalingam *et al.*). However, in most insects this is a very inefficient approach.

D. Control of Transgene Expression in Non-drosophilids

Several of the reports in this Workshop addressed the efficiency and control of transgene expression in non-drosophilids. In the short term, as we have described above, this is important for the efficient expression of selectable marker genes following transformation. However, in the longer term there will clearly be a need to control (spatially and/or temporally) the expression of medically or commercially relevant transgenes which are introduced into insect genomes. Most of the current work, both in insect cell cultures and in living insects, has involved the use of *Drosophila* promoters to control transgene expression. These have included hsp70 (Lycett *et al.*; Comley *et al.*; Morris and James; Mathi *et al.*), actin 5C and E74 (Lycett *et al.*), copia (Mathi *et al.*) and metallothionein (Kovach *et al.*). Other interesting *Drosophila* promoters include alleles of the lysozyme locus (Pereira), several of which were shown to be expressed in a tissue-specific fashion.

In the case of the inducible promoters, such as hsp70, several research groups had attempted to exploit those conditions which, in *Drosophila*, yielded high levels of gene expression. This approach is not necessarily successful, as shown by the experiments of Lycett *et al.* and Comley *et al.*, where conditions for optimal expression from this promoter were determined in *Aedes aegypti* cells and embryos respectively. Such experiments show clearly that optimal expression from hsp70 in *Aedes aegypti* is at 41°C and not 37°C as in *Drosophila*. This may reflect differences in the physiology of the organisms and also serves to show how cell cultures can usefully be employed to optimize such basic criteria for subsequent use in embryos.

It may be the case, however, that endogenous promoters will provide the most efficient control of transgene expression and several groups are now working in this area. Clearly, in the mosquito at least, promoters which drive salivary-gland- or gut-specific gene expression are of interest and are actively being sought.

E. Conclusions

In conclusion, this Workshop highlighted the growing international interest in molecular entomology. If the role model of *Drosophila* can be emulated, the rate of progress of both basic and applied molecular research in insects of major medical and commercial importance should now begin to accelerate. The next five to ten years will be an exciting time for those of us lucky enough to be involved.

Index

A

Abecins, 130
Ac, 44
Acetylcholinesterase
  Anopheles stephensi, 176
  Drosophila, 176
Acheta
  achetakinins, 209
  AKH, 207
Achetakinins, 209
  N terminal modifications, 220
  receptor, 216
  relationship with AKH, 218
  secondary structure predictions, 210
Acetylcholinesterase
  insecticide resistance, 175
Acyrthosiphon pisum
  sex pheromone, 166
Adipokinetic hormone (AKH), 187, 206, 212
  family, 212
  functions, 191
  inactivation, 191
  localisation, 189
  metabolism, 191
  N-terminal modifications, 218
  precursor biosynthesis, 194
  processing
    in vivo, 198
    in vitro, 198
  prohormone, 194
  receptor binding model, 212
  release, 191
  site of synthesis, 189
Aedes aegypti
  CAT, 252, 254
  eclosion hormone, 230
  gene expression, 252
  HSP70, 248
  microinjection of embryos, 7
  reporter gene expression, 247
  retrotransposons, 244

  sex determination, 82
  transposable element, 244
Aedes albopictus
  genome organisation, 7
Aedes triseriatus
  genetic map, 54
  genome organisation, 7
Aggregation pheromone
  Anthonomus grandis, 142
  Sitona lineatus, 142
Ajugarin I, 146
AKH *see* Adipokinetic hormone
Alarm pheromone, 169
  aphids, 142
Andropin gene, 132
Anopheles albimanus
  genetic map, 54
Anopheles culicifacies
  sex determination, 82
Anopheles gambiae
  microdissection libraries, 59
  microinjection of, 10
  Plasmodium cynomolgi, 55
  rDNA, 12
  retrotransposons in, 12, 245
  sex determination, 82
  transposable element, 245
Anopheles quadrimaculatus
  genome organisation, 7
Anopheles stephensi
  acetylcholinesterase, 176
Antheraea pernyi
  cecropin, 128
  genome organisation, 7
Anthonomus grandis
  aggregation pheromone, 142
  ajugarin I, 146
  antifeedants, 146
Antibacterial proteins
  Attacin-like molecules, 129
  Cecropins, 128
  Lysozyme, 128
Antifeedants, 145

Index

ajugarin I, 146
Aphid sex pheromone, 142, 165
Aphis fabae
  alarm pheromone, 169
  semiochemical receptors, 160
  sex pheromone, 166
Apidaecins, 130
Apis apis
  antibacterial proteins, 130
Apis cerana indica
  sex determination, 84
Apis mellifera
  genome organisation, 7
  sex determination, 84
Athalia rosae ruficornis
  sex determination, 85, 86
Attacin-like molecules, 129

B

Blattella germanica
  sex pheromone, 146
Bombyx mori
  cecropin, 128
  chorion genes, 255
  embryo transformation, 250
  genetic maps, 52
  genome organisation, 7
  lysozyme, 128
  PTTH, 227
  semiochemical perception, 155
  transcription factors, 255
  zinc fingers, 255
Bracon hebetor
  sex determination, 84
Brevicoryne brassicae
  alarm pheromone, 169
  semiochemical receptors, 160

C

Caenorhabditis elegans, 24
  cosmid map, 67
  rDNA, 24
  sex determination, 98
Calliphora
  sex determination, 82
  yolk protein genes, 97

CAT *see* Chloramphenicol acetyl transferase
Carausius
  AKH, 207
Cecropins, 128
  Drosophila, 132
  regulation of expression, 134
Cell–cell communication
  segment polarity gene products, 116
Ceutorhynchus assimilis
  perception of volatiles, 160
  pheromones, 148
Ceratitis capitata
  genetic manipulation of, 256
Chemical ecology
  behavioural studies, 164
Chironomus tentans
  genome organisation, 7
  sex determination, 82
Chironomus thummi
  sex determination, 95
Chloramphenicol acetyl transferase (CAT)
  Locusta migratoria, 253
  mosquitoes, 13, 247, 254
  mosquito cells, 247, 252, 254
  reporter gene, 247
  sinbis virus vector, 254
Chorion genes
  Bombyx mori, 255
Classification
  insects, 24ff
  transposable genetic elements, 36
Corpora allata
  peptide regulation of, 228
Cosmid mapping, 67
Culex pipiens
  esterases, 180
  genome organisation, 7
  sex determination, 82
Culex quinquefasciatus
  oviposition pheromone, 156
  sex pheromone, 143
Cytochrome P450
  insecticide resistance, 179

D

Dendroctonus spp
  pheromone perception, 157
Defensins, 130

Index 265

Diptera
    sex determination, 80
Diptericins, 129
Drosophila melanogaster, 133
Diuretic peptides, 221
Drosophila hawaiiensis
    microinjection of, 43
Drosophila mauritiana
    mariner, 45
Drosophila melanogaster
    Andropin, 132
    ecdysteroid receptors, 232
    eclosion hormone, 230
    HSP70 promoter, 248, 253
    immune system, 131
        cecropin locus, 132
        diptericin gene, 133
        lysozyme locus, 134, 256
        cecropin gene regulation, 134
    insecticide resistance, 175
        acetylcholinesterase, 176
        cytochrome P450, 180
        esterases, 180
        GABA receptor, 175
        Glutathione S-transferase, 179
        sodium channel, 176
    juvenile hormone receptors, 234
    mapping, 51
    metallothionein promoter, 248
    microdissection library, 59
    pattern formation, 102, 112
    polytene chromosomes, 59
    RNA processing, 91
    sex determination, 77, 81
    YAC map, 66
Drosophila pseudoobscura
    pattern formation, 114
Drosophila simulans
    microinjection of, 43
Drosophila virilis
    pattern formation, 114
Drosophila willistoni, 40

E

Ecdysteroids
    development, 233
    receptors, 232
    transcription factors, 233

Eclosion hormone, 228
    Aedes aegypti, 230
    Bombyx, 230
    Drosophila, 230
    Manduca, 230
Economic insects
    transgenic technology, 15
Endocrinology
    molecular approaches to, 226
Enhancers
    stage specific, 13
    tissue specific, 13
Enhancer traps, 43, 44
Esterases
    insecticide resistance, 180
Evolution
    DNA analysis, 21
    genes and genomes, 39
    rDNA, 24

F

Fat body
    neuropeptides, 206
FB elements, 38
Filarial susceptibility
    f^m locus, 16
FLP-mediated recombination
    Aedes aegypti, 249

G

GABA receptor, 174
    chloride channel complex, 174
    Drosophila, 175
Galleria mellonella
    lysozyme, 128
Gene amplification
    esterases, 181
    insecticide resistance, 181
Genetic manipulation, 5ff, 256
Genetic maps, 51
Genome complexity, 6
Genome evolution, 22, 39
Genome mapping, 64, 257
Genome organisation, 6ff, 22
    Aedes aegypti, 7
    Aedes albopictus, 7

Aedes triseriatus, 7
Anopheles quadrimaculatus, 7
Antherea pernyi, 7
Apis mellifera, 7
Bombyx mori, 7
Chironomus tentanus, 7
Culex pipiens, 7
Lucilia cuprina, 7
long period interspersion, 6
Musca domestica, 7
Sarcophaga bullata, 7
short period interspersion, 6
Glutathione S-transferase
  insecticide resistance, 179
Gryllus
  AKH, 207

H

Haemolin, 130
Haplodiploid
  Hymenoptera, 84
  sex determination in, 84
Heliothis
  AKH, 188, 207
  juvenile hormone esterase, 231
Heteropza
  sex determination, 87
hobo, 44
Homeodomain proteins
  structure–function relationship, 120
Homeotic genes, 112
  regulatory targets, 117
Horizontal transfer
  genetic information, 40
HSP70
  Aedes aegypti, 14
  Drosophila melanogaster, 14
HSP70 promoter
  Aedes aegypti, 248
  Locusta migratoria, 253
Hyalophora cecropia
  attacins, 129
  cecropins, 128
  immune proteins, 127
  lysozyme, 134
Hyalopterus pruni
  sex pheromone, 166
Hybrid dysgenesis, 38

Hymenoptera
  haplodiploid, 84
  sex determination, 84

I

I element, 36
Immune recognition molecules
  haemolin, 130
  lectins, 131
Immune system
  Drosophila melanogaster, 132
  cellular, 126
  humoral, 127
  immune proteins
    antibacterial proteins, 128
    Lysozyme, 128
    Cecropins, 128
    Attacin-like molecules, 129
  recognition molecules, 130
    haemolin, 130
    lectins, 131
Insecticide resistance, 173
  acetylcholinesterase, 175
  cytochrome P450, 179
  esterases, 180
  GABA receptor, 174
  Glutathione S-transferase, 179
  metabolic resistance, 179
  molecular biology of, 180
  monooxygenases, 179
  sodium channel, 176
  target site resistance, 174
Insertional mutagenesis, 41
in situ hybridisation
  genome mapping, 257
  dipterans, 257
Integration
  control of, 249
Ips paraconfusus
  pheromone receptors, 157
Ips pini
  pheromone receptors, 157

J

Juvenile hormone, 234

Index

K

Kairomones, 158

L

Lectins, 131
Leptinotarsa decemlineata
  perception of volatiles, 159
Leucophaea
  achetakinins, 209
Lilium henryi
  retroviral element, 45
Lipaphis erysimi
  alarm pheromones, 169
  semiochemical receptors, 160
Locusta
  AKH, 207
  diuretic peptides, 221
Lucilia cuprina
  eye colour genes, 246
  genetic map, 52
  genome organisation, 7
  transposable elements, 243
  sequence homology with *Drosophila*, 243
Lysozyme, 128
  Drosophila, 256
  gene organisation, 256
  gene expression, 256

M

Mamestra brassicae
  sex pheromones, 164
Manduca sexta
  AKH, 188, 207
  cecropin, 128
  diuretic hormone, 231
  diuretic peptides, 222
  ecdysteroid receptors, 232
  eclosion hormone, 230
  juvenile hormone receptors, 234
  PTTH, 227
Mapping
  cosmid, 67
  genetic, 51ff
  genome, 51ff, 64
  mini-satellite, 62

  micro-satellite, 62
  multilocus, 53
  P1, 66
  polytene chromosomes, 56
  restriction fragment length polymorphisms (RFLP), 60
  YAC, 66
Mariner, 45
Megoura viciae
  sex pheromone, 142, 166
Meiotic drive, 17
Metamorphosis
  hormones, 231
Microdissection
  libraries, 59
Microinjection of embryos, 7ff, 250
Micro-satellite mapping, 62
Mini-satellite mapping, 62
Monooxygenases
  insecticide resistance, 179
Moulting
  regulation of, 231
M strains 37, 39
Multilocus mapping, 53
Musca domestica
  genome organisation, 7
  sex determination, 80, 83
Myzus nicotianae
  esterases, 181
Myzus persicae
  alarm pheromones, 169
  esterases, 181
  pheromone attractants, 145

N

Neuropeptides
  fat body metabolism, 206

O

Odours
  electrophysiology, 154
  perception, 152
Onchocerca volvulus
  transmission of, 32

P

P1 mapping, 52, 66
Pattern formation
  Drosophila, 102ff
  transcription patterns, 113
PCR *see* Polymerase Chain Reaction
Pectinophora gossypiella
  sex pheromones, 143
P element, 41, 175
  mosquitoes, 8ff
Periplaneta americana
  AKH, 188, 207
  sex pheromone, 142
Pheromones, 141
  aggregation, 142
  alarm, 142, 169
  antifeedants, 145
  oviposition, 156
  perception of, 152, 155
  receptors, 157
  sex, 142
Phormia terranovae
  attacin-like molecules, 129
  defensins, 130
  diptericin, 133
Pheromone receptors, 157
Phorodon humuli
  alarm pheromone, 169
  semiochemical receptors, 160
  sex pheromone, 149, 158, 167
Phylogenetic relationships, 25
Plant volatiles
  kairomones, 158
Plasmodium cynomolgi
  Anopheles gambiae, 55
Plutella xylostella
  sex pheromone, 143
P–M Dysgenesis, 39
Polarity genes
  Drosophila, 104, 112
Polymerase Chain Reaction (PCR)
  acetylcholinesterase, 176
  eclosion hormone, 230
  GABA receptor, 175
  neuropeptides, 227
  retrotransposon-like elements, 12
  Simulium damnosum rDNA, 28
  sodium channel, 178
Polytene chromosomes

mapping, 56, 59
Proctolaelaps regalis
  P elements, 41
Promoters
  Drosophila, 253
  metallothionine, 248
  stage specific, 13
  tissue specific, 13
Prothoracicotropic hormone (PTTH), 227
  Bombyx mori, 227
  Manduca sexta, 227
  Samia cynthiaricini, 227
P strains, 39
Psylliodes chrysocephala
  perception of volatiles, 160
  pheromones, 148
PTTH *see* Prothoracicotropic hormone

Q

Quadraspidiotus perniciosus
  pheromones, 142

R

Receptors
  achetakinins, 216
  AKH, 212
  ecdysteroid, 232
  hormones, 231, 234
  juvenile hormone, 234
Repetitive DNA, 30
Reporter genes
  mosquitoes, 244
Restriction fragment length polymorphisms, 53, 60
Retrotransposons
  mosquito, 244
  Anopheles gambiae, 12
Reverse transcriptase
  mosquitoes, 244
  retrotransposon, 244
RFLP *see* Restriction Fragment Length Polymorphism
Rhodnius prolixus
  juvenile hormone receptors, 234
Rhopalosiphum padi
  alarm pheromones, 169

Ribosomal RNA
  evolution of, 24
  systematics, 24
Ribosomal DNA
  Simulium damnosum, 28
RNA processing
  sex specific, 91
Romalea
  AKH, 207

S

Saccharomyces cerevisiae
  Ty1, 36
Samia cynthia ricini
  PTTH, 227
Sarcophaga bullata
  genome organisation, 7
Sarcophaga peregrina
  attacin-like molecules, 129
  cecropin, 128
  defensins, 130
  sarcotoxins, 132
Sarcotoxins, 129, 132
Satellite DNA, 22
Schistocerca
  AKH, 188, 207
Schizaphis graminum
  sex pheromones, 142, 166
Scolytus scolytus
  pheromones, 157, 159
Segmentation genes, 104
Segment polarity gene products, 116
Selectable markers
  eye colour genes, 13, 246
  selection of transformants, 9, 12
Selectable markers, 9, 12, 246
Semiochemicals, 142
  activity, 142
  analogues, 146
  attractants, 145
  behaviour mediated by, 163
  biosynthesis, 147
  interaction, 144
  molecular determinants of, 141, 147
  perception, 152
  receptors, 160
Sex determination
  genetics of, 80

  model of, 85, 89
  molecular analysis, 95
  molecular basis, 89
Sex pheromones
  Acyrthosiphon pisum, 166
  aphid, 142, 165
  Aphis fabae, 166
  behavioural response to, 165
  Blattella germanica, 146
  Culex quinquefasciatus, 143
  Hyalopterus pruni, 166
  Mamestra brassicae, 164
  Megoura viciae, 166
  Pectinophora gossypiella, 143
  perception, 155
  Periplaneta americana, 142
  Phorodon humuli, 149, 158
  Plutella xylostella, 143
  Schizaphis graminum, 166
  Sitobion spp, 166
  Spodoptera littoralis, 149
Simulium damnosum
  complex, 32
  rDNA, 28
Sitobion spp
  alarm pheromone, 169
  sex pheromone, 166
Sitona lineatus
  aggregation pheromone, 142, 144, 157
Solenopsis invicta
  sex determination, 84
Speciation
  transposable elements, 41
Spodoptera littoralis
  sex pheromone, 149
Systematics
  rRNA, 24

T

Target genes
  honeybee, 15
  mosquitoes, 15
  silk moth, 15
Taxonomy
  DNA analysis, 21
Tetranychus urticae
  alarm pheromone, 142
Transcription factors

Bombyx mori, 255
  chorion genes, 255
  hormones, 231
Transfection of cells, 13, 252
  calcium phosphate precipitation, 13
  electroporation, 13
  lipofection, 13
  polybrene, 13
Transformation techniques, 250
  cultured cells, 13, 247, 252
  honeybee, 251
  mosquitoes, 7ff
  non-drosophilids, 241
  silkmoth, 250
  sperm mediated, 251
Transformation
  germline, 42
Transformation vectors, 8, 11
  control of integration, 249
  promoters, 247
  selectable markers, 9, 12, 246
Transgenic mosquitoes, 10
  natural populations, 17
Transgenic technology, 4ff
  economic insects, 15
  functional cloning, 14
  transposon tagging, 14
Transmission blocking vaccines
  malaria, 16
  mosquitoes, 16
Transposable genetic elements
  classification of, 36
  Drosophila melanogaster, 41
  exploitation of, 41
  horizontal transfer, 40
  Lucilia cuprina, 243
  mosquitoes, 11, 244
  mutagens, 38
  search for, 11, 44
  speciation, 41
Transposition
  biological consequences, 35ff, 38, 40
Transposon tagging, 41
  mosquito, 14
  Drosophila melanogaster, 41
Ty1, 36

V

Volatiles
  host-plant, 167

Y

YAC mapping, 52, 65
  Drosophila, 66
Yolk protein genes
  Calliphora, 97

Z

Zinc-finger
  silk moth, 255
  chorion genes, 255